FLUIDICS

Components and Circuits

FLUIDICS
Components and Circuits

K. Foster
*Reader, Department of Mechanical Engineering
University of Birmingham, England*

G. A. Parker
*Reader, Department of Mechanical Engineering
University of Surrey, Guildford, England*

WILEY–INTERSCIENCE
a division of John Wiley & Sons Ltd
London · New York · Sydney · Toronto

Copyright © 1970 John Wiley & Sons Ltd. All Rights Reserved. No part of this publication may be reproduced, stored in a retrieval system, or transmitted, in any form or by any means, electronic, mechanical photo-copying, recording or otherwise, without the prior written permission of the Copyright owner.

Library of Congress catalog card number 76-116650

ISBN 0 471 26770 8

Printed by offset in Great Britain by
William Clowes and Sons, Limited,
London, Beccles and Colchester

Preface

Fluidics has developed in a rather unbalanced way since it crystallized as a new branch of Control Systems in the late 1950's. Some areas have received considerable attention both in North America and Europe, particularly in relation to aerospace applications. However, in the purely industrial field little advance has been made in the application of fluidic–pneumatic control systems until comparatively recently due to the lack of commercial viability of the majority of devices available. One exception to this has been the widespread use of miniaturized moving-part pneumatic devices in Eastern Europe mainly for Process Control applications.

It is clear that a considerable amount of practical knowledge exists on both individual device performance and also circuits. Although much of this has arisen from aerospace applications, with consequent rather restricted information dissemination, it is gradually passing into the industrial field. Many introductory articles and research papers on fluidics have appeared in the literature, which have tended to produce a gap in understanding between the specialist and the non-specialist. We have attempted to bridge this gap by writing a book which we hope will appeal both to the specialist and layman alike.

The original material for the book was written for a Fluidics course organized by the authors at Birmingham University (Easter 1967) and used again in slightly modified form for a similar course in 1968. Since that time substantial revisions and additions to the material have been made both as a result of experience gained in running the courses and also new research in the field. One of the difficulties experienced has been writing an up-to-date text at a time of rapid technological change in Fluidics.

Although some of the topics dealt with in the book are not directly related, we have tried to follow a logical development of the subject. Chapter 1 is an introduction to all the well-known non-moving part devices and sensors which may be found in fluidic systems. The explanation is simple and essentially non-mathematical so that the non-specialist may easily grasp the main features. Many of the devices are described in more detail in later chapters. In Chapters 2 and 3 the fluid mechanics of systems components using a lumped and distributed parameter representation is developed on a general basis together with the mechanism of fluid flow within fluidic devices.

At this point the book divides into two distinct sections. Chapters 4 and 5 deal exclusively with pure fluid analogue devices and methods of signal shaping for fluidic circuits. Chapters 6 to 14 deal with many aspects of both moving and non-moving part digital fluidic devices. Particular attention is paid to the description of standard system sub-assemblies (such as counters, encoders, etc.) and also system organization. Finally, Chapter 15 deals briefly with some of the graphic techniques available for obtaining system operating-conditions in both digital and analogue circuits.

A difficulty in writing a compendium of material of this kind is that notation differs widely between authors; symbols for fluidic devices are by no means standardized. Some attempt has been made to rationalize both of these, simply in order to avoid confusion in any one chapter, but neither notation nor symbols have been standardized completely and the authors hope that the compromise will be acceptable to the reader.

A further difficulty is that it is impossible to cover every reference in the vast amount of literature available on fluidics and the choice of papers to which reference is made inevitably reflects the interest of the authors. There will doubtless be many excellent references not discussed.

The authors would like to acknowledge the generous support of the Science Research Council for a considerable proportion of their research work, the results from which constantly appear throughout the book. Thanks are also due to the National Coal Board and the Admiralty for contracts which provided further impetus to the research.

It is impossible to end the preface without acknowledging the efforts of those concerned with carrying out the research projects and responsible for many of the results quoted. Drs. N. S. Jones, D. G. Mitchell, B. Jones, and D. A. Retallick were working on the projects while most of the present material was being gathered together, and most thanks are due to them. In the latter stages, discussions with M. K. Addy, P. J. Cleife, R. M. H. Cheng and P. Drazan have been invaluable and we extend our thanks to them also. No book can be written without substantial secretarial aid and we are indebted to Mrs. M. Plant for most of the early work and more recently to Miss R. Croger, Miss J. Nightingale and Mrs. M. Eardley.

K. Foster
G. A. Parker

Glossary of Symbols

If a symbol has an alternative use in a specific chapter the chapter is quoted after the definition in parentheses.

A	area
a	local propagation velocity
b	channel width
C	fluid capacitance
C_{12}	$= N_{12} \cdot (P_1/P_2)$
C_D	coefficient of discharge
c	absolute propagation velocity
D	diameter
d	depth, setback (Chapter 7)
E	ratio of dynamic to total pressure
F_w	pressure force on wall
f	frequency or friction coefficient
f_1	non-linear static characteristics of analogue amplifier 1
G	gain
$G(s)$	transfer function
g	gravitational constant
H	fluid inductance or inertance
h	channel height
I	rotary inertia
J	fluid momentum
j	$= \sqrt{-1}$
K	compressible orifice flow constant
K_{12}	$= \dfrac{(P_1/P_2)}{N_{12}} \cdot \dfrac{\partial N_{12}}{\partial (P_1/P_2)}$
L	length
$L_{f,i}$	free and impacting jet potential flow core length respectively
M	weight of fluid (Chapter 2), mass flow rate (Chapter 4), magnitude ratio (Chapter 5)
m	incremental mass flow rate
N	gear ratio
N_{12}	$= W_{12}/W_{\mathrm{crit}}$, n_{12} incremental value
n	polytropic constant
P, p	steady state and incremental pressure, total and static pressures respectively (Chapter 4)
ΔP	pressure difference

P_m	mean pressure
Q, q	steady state and incremental volume flow rate
R, r	steady state and incremental resistance
$\mathbf{R}$	gas constant
R_e	Reynolds number
r	radius (Chapter 4)
S	wall offset
$\mathscr{S}$	LaPlace transform, specific weight (pp. 49, 50 only)
T	absolute temperature
t	time
U, u	particle velocity in the x direction
$\bar{u}$	sectional mean velocity
V	volume
v	particle velocity in the y direction
W, w	steady state and incremental weight flow rate, nozzle width (Chapter 1)
$\mathscr{W}$	power
X, x	steady state and incremental distance
$Y(j\omega)$	admittance
$Z(j\omega)$	impedance
Z_{ch}	characteristic impedance
α	wall angle
β	jet deflexion due to control momentum and static pressure forces
β_m	jet deflexion due to control momentum alone
γ	ratio of specific heats ($= 1\cdot 4$ for air)
θ	total jet deflexion
μ	absolute viscosity
ν	kinematic viscosity
ξ	damping ratio
ρ	density
τ	time constant
ϕ	phase angle, jet spread angle (Chapter 4)
ω	angular frequency
ω_r	resonant frequency

Suffices

c	control or input
col	collector
e	emitter
o	output
s	supply

Chapter 3

B	width of nozzle
b	width of jet at distance x from nozzle or slit
$b_{\frac{1}{2}}$	distance between $\frac{1}{2}U$ velocity points in jet
b'	y at point of inflection of velocity profile of jet
C'_p	$(P_\infty - P_b)/(P_\infty - P'_b)$
C_Q, C_r	source and vortex strengths
c	$c_r + jc_i = \beta/\alpha$ for laminar jet

GLOSSARY OF SYMBOLS

D	setback
d	diameter of circular nozzle, distance apart of vortex rows
E	entrainment parameter—subscripts 1, 2 denote inner and outer edges of jet respectively
G	value set for $\bar{W}_c^*$ in vortex valve
h	distance of reattachment point from nozzle centre line, also nozzle-edge distance, also height of vortex chamber
h_{c_i}, h_{c_o}	distance of inner and outer nozzle edges from jet centre line for curved flow
J	momentum of jet
K	$= J/\rho$ also constant
K_{e_c}, K_{e_0}	calculated and actual K at nozzle
l	mixing length, also distance of reattachment point from nozzle in x direction
m	$= C\alpha/2\beta$ in vortex flow
N	radial Re, also an integer
P_b	average bubble pressure
P_b'	pressure along boundary close to nozzle
$\bar{P}_{cc}$	non-dimensional cut-off control pressure for vortex valve
P_∞	static pressure outside bubble
P_∞'	static pressure downstream of reattachment
$R_{e,o}$	inner and outer radii of vortex chamber
t	a parameter $\tanh\left(\dfrac{\sigma y}{x + x_0}\right)$
U	centre line velocity in jet analyses
U_m	maximum velocity of mean flow
U_R	maximum velocity of reverse flow in bubble
u_v	velocity of propagation of vortices in jet
X	$= r/R_0$ for vortex analysis
x_r	reattachment distance along the wall
α	reciprocal of wavelength of disturbances in jet ($= \alpha_r + j\alpha_i$), a wall angle, angle flow makes with radius in vortex flow analysis, a parameter
β	speed of propagation of disturbance in jet ($= \beta_r + j\beta_i$), a parameter
ε	eddy viscosity
η	$\sigma y/x$; also a co-ordinate
θ	angle between main jet and control jet, also angle subtended by bubble at centre of curvature
κ	a constant in expression of eddy viscosity
ξ^*	distance from one edge of shear layer to where velocity is $\tfrac{1}{2}U$
σ	turbulent mixing coefficient
ϕ	jet deflection angle

Notation for Analogue Devices

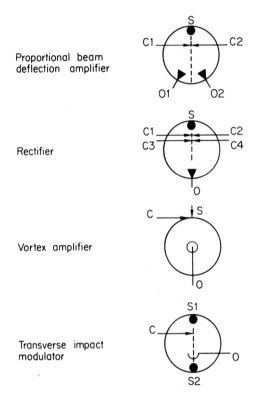

Proportional beam deflection amplifier

Rectifier

Vortex amplifier

Transverse impact modulator

GLOSSARY OF SYMBOLS

Notation for Digital Devices

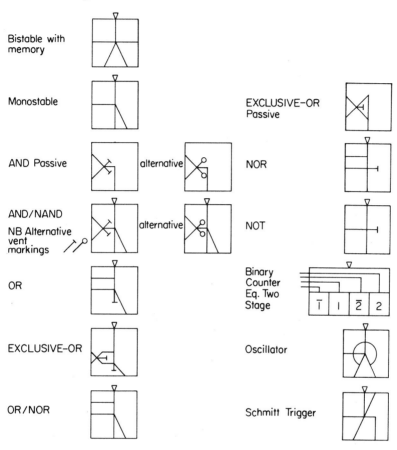

Contents

1. **Introduction to Pure Fluid Devices and Sensors**
   - 1.1 General 1
   - 1.2 Classification of Element Types 2
   - 1.3 Wall Re-attachment Digital Devices 5
   - 1.4 Momentum Interaction Digital Devices 13
   - 1.5 Combined Wall Re-attachment and Momentum Effects 17
   - 1.6 Transition from Laminar to Turbulent Flow . . . 21
   - 1.7 Miscellaneous Digital Devices 25
   - 1.8 Momentum Interaction Analogue Devices . . . 29
   - 1.9 Momentum Interaction and Boundary Layer Analogue Devices 34
   - 1.10 Modulated Pressure Field Analogue Devices . . . 38
   - 1.11 Position Sensors 39

2. **Lumped and Distributed Parameter Fluid System Components**
   - 2.1 General 46
   - 2.2 Basic System Parameters 47
   - 2.3 Linearization of System Parameters 60
   - 2.4 Lumped Parameter Fluid Networks 64
   - 2.5 Fluid Transmission Lines 76

3. **The Fluid Mechanics of Jets, Wall Attachment and Vortices**
   - 3.1 The Laminar Jet 93
   - 3.2 The Transition from Laminar to Turbulence of a Jet 95
   - 3.3 Noise Associated with a Laminar Jet. Edgetones . . 110
   - 3.4 The Turbulent Jet 117
   - 3.5 Power Consumption of a Jet 127
   - 3.6 The Deflexion of a Jet by Another Transverse Jet . 129
   - 3.7 The Turbulent Re-attachment Process 131
   - 3.8 The Trapped Vortex 161
   - 3.9 Theoretical Approaches to the Problem of Vortex Flow 163

4. Pure Fluid Analog Amplifiers
- 4.1 General 178
- 4.2 Beam-deflexion Amplifier 181
- 4.3 Impact Modulator 195
- 4.4 Vortex Amplifier 206

5. Analog Signal Control Techniques
- 5.1 High-gain Amplifiers 224
- 5.2 Low-gain Amplifier Signal Shaping 237
- 5.3 Frequency Modulation and Phase Discrimination . 252
- 5.4 Pulse-width Modulation 265

6. Boolean Algebra Theorems and Some Methods of Manipulation of Switching Functions
- 6.1 Introduction 274
- 6.2 Postulates of Boolean Algebra 277
- 6.3 Definitions in Boolean Algebra 278
- 6.4 Theorems of Boolean Algebra 280
- 6.5 The Use of Duality in Simplification of Switching Functions 285
- 6.6 Canonical Forms of Boolean Functions . . . 287
- 6.7 Simplification of Switching Functions: Tabulation Method 288
- 6.8 Simplification of Switching Functions: Karnaugh Maps 291

7. The Design of Pure Fluid Digital Elements
- 7.1 General 299
- 7.2 The Turbulence Amplifier 301
- 7.3 The Wall Attachment Amplifier 308
- 7.4 Monostable Devices 328
- 7.5 The Switching Action of a Bistable Wall Attachment Element 329

8. Moving-Part Logic Devices
- 8.1 Introduction 338
- 8.2 Diverter Valves 339
- 8.3 Devices Working on a Back-Pressure Principle . . 352
- 8.4 A Diaphragm 'Threshold Logic' Device . . . 365
- 8.5 Speed of Operation of Elements 367
- 8.6 Valves for Power Amplification 370
- 8.7 Conclusions 372

CONTENTS

9. The Design of Sequential Systems and Counters, with a Note on Number Codes
- 9.1 Introduction 374
- 9.2 State Diagrams and Flow Tables 376
- 9.3 Secondary and Output Excitation 379
- 9.4 Pneumatic Ladder—Type Sequence Circuits . . . 384
- 9.5 Sequential Logic and Some Ideas on a Universal Logic Block 389
- 9.6 Number Codes 395
- 9.7 Counting Techniques 402

10. Fluidics Applied to Sequence Circuits and Simple N.C. Systems
- 10.1 Introduction 421
- 10.2 Simple Asynchronous Circuits 425
- 10.3 Counter Controlled Circuits 429
- 10.4 A Punched Card Controlled Sequence 433
- 10.5 Air Consumption of Pure Fluid Devices and Moving Part Circuits 436
- 10.6 Point-to-point Position Control 439

11. Card Readers, Tape Readers, Decoding and Visual Displays
- 11.1 Plug Boards and Card Inputs 447
- 11.2 Codes, Tapes and Tape Readers 450
- 11.3 Decoding 453
- 11.4 Visual Displays 462

12. Shift Registers, Ring Counters and Binary Counters
- 12.1 Introduction 469
- 12.2 The Shift Register 470
- 12.3 Ring Counters 481
- 12.4 Binary Counters. The 'T' Flip-Flop 485
- 12.5 Up-and-Down Counters 491
- 12.6 Pulse Shapers 492
- 12.7 Binary-Coded Decimal Counters 495
- 12.8 Unit-Distance Binary-Coded Decimal Counters . . 496

13. Digital System Encoders
- 13.1 Introduction 500
- 13.2 Classification of Encoders 501
- 13.3 Incremental Pattern Encoders 502
- 13.4 Coded Pattern Encoders 504
- 13.5 Practical Angular Encoders 512
- 13.6 Practical Linear Encoders 518

14. Summing Junctions, Converters and Signal Hazards in Digital Systems
- 14.1 Introduction 521
- 14.2 Adder Logic 522
- 14.3 Fluidic Adder Circuits 526
- 14.4 Subtractor Logic 532
- 14.5 Fluidic Subtractor Circuits 535
- 14.6 Comparator Logic 537
- 14.7 Fluidic Comparators 538
- 14.8 Digital-to-Analogue Converters 544
- 14.9 Static Hazards in Fluidic Circuits 555

15. Graphical Techniques Applied to Elements and Circuits
- 15.1 General 572
- 15.2 Digital Element Input Characteristics 572
- 15.3 Digital Element Transfer Characteristics . . . 576
- 15.4 Digital Element Output Characteristics . . . 581
- 15.5 Combined Digital Characteristics 583
- 15.6 Two Circuit Analysis Examples 584
- 15.7 Interstage Static Matching of Analogue Beam-deflexion Amplifiers 589

Subject Index 593

1
Introduction to Pure Fluid Devices and Sensors

1.1 General

What is Fluidics? Fluidics is a new technology arising from a re-appraisal of a very old technology, namely fluid power and its control. The basic distinction to be drawn between conventional fluid flow engineering and fluidics is that in the former case we are normally concerned only with harnessing fluid power in an economic manner, whereas with fluidics we are thinking in terms of processing information through a fluid medium as well as transmitting power. For this reason the attitude towards the development of fluidic devices and circuits clearly owes as much to the well-established technology of electronics as it does to fluid mechanics. Of course we have been able to control fluids in very sophisticated ways for a long time; but generally the control systems have tended to be limited in performance by the mechanical nature of the control devices, such as spool valves, diaphragms, springs, etc., used in the system. The philosophy of fluidics, on the other hand, has been to examine new devices which use extremely simple mechanical shapes or alternatively rely on fluid mechanic effects within the fluid itself, so that moving parts are eliminated. Such elements are considerably faster than conventional mechanical, electro-mechanical, pneumatic and hydraulic components, and potentially have high reliability. It is these factors which make it technically and economically attractive to process information in a fluid medium for certain applications.

Furthermore, fluidic circuits can operate under severe environmental conditions, when made from suitable materials, without loss of reliability. This increases the possible area of fluidic applications considerably, particularly as electronic components and systems are at a disadvantage in these conditions. Generally fluidic control devices are never expected to compete with electronics where speed of operation is important, such as in general purpose computing; nevertheless, there are many applications where this speed is not required but rather a premium is set on cheapness, reliability and an ability to function in adverse environments. In these circumstances fluidic devices show up in a more favourable light and may be regarded as complementing the existing range of control system components.

Generally, any fluid may be used in fluidic control elements provided the designs conform to the fluid mechanic principles involved. However, as far as industrial fluidics is concerned, the usual working fluid is air because of its relative cheapness, availability and ease of use. Special applications using other common fluids such as hydraulic oil and water are mainly still in the experimental stage.

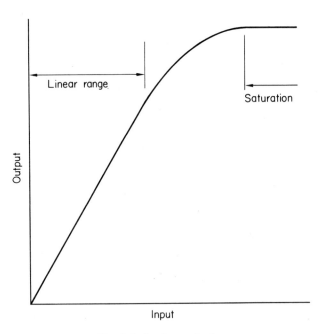

Fig. 1.1 Analogue device

1.2 Classification of Element Types

Broadly speaking, elements may be divided into two categories, depending on whether they are to be used for analogue or digital circuit applications. In analogue circuits continuous change or modulation of the

[1.2] CLASSIFICATION OF ELEMENT TYPES

signals occur in the system, with each element in the circuit bearing a direct relationship between the input and output signals. Fig. 1.1 shows a typical characteristic for analogue elements which, in practical designs, always exhibit a limited linear or proportional range and a corresponding non-linear range associated with the overloaded or saturated condition.

A digital or switching device operates on the basis of discrete signal changes; that is, only one output signal change is produced when the input signal magnitude is raised above a minimum level or switching point, as shown in Fig. 1.2. It is seen that such elements respond only to two signal

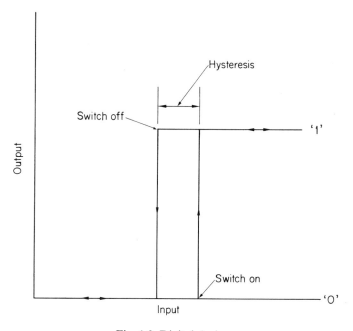

Fig. 1.2 Digital device

levels (usually designated '1' and '0') and, unlike analogue elements, are ideally immune from intermediate signal levels. The resulting switching action may, in practice, also produce a switching deadband or hysteresis loop signifying a different switching 'on' point to the switching 'off' point due to internal energy losses within the device.

Further classifications for elements may be made on the basis of whether a separate energy supply is required. If no energy is added, the device is termed passive as distinct from an active device which does require a separate supply. A usual requirement for an active device is that the output signal should be amplified compared with the input or control signal.

Although the intention is that all fluidic elements should be inherently highly reliable and fast operating, this does not exclude the use of moving mechanical components of simple geometry within the device. This class of device is almost exclusively digital and is described fully in Chapter 8.

Before describing the various digital or switching devices it is of value to define some of the characteristics they share in common. These are:

(a) *Switching gain*. Is broadly defined as the ratio of the change in output signal at switching to the change of input signal causing it. The difficulty arises in practice in determining the smallest input signal change which will cause switching as, clearly, an infinite number of input signal changes larger than this value is possible. In most fluidic devices we are concerned with pressure gain although a corresponding gain based on flow is equally valid. Roughly speaking a device designed to have high pressure gain has low flow gain and vice versa.

We may write gains as

$$\text{pressure switching gain} = \frac{\text{change in output pressure from '0' to '1' (or vice versa)}}{\text{smallest control pressure change to cause switching}}$$

$$\text{flow switching gain} = \frac{\text{change in output flow from '0' to '1' (or vice versa)}}{\text{smallest control flow change to cause switching}}$$

As the switching gain in most fluidic devices varies with load, supply pressure, etc., it is essential to specify the operating conditions.

(b) *Fan-in*. This quantity is the maximum number of control input connexions which may be made to the device without causing it to malfunction.

(c) *Fan-out*. When the output of a digital device is connected to the input of another device there is a reduction in the maximum output signal amplitude. The fan-out defines the maximum number of identical elements which may be connected in parallel to the output of a similar element and still be switched when the output of the first element changes.

Many fluid mechanic phenomena are exploited in pure fluid elements, such as jet wall re-attachment, momentum interaction of two or more jets, the transition of a jet from laminar to turbulent flow and so on. Practical devices may be dominated by one of these effects or may rely on a combination of effects. In this chapter different elements are classified according to the most significant effect within the device.

1.3 Wall Re-attachment Digital Devices

1.3.1 Bistable Amplifier

The mechanism of wall re-attachment by a fluid jet is discussed in detail in Chapter 3 so that only a brief description is included in this chapter.

Wall re-attachment of a fluid jet occurs when a boundary wall is placed in close proximity to it, causing the jet to bend and adhere or latch on to the wall. Normally, a jet of fluid issuing from a rectangular or circular nozzle into a stationary fluid spreads outwards as it moves undeflected in a downstream direction. This is due to entrainment of the stationary fluid

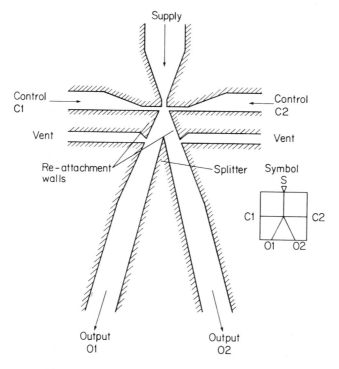

Fig. 1.3 Bistable wall re-attachment amplifier

into the moving jet in a momentum exchange process. However, if the entrainment on one side of the jet is reduced due to an adjacent boundary, then the local pressure on that side of the jet is also reduced, causing the jet to be deflected towards the boundary. With suitable positioning of the boundary wall the jet is deflected so far that it re-attaches to the wall enclosing a low pressure region, known as the re-attachment bubble, and is little affected by disturbances to the flow downstream. Many pure fluid digital devices rely on this effect for their operation and achieve control of the switching action by introducing a small flow into the re-attachment bubble. A good example of this type of element is the well known bistable wall re-attachment amplifier.

Fig. 1.3 shows the geometry of the device with two boundary walls positioned symmetrically downstream of a supply nozzle. Normally such devices are two-dimensional; that is the channels are rectangular in section, being formed by sandwiching the profile between top and bottom cover plates. Consider the supply jet re-attachment to the left boundary wall. Under these conditions essentially all the flow passes down the left

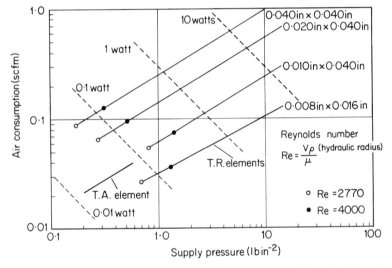

Fig. 1.4 Nozzle flow characteristics for turbulent amplifier (T.A.) and turbulent re-attachment amplifiers (T.R.) (Parker, *7th Int. Machine Tool Design Res. Conf.*, 587, 1966

output channel, marked O1, into the next element. Because control channel C1 is connected to the re-attachment bubble, a small control flow through it is sufficient to raise the pressure inside the bubble to the point where the jet unlatches from the wall and deflects over to the other wall. The supply flow now passes along output channel O2 instead of O1 even when the control flow C1 is removed. The only way of returning the jet to its original position is to apply a similar control flow to the other control channel C2. Thus it is seen that the device acts like a switch with two stable states, giving it the important property of memory as it 'remembers' the

[1.3] WALL RE-ATTACHMENT DIGITAL DEVICES 7

state of a previously applied signal in C1 or C2. It is also an amplifier of the control signal, as the jet switching action is very sensitive. Finally, vent channels in the amplifier allow the output channel to pass the required flow for the next element, any excess being spilled away through the vent channel rather than through the unused output channel.

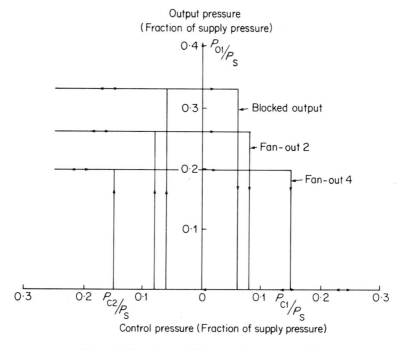

Fig. 1.5 Bistable amplifier transfer characteristic

The characteristics of digital fluidic devices may be completely represented by three standard curves; supply pressure-flow curve, the transfer characteristic between control and output signals, and the output pressure-flow curve with external load lines superimposed on it. The first of these, the supply characteristic, is necessary to determine the device fluid consumption at the design supply pressure for the system. A comparison of the air consumption of various turbulent re-attachment elements with a turbulence amplifier (see Section 1.6.1) is shown on a logarithmic scaled graph in Fig. 1.4. The rectangular nozzle dimensions for the turbulent re-attachment elements (T.R.) are quoted as width × depth in inch units. The turbulence amplifier (T.A.) is a standard tubular type. Also shown on the curve are lines of constant power which indicate that considerable care must be exercised in choosing the supply pressure for a complex system of T.R. elements if excessive air consumption is not to result.

The transfer characteristic for wall re-attachment devices is usually considered between the control and output pressures over the whole switching cycle as the control pressure is varied between maximum and minimum values.

Fig. 1.5 shows a typical non-dimensional transfer characteristic for a bistable amplifier which provides information on the variation of switching points and hysteresis with load. An essential feature of bistability is that the hysteresis loop extends across the line of zero control pressure so that either control signal may be set to zero without destroying the memory property.

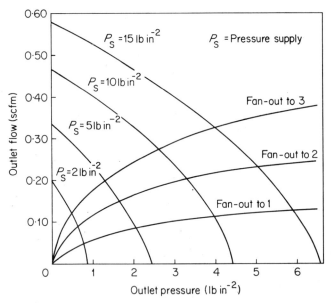

Fig. 1.6 Output characteristic with load lines (*Aviation Electric sales brochure*)

The output characteristic of the wall re-attachment amplifier, and many other fluidic devices, is dependent on the shape of the diffuser geometry in the receiver channels. A typical characteristic is shown in Fig. 1.6 although other shapes are possible. It provides information concerning the maximum pressure and flow recoveries possible at different constant supply pressures. Superimposed on the same graph are load lines, representing the effect of increasing the fan-out from one to three, which provide the operating pressure and flow value where they intersect the appropriate supply pressure curve. The maximum pressure recovered in the output channel usually varies between 35 and 50 per cent of the supply pressure and the corresponding maximum flow recovered is typically 90 to 110 per cent of the supply flow, the increased flow being due to atmospheric entrainment.

The switching speed of a wall re-attachment element is dependent largely on the geometric size of the element profile, the fluid used and the jet velocity. Operating with air and using supply nozzles in the range 0·010 in. to 0·020 in. wide and 0.040 in. deep the switching time is typically 0·5 msec for this type of device. Denser fluids and larger devices produce slower switching.

1.3.2 OR/NOR Logic Element

Another commonly used wall re-attachment element is the OR/NOR device, which is illustrated for a two input configuration in Fig. 1.7.

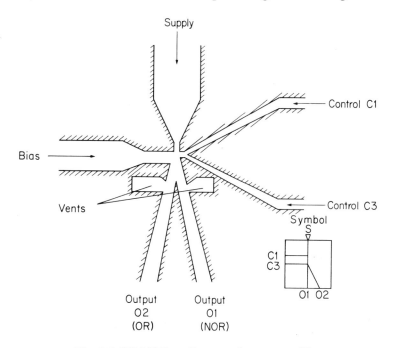

Fig. 1.7 OR/NOR wall re-attachment amplifier

The basic bistable element configuration is modified so that the left-hand control channel, known as the bias port, is enlarged and usually opened to the atmosphere when the device is operated using air. This makes the jet stable only on the right-hand wall, thus converting the device from a bistable to a monostable mode of operation. Between two and four input channels may be connected into the re-attachment bubble so that when any control signal is present it is sufficient to switch the jet causing a flow to appear in the output OR channel. Conversely, when all the control signals are absent a flow is present in the NOR channel. The logic functions are explained in greater detail in Chapter 6.

The characteristics of an OR/NOR element are very similar to those of the bistable amplifier except that a monostable device has a modified transfer characteristic. Fig. 1.8 shows the usual form of this characteristic.

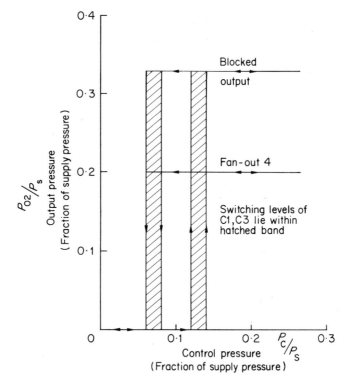

Fig. 1.8 OR/NOR element transfer characteristic

Inherently a monostable action requires the hysteresis loop not to cut the zero control pressure axis and also, for good design, the loop itself should be small so that good signal discrimination is achieved between the 'on' and 'off' states. Again the fan-out load will affect the output pressure recovery and may also modify the switching points and hysteresis band[1].

[1.3] WALL RE-ATTACHMENT DIGITAL DEVICES

Summarizing the common characteristics of both bistable and OR/NOR switching devices operating using air:

Supply pressure range	0·5 to 15 lb in^{-2}
Supply nozzle size	0·010 in. to 0·020 in. wide by 0·040 in. deep
Supply flow	0·05 to 0·25 ft^3 min^{-1}
Maximum output	35% to 50% of supply pressure
Maximum output flow	90% to 110% of supply flow
Maximum number of similar elements that can be switched (fan-out)	2 to 4
Maximum number of input connexions (fan-in)	2 to 4 (can be extended greatly by using foil isolators)
Pressure gain at switching	5 to 15
Switching speed	0·5 msec

The main advantages of the wall attachment elements may be summarized as:

(a) Fast switching and good signal discrimination in switching circuits.
(b) Capable of integration and miniaturization.

The corresponding disadvantages are:

(a) Poor isolation between input channels.
(b) Switching characteristics very sensitive to small changes in geometry in the re-attachment region.
(c) Poor fan-out.

Some of the difficulties of input isolation of the OR/NOR device may be overcome by using free foil diode isolators in each input channel as discussed by Foster and Retallick[2]. Foil devices are described in detail in

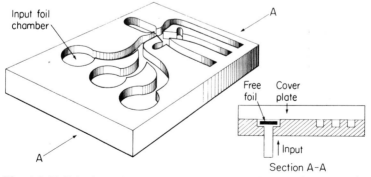

Fig. 1.9 Foil isolated inputs for an OR/NOR element (Foster and Retallick, *Ref. 2*, Fig. 3.3)

Chapter 8. Fig. 1.9 shows the arrangement of foil chambers and wall attachment element for a three input OR/NOR. If any control signal is present it lifts its foil allowing a free flow of fluid to the wall attachment element but also isolates the other inputs. This is due to the remaining foils being held shut by the back pressure. In a planar construction this arrangement has the further advantage that only one control channel need be provided into the re-attachment zone/thereby considerably reducing design difficulties. Designs of OR/NOR elements with a fan-in of greater than 20 have been successfully tested which operated satisfactorily at pulse frequencies greater than 200 Hz. Generally the foil dimensions are sufficiently small to give a response compatible with pure fluid devices.

1.3.3 AND–NAND Logic Element

An active AND logic device using a single wall re-attachment element has not yet been successfully developed. In general it is possible to logically invert signal senses so that, for instance, OR/NOR can be converted to AND/NAND logic. Consider a simple example of two input signals A and B entering an OR/NOR element; then the outputs may be written (see Chapter 6) as

$$\text{OR output} = A + B$$
$$\text{NOR output} = \overline{A + B}$$

If the input signals are now inverted, the signal in the output channel which was formerly the OR would be $\overline{A} + \overline{B}$ which is equivalent to $\overline{A.B}$ forming the NAND function. Similarly the former NOR channel would have an output of $\overline{\overline{A} + \overline{B}}$ which is equivalent to the AND function A.B.

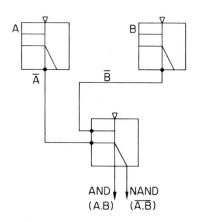

Fig. 1.10 Active AND/NAND element

[1.3] WALL RE-ATTACHMENT DIGITAL DEVICES

Thus the use of OR/NOR and AND/NAND logic are interchangeable with the appropriate change in signal convention. Normally, a positive fluid pressure or flow level is taken as the logical '1' state which is the convention used throughout this book.

If more than one active wall attachment element is acceptable in terms of increased power consumption and increased switching time then an active AND–NAND may be constructed by using monostable amplifiers to invert the input signals.] Fig. 1.10 shows the circuit arrangement required to form a two input AND–NAND function using three OR/NOR devices.

1.4 Momentum Interaction Digital Devices

1.4.1 AND Logic Element

The second fluid mechanic effect of importance in fluidic devices is the momentum interaction of jets. When a jet impinges on, say, a flat plate positioned perpendicularly to the flow, the force exerted on the plate by the jet may be expressed as the time rate change of jet momentum (see Chapter 3). If instead of a plate another jet interacts with the original jet, the forces exerted on the resulting jet now depend on both jet momenta.

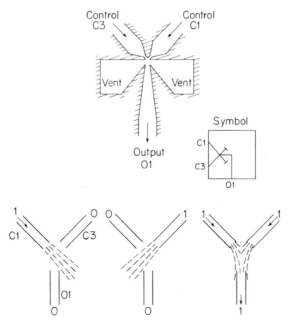

Fig. 1.11 Passive AND with signal combinations

Usually a simple triangle of forces is sufficient to determine the magnitude and direction of the mixed jet.

The simplest arrangement possible is a planar two input passive AND element consisting of two input nozzles set mutually at right angles with an output channel positioned symmetrically to receive the resulting jet. Fig. 1.11 shows schematically the basic operation of the element.

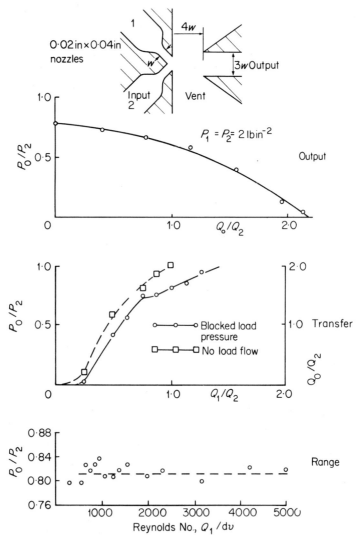

Fig. 1.12 Momentum interaction AND element characteristics (Parker and Jones, *Ref. 3*, Fig. 9)

[1.4] MOMENTUM INTERACTION DIGITAL DEVICES

When input signals C1 and C2 are present they impinge on each other giving a resultant jet which is directed into a receiver channel O1 provided that the jet momenta of the two input jets are approximately equal. If either signal C1 or C2 is absent, the other signal jet is undeflected and enters a large vent region, without touching the output channel. In the absence of both signals there is no possibility of output signals since the device has no external energy supplied to it. In common with all passive elements no amplification of the input signals can occur so that the objective of a good design is to minimize losses.

Fig. 1.12 gives the dimensions and characteristics of a typical AND element design used by Parker and Jones[3]. Of particular interest is the wide range of operation for the input jet flows signifying a pressure recovery better than 80 per cent of the input signal pressure when operating in both laminar and turbulent flow regimes. This makes it a very adaptable element, particularly as it also has good dynamic characteristics.

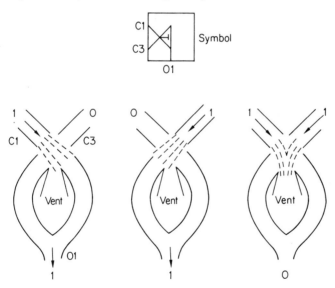

Fig. 1.13 Signal combinations for a passive EXCLUSIVE–OR

1.4.2 EXCLUSIVE–OR Logic Element

The EXCLUSIVE–OR function differs from an OR function in that it provides an output signal only when one input signal is present (see Chapter 6). For a logic element of this type having two inputs A and B the output would be $A\bar{B} + \bar{A}B$. Again the simple two perpendicular nozzle arrangement may be used but in this case there are two receiver channels which must be joined together. Fig. 1.13 shows the operation of the device for the different input signal combinations.

When input C1 alone is present a jet passes across the interaction region into a receiver aligned with the nozzle. Flow then passes out of the device through a Y junction which ideally does not allow reverse flow down the opposite channel. A similar action results if input C3 is present by itself. When both signals are present the jets impinge and are vented into a central area between the output channels.

In order to maintain uniform pressure conditions in the interaction region practical designs often include side vents so that momentum interaction is not modified by a non-uniform pressure field due to adjacent boundaries. Most designs exhibit poor static pressure and flow recovery[3] often caused by reverse flow down the inactive output channel into the jet interaction region. Good discrimination between signal combinations is also difficult to achieve particularly when the input signals are pulsed. There is no published information on a pure momentum interaction device of this type which satisfactorily overcomes all of these problems (see Chapter 7).

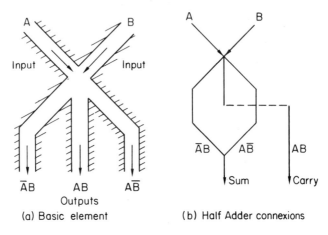

Fig. 1.14 Half adder after Greenwood (*Ref. 4*)

1.4.3 Half Adder

A Half Adder logic element is required in binary addition and subtraction circuits (see Chapter 14). It accepts two input signals and provides two output signals which are the AND and EXCLUSIVE–OR functions, normally denoted as the Carry and Sum signals respectively. It is possible to generate a Half Adder function using a combination of fluidic elements

[1.4] MOMENTUM INTERACTION DIGITAL DEVICES

but two single passive element designs have been suggested by Greenwood[4,5] and Gray and Stern[6]. The former design is a combination of momentum interaction AND and EXCLUSIVE–OR elements described above while the latter relies on both wall attachment and momentum interaction effects and is discussed later in the chapter.

The Greenwood Half Adder is shown in basic form in Fig. 1.14. The difficulties of optimizing the geometry in this device are obviously more severe than for the EXCLUSIVE–OR alone as the central output channel may no longer be regarded as a vent. It is probable that a design relying purely on momentum interaction would be very difficult to achieve, particularly when the outputs are loaded and the device is operating under dynamic conditions. An unvented design might be more practical but no published information is available.

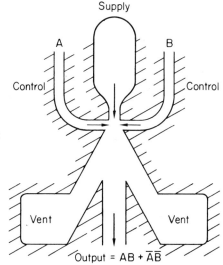

Fig. 1.15 EQUIVALENCE element (Parker and Jones, *Ref. 3*, Fig. 1B)

1.5 Combined Wall Re-attachment and Momentum Effects

1.5.1 EQUIVALENCE Logic Element

Logically this is a device which gives an output signal when two input signals are identical, either both '0' or both '1'. An active fluidic element which gives this function may be constructed by combining momentum interaction and wall attachment effects as shown in Fig. 1.15.

The geometry shows the boundary walls to be inclined back more steeply than a wall re-attachment device so that, by itself, the supply jet is not deflected and passes out through a symetrically positioned output channel. However, if an input signal A is introduced a small deflexion of the supply jet results due to momentum interaction, thereby venting the flow to the right rather than being collected in the output receiver. A similar deflexion to the left vent occurs if input B alone is present. When both A and B are present, their momenta oppose each other and the supply jet remains undeflected giving an identical output signal to the case when both A and B are absent. Thus the output channel gives the EQUIVALENCE function, i.e. provides an output signal whenever A and B are equal or equivalent (both '0's or '1's). It is to be noted that the geometrical arrangement may be such as to induce temporary wall attachment when the flow is being vented in order to ensure that all the flow enters the vent cleanly. Removal of the control signal automatically stops wall attachment thus returning the supply jet to the central position.

Mitchell[7] has briefly described the element and it has been used by Glaettli[8] in a binary adder circuit and also by Barker[9] in a comparator circuit. Difficulties have been reported with the receiver noise level and also marked differences in the output signal when both signals are present and when they are both absent. However, Glaettli has described its use in complex circuits so that it may be inferred that satisfactory performance is possible. No detailed characteristics have been published.

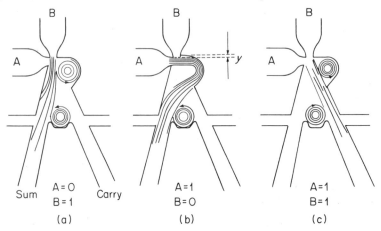

Fig. 1.16 Cusp Half Adder (Parker and Jones, *Ref. 10*, Fig. 25)

1.5.2 Half Adder Logic Element

A suitably shaped passive element to generate the Half Adder function using momentum interaction and wall re-attachment effects is shown in Fig. 1.16. For the case shown in (*a*), a sum output is generated by the wall

[1.5] COMBINED WALL RE-ATTACHMENT AND MOMENTUM EFFECTS 19

re-attachment of a turbulent jet produced by an input at B. The more complex and difficult sum output to generate is shown in (b) where a turbulent jet issuing from A is deflected through a large angle by a cusp into the receiver with no wall attachment effect. The carry signal is produced more easily by the momentum interaction of the turbulent jets issuing from A and B illustrated in (c).

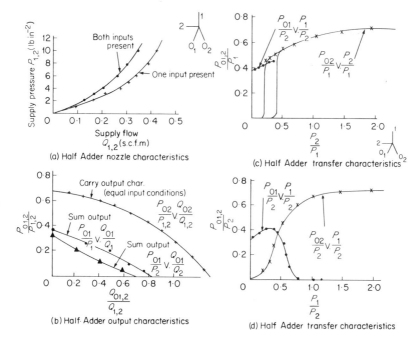

Fig. 1.17 Static characteristics of a cusp Half Adder (Parker and Jones, *Ref. 10*, Figs. 2a, b, c, d)

For stable operation the device relies upon the formation of vortices generated by the two cusps, one located adjacent to nozzle B and the other on the flow splitter. The vortex generated by the splitter cusp, when the sum output is formed, provides a stabilizing effect particularly when the output is produced by an input at A. This stabilizing vortex virtually eliminates leakage flow down the carry output under these conditions and reduces the load sensitivity of the device. When a carry signal is generated by the introduction of the two jets, the direction of this vortex is reversed. In this instance the vortex generated by the cusp close to nozzle B combines to produce a stabilizing effect on both sides of the carry output jet.

Parker and Jones[10] have successfully used an element of this type in

Adder and Subtractor circuits. They used an element with the following critical dimensions:

> For the nozzles $W_A = W_B$
> Setback of re-attachment wall $= 0 \cdot 25 W_A$
> Wall angle $= 12°$
> Cusp setback, $y = 0 \cdot 25 W_A$
> Splitter distance $= 14 W_A$

Signal nozzle dimensions were 0·020 in width × 0·040 in depth. The static characteristics obtained are shown in Fig. 1.17 and indicate that large differences in characteristics are to be expected between the sum and carry output signals. Both sum output signal combinations give a poor pressure and flow recovery with the output resulting from the input signal 2 being predictably the worst due to the complicated signal path around the cusp. Although it is desirable to have matched output characteristics they are not critical in the normal mode of operation in adder circuits (see Chapter 14).

Parker and Jones found that the cusp setback (marked 'y' in Fig. 1.16) was critical in the design described above for obtaining good dynamic characteristics, the maximum frequency at which the Half Adder could

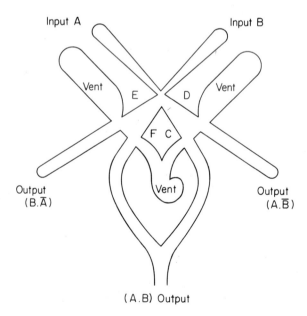

Fig. 1.18 Cross-over AND element (Hellbaum, *Ref. 11*, Fig. 3)

be operated without exhibiting poor signal definition being approximately 500 Hz.

1.5.3 AND Logic Element

A further passive AND device may be constructed by overlapping two identical wall re-attachment element profiles. Hellbaum[11] has described such an element which is shown schematically in Fig. 1.18. The wall re-attachment geometries provide monostable action so that when inputs A or B alone are present the jets re-attach to walls D and E respectively and are vented. Momentum interaction occurs when both signals are present so that the combined jets either attach to wall C or F. The AND function is then generated by joining the two output channels adjacent to C and F in a Y connexion. Two attractive features of this design would appear to be, firstly, that the device can still operate when either input signal is considerably weaker than the other and, secondly, provision is made for venting so that load sensitivity should be reduced. It should also be possible to join the two vent channels together to form the EXCLUSIVE–OR and hence use the element for Half Adder applications. However, no information is available on this mode of operation.

1.6 Transition from Laminar to Turbulent Flow

1.6.1 Turbulence Amplifier

Fluid flowing through a pipe or channel is subjected to viscous shear stresses which are normally proportional to the velocity gradient across the channel section. At some distance downstream from the pipe entrance, the flow profile becomes fully developed and is characterized by smooth continuous flow patterns. Such conditions may be analysed and are called laminar flow. Normally laminar flow is associated with low fluid velocities, which, when increased, pass through a transition into a turbulent flow regime. The flow is no longer smooth; the average level of shear stress is greatly increased, and localized shear stresses may vary considerably.

Fortunately, the dimensionless Reynolds Number, Re, gives a good indication of the transition from laminar to turbulent flow. For narrow, deep channels the characteristic length is taken to be the channel width, while for tubes the diameter is used for the same purpose. As a rough rule, the transition to turbulent flow in pipes occurs at Re $\geqslant$ 2000 (see Chapter 2).

A laminar jet (that is one operating at low Reynolds Numbers) issuing from a nozzle does not normally spread as rapidly as a turbulent jet. If a receiver is placed axially in line with the supply jet, as shown in Fig. 1.19, the pressure recovery is good provided the jet remains laminar up to the

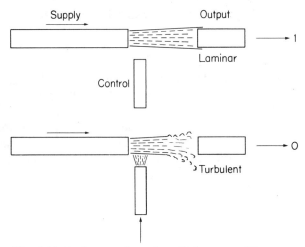

Fig. 1.19 Basic operation of a turbulence amplifier

receiver. On the other hand, the receiver pressure is low when the jet becomes turbulent due to the increased spread of the jet in this flow regime. The pressure at the supply nozzle for a practical device is adjusted so that the jet becomes turbulent just beyond the receiver giving extremely sensitive conditions to small disturbances to the flow.

A control flow channel positioned close to the supply nozzle introduces sufficient flow into the jet to cause a transition from laminar to turbulent conditions before it reaches the receiver, thus drastically lowering the receiver pressure. Auger[12] has described a practical three-dimensional device, Fig. 1.20, with multiple control nozzles, any of which may cause the output to switch off when a control signal is present. It provides the NOR logic function and has the advantage of better isolation between input and

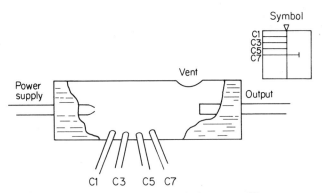

Fig. 1.20 Axi-symmetric turbulence amplifier

[1.6] TRANSITION FROM LAMINAR TO TURBULENT FLOW

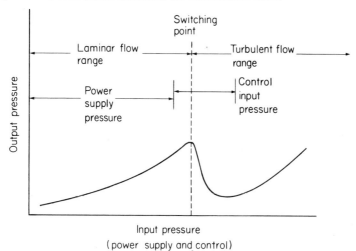

Fig. 1.21 Turbulence amplifier supply-output characteristic

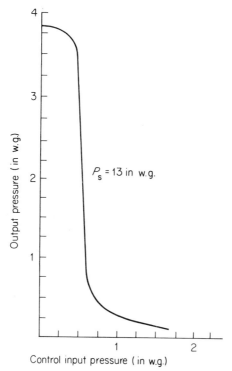

Fig. 1.22 Turbulence amplifier control-output characteristic

output signals compared with the wall re-attachment device. However, this tends to be offset by a rather slower operating speed.

Typical supply pressure–output pressure characteristics for a turbulence amplifier are shown in Fig. 1.21 indicating increasing pressure recovery in the laminar region as the supply pressure and flow are increased. The transition to turbulence is characterized by an irregular and rapid decrease in output pressure recovery. Fig. 1.22, showing the output pressure against input pressure characteristic for no output flow conditions, indicates the very large switching pressure amplification which may be achieved. However, an essential feature of turbulence amplifiers is their small supply flow velocities compared with many wall re-attachment devices. This results in a very low power supply consumption for a turbulence amplifier compared with many other pure fluid active devices which may be an important advantage in some complex circuits (see Fig. 1.4).

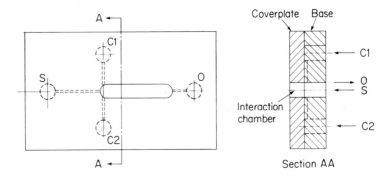

Fig. 1.23 Planar turbulence amplifier (Verhelst, *Ref. 13*, Fig. 11)

Auger has concluded that a nozzle diameter of 0·030 in gives a suitable compromise between good power gain and miniaturization. Under these conditions, a nozzle–output distance of 0·7 to 1 in is appropriate for logic operation. By increasing this distance to greater than 1·3 in, the jet becomes sound sensitive and breaks into turbulence in the presence of high frequency sound waves. He also suggests that careful control of upstream supply conditions is necessary for the long straight inlet pipes.

Recently Verhelst[13] has described a planar turbulence amplifier which is more amenable to mass production manufacturing techniques. Fig. 1.23 shows the geometry of the device which has two control input channels C1 and C2 positioned normal to the supply S and the output O. The design although having a limited fan-in compared to the Auger design could probably be extended to at least 4. Fig. 1.24 shows the static characteristics of the planar arrangement which may be very conveniently presented in graphical form due to the good isolation between input and output conditions.

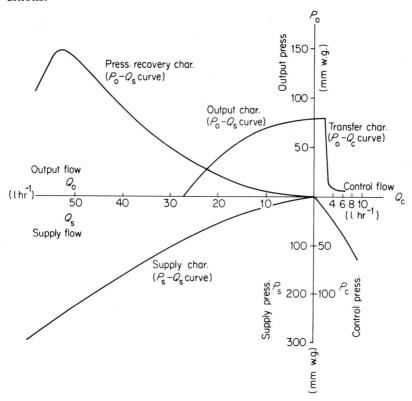

Fig. 1.24 Planar turbulence amplifier static characteristics
(Verhelst, Ref. 13, Fig. 5)

1.7 Miscellaneous Digital Devices

1.7.1 Induction AND Logic Element

Swartz[14] has given information on a passive AND device which relies on the formation of low pressure regions in the wall re-attachment process to provide interaction between the two input signal flows. Fig. 1.25 illustrates schematically the arrangement of the device which is basically a combination of two wall re-attachment amplifiers with a common output channel and a common control channel. Both outer walls are positioned to cause strong wall re-attachment of the input jets when only one of them is present.

Consider the action of one input signal alone. As the stream leaves the input nozzle the pressures in the common control channel and the entrainment passage are identical so that the jet will latch onto the outer wall and flow out of a vent. In this condition the common control channel pressure is reduced below atmospheric as there is entrainment from the atmosphere along a small long channel. If a second input is now introduced, the second

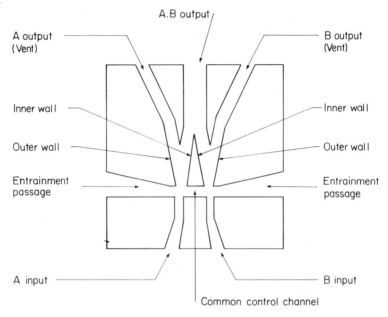

Fig. 1.25 Passive induction AND element (Swartz, *Ref. 14*, Fig. 1)

stream immediately attaches to the inner wall due to the low pressure existing in the common control channel and passes out through the AND output. The common control channel is now restricted at both ends and rapidly reduces its pressure still further thereby causing the first jet to also attach to the inner wall and completing the capture of all the flow into the AND output. Removal of one of the inputs allows the wall re-attachment effect to dominate the other jet again causing it to deflect into the venting position.

It is claimed that this arrangement provides a more efficient design of AND element as there are no sharp turns of the stream after they interact, although no published information is available to substantiate this claim. The author reports that there are several critical dimensions in the profile, the most important being the setback of the inner and outer walls.

[1.7] MISCELLANEOUS DIGITAL DEVICES

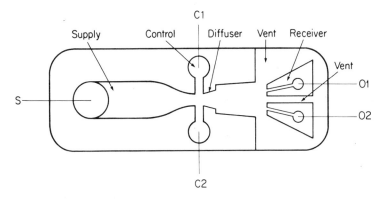

Fig. 1.26 Diffuser switch profile and connexions (König, Ref. 15, 56, Fig. 1)

1.7.2 Diffuser Switch Logic Elements

A special use of wall attachment effects may be made by positioning the jet boundary walls without setback thereby forming the supply nozzle into a diffuser. König[15] has described mono and bistable amplifiers of this type operating at low supply pressures using a planar configuration, which is shown in basic form in Fig. 1.26. Two control channels C1 and C2 are situated at the throat of the nozzle in a low pressure region of the diffuser which control the separation of the jet from one or other of the walls of the diffuser. Using a diffuser at an angle of 15° the device exhibits bistability over a wide range of supply pressures (0·25–10 lb in^{-2}). In the diagram S is the supply port and O1 and O2 are the receivers, which have

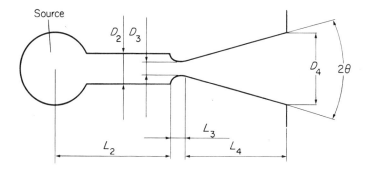

Fig. 1.27 Dimensions of a diffuser switch (König, Ref. 15, Fig. 8)

narrow channels to achieve good pressure recovery. Typical diffuser dimensions are shown in Fig. 1.27 which have the following values:

> Contraction ratio $D_2^2/D_3^2 \simeq 6$
> Expansion ratio $D_4^2/D_3^2 \simeq 7$
> Supply channel length $L_2/D_2 \simeq 3$
> Nozzle length $L_3/D_3 \simeq 1$

With an operating supply pressure of 0·5 lb in^{-2} a control switching pressure of about 0·015 lb in^{-2} is required to give an output signal change of 0·15 lb in^{-2}. A fan-out to between 4 and 6 similar elements is possible.

The diffuser switch has found particular use in conjunction with the Dreloba diaphragm logic elements (see Chapter 8) for applications in which pressures below the range 3–15 lb in^{-2} are used.

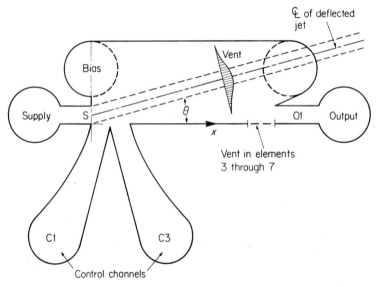

Fig. 1.28 Laminar NOR logic element (Walker and Trask, *Ref. 16*, Fig. 1)

1.7.3 Laminar Flow Logic Elements

Recent investigations into new designs of digital elements have shown a trend towards reducing power consumption. One approach is to use laminar instead of turbulent jets although this means that jet wall re-attachment is largely impracticable (see Chapter 3). A solution to this problem has been suggested by Walker and Trask[16] using a two input NOR element as shown in Fig. 1.28. A laminar jet issuing from supply nozzle S is guided by the straight wall to pass without deflexion into the

[1.7] MISCELLANEOUS DIGITAL DEVICES

output channel O1 in the absence of a signal in either control channel marked C1 and C3. The supply jet is deflected through an angle θ into a vent region when either or both control signals are present due to the increased static pressure in the control channel. Momentum effects are assumed to be small due to the low velocities involved. The authors used an element with a supply nozzle width of 0·51 mm and aspect ratio of approximately unity and achieved fan-out ratios of about 3 at supply pressures less than 10 cm w.g. This corresponded to power consumption levels of approximately 2 per cent of present wall attachment devices.

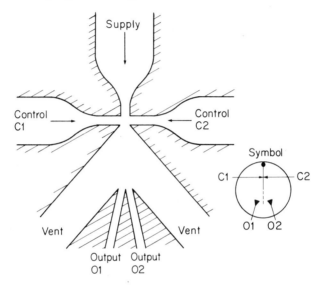

Fig. 1.29 Proportional beam deflexion amplifier

1.8 Momentum Interaction Analogue Devices

1.8.1 Beam Deflexion Amplifier

Beam deflexion amplifiers use a superficially similar arrangement to wall attachment devices in that a supply nozzle, two output and two control channels are used in a symmetrical arrangement, Fig. 1.29. The geometries differ radically just downstream of the supply jet because, in order to avoid wall re-attachment effects, the beam deflexion amplifier boundaries are removed a considerable distance in a transverse direction. This allows the supply jet to flow undeflected across the interaction chamber and divide equally across the output splitter. In the absence of control flow from C1 or C2, equal flows issue from each output port O1 and O2. The output channels may be separated by a splitter or alternatively a further vent may be used at this point to improve the output characteristics. The position and shape of these channels is a compromise between being far enough downstream to accept reasonable deflexions of the supply jet and also being far enough upstream to recover a significant proportion of the supply jet pressure. Any flow not required by the output channels passes out through the vents.

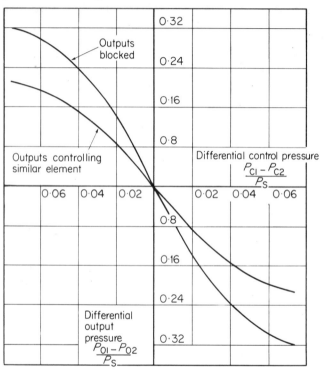

Fig. 1.30 Beam deflexion amplifier pressure transfer characteristics (*Aviation Electric sales brochure*)

The control signals are directed into the interaction region from opposing nozzles perpendicular to the supply jet. The momentum flux of the supply jet and the forces acting on it form the control flows which determine the resultant direction of the main jet as it leaves the interaction region. That is to say, as flow is introduced into either control channel, it interacts with the supply jet and deflects it to the opposite output channel. The action is proportional and amplification is obtained. As two output channels are available they may be used singly or alternatively together as a positive and negative valued differential signal as shown in Fig. 1.30.

The slope of this characteristic is known as the gain and is a measure of the amplification achieved by the device. The pressure gain (that is the gain of the control–output pressure characteristic under specified load conditions) for the beam deflexion configuration shown in Fig. 1.29 is usually in the region of 6–8 in the unloaded condition. It has a tendency to unstable operation with loads approaching the blocked load condition, that is with small load flows. Two main approaches have been used to overcome this limitation: firstly, a central vent may be inserted between the two output channels which achieves blocked load stability at the expense of a reduced pressure gain and, secondly, vortex vents placed in the output channels which have similar effects. Both profiles are shown in Fig. 1.31.

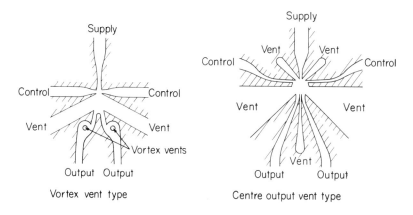

Fig. 1.31 Matching vent configurations for beam deflexion amplifiers

The dynamic characteristics of beam deflexion amplifiers, as with other analogue amplifiers, vary considerably with the loading and geometric size of the device. Upper operating limits for amplifiers with 0·010 in wide supply nozzles are reported to be about 3 kHz, although 1 kHz would be the maximum operating frequency for many applications. Because of the high frequency response and versatility of the beam deflexion amplifier it is suited to a wide range of low pressure pneumatic control systems.

1.8.2 Impact Modulators

This device employs two axially opposed circular nozzles, through which supply fluid flows to cause direct impingement of the two jets. If the momenta of the two supply jets are equal, the impact or balance point occurs midway between the nozzles and is very sensitive to small changes in axial momentum. A wide range of operating conditions are possible using both incompressible and compressible jet flow. The two practical devices operating on this principle are known as the Transverse Impact Modulator (TIM) and the Direct Impact Modulator (DIM). Fig. 1.32 illustrates the Transverse type which has an annular chamber surrounding one of the supply tubes, known as the collector, which senses the output flow and pressure. The other supply tube is called the emitter and reacts with control flow from a nozzle situated perpendicular to the supply flow axis. Normally with no control flow present, the momentum of the emitter jet is arranged to be slightly greater than that of the collector jet

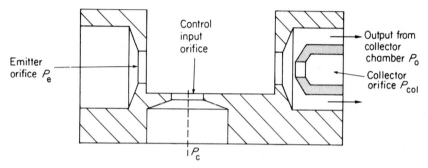

Fig. 1.32 Transverse Impact Modulator (TIM) (*N.A.S.A.*, CR–245, 4–22, Fig. 4.3.2–2)

so that the balance point occurs just outside the collector orifice. This produces the maximum output pressure or flow signal. When control flow occurs, the impingement of the control and emitter jets is similar to the action occurring in a beam deflexion amplifier with momentum, pressure and centrifugal forces modifying the jet deflexion. The flow field then becomes asymmetric, allowing an increased flow and momentum condition to be established in the collector tube. The slight reduction in the axial momentum of the emitter nozzle and the corresponding rise in the collector nozzle results in a shift of balance point away from the collector thereby reducing the output signal level. The process is very sensitive to changes in control flow and has a proportional action resulting in an amplifier of high negative gain, that is, a device giving a reduced output signal as the input is increased.

The second arrangement of this type of device is known as the Direct Impact Modulator, which is shown in Fig. 1.33. The arrangement differs from the transverse type in the method of controlling the changes in axial momentum. Instead of a control nozzle perpendicular to the jet axis, an annular control chamber, concentric with the emitter nozzle, is used. As an increase in control flow increases the jet momentum from left to right, the balance point is initially arranged to be away from the output chamber

[1.8] MOMENTUM INTERACTION ANALOGUE DEVICES

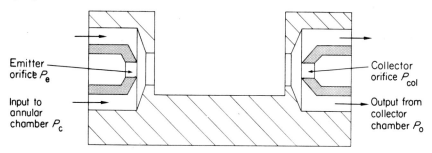

Fig. 1.33 Direct Impact Modulator (DIM) (Bjornsen, *H.D.L.* (2), **II**, Fig. 10)

so that the point moves towards the output with increasing control flow. This results in a positive gain amplifier with similar sensitivity to the transverse type. One difference of interest is that the direct type control chamber has a pressure–flow characteristic indicating a region of operation involving very small constant control flow as shown in Fig. 1.34. This

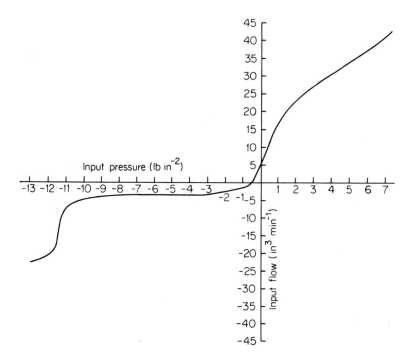

Fig. 1.34 Control characteristics of a DIM (Bjornsen, *H.D.L.* (2), **II**, 28, Fig. 11 (1964))

would suggest that the emitter jet is almost entirely pressure controlled. Frequency response tests on a commercially available unit using 0·016 in diameter supply nozzles are reported to give a characteristic with a break frequency close to 50 Hz, although noise becomes a problem above 100 Hz. A smaller device may be expected to give an improved dynamic performance.

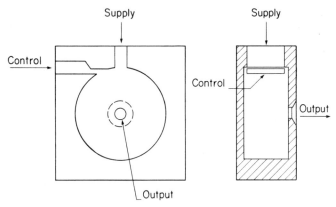

Fig. 1.35 Basic vortex amplifier configuration

1.9 Momentum Interaction and Boundary Layer Analogue Devices

1.9.1 Vortex Amplifier

The vortex amplifier, together with the beam deflexion amplifier, are the two most widely used fluidic analogue amplifiers at the present state of the technology. Their characteristics are dissimilar resulting in significantly different areas of application. Essentially the vortex amplifier is a flow control device which will operate over a wide range of pressures (greater than 1000 lb in^{-2}) using liquids or gases as the operating medium. It may be constructed in a variety of forms to facilitate signal summing or shaping with multiple inputs. To be balanced against these advantages, it is difficult to miniaturize to the point where the amplifier dynamic response is satisfactory for a wide range of applications. Generally the response is an order of magnitude worse than the beam deflexion amplifier.

Fig. 1.35 shows the basic form of the vortex amplifier which consists of a shallow circular chamber which is arranged to admit fluid in a radial direction towards the centre. An outlet tube, positioned at the chamber centre, allows the flow to pass out in a direction perpendicular to the plane of the chamber. The amplifier has one or more control nozzles positioned

[1.9] MOMENTUM INTERACTION AND BOUNDARY LAYER 35

so that they are tangential to the periphery of the vortex chamber. In the absence of a control signal the supply flows radially inwards through the chamber and leaves through the outlet tube with a minimum of losses. Control flow deflects the supply flow by momentum interaction thereby causing a spiral motion to the chamber centre. The tangential velocity component imparted to the flow increases as it moves towards the outlet due to the conservation of angular momentum. This tangential velocity amplification is the essential feature of the vortex amplifier, resulting in a centrifugal pressure build-up radially across the chamber. If the supply is pressure regulated, increased chamber pressure drop causes flow throttling. When the control pressure and flow are sufficiently high to reduce the supply flow to zero in order to maintain a constant supply pressure, the output is provided entirely by the control flow.

Practical vortex amplifiers have flow gains in the region of 20 and can be modified to form pressure amplifiers with gains of approximately 10. However, the linearity of the amplifier is not as good as the beam deflexion type but does tend to improve with increasing supply pressure. A further disadvantage in practice is that the control signal pressure must exceed the supply pressure in order to control the amplifier.

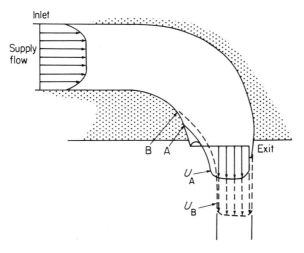

Fig. 1.36 Velocity profiles in a curved channel

1.9.2 Double Leg Amplifier

The double leg amplifier has a complex geometry designed to provide controlled separation of flow from a convex wall. Fig. 1.36 shows typical velocity profiles through a curved channel. When the radius of curvature is small enough, the flow separates from the inner wall and alters the velocity profile. The figure indicates roughly how the profile changes as the separation point is moved from A back to B. It is the change in profile shape which significantly alters the momentum flux passing through the output region.

Zisfein and Curtiss[17] describe a practical flow amplifier based on this principle. The main feature is that the point of separation is carefully controlled by introducing flow into the boundary layer of the separation wall in an opposite direction to the main flow. Fig. 1.37 shows the double leg amplifier and the method of introducing the control flow. Apart from

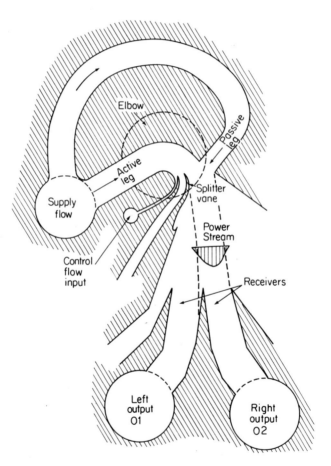

Fig. 1.37 The double leg amplifier configuration (*N.A.S.A.*, CR 245, 4–17, Fig. 4.2.5–2.)

[1.9] MOMENTUM INTERACTION AND BOUNDARY LAYER 37

the controlled separation, momentum interaction is also important between the main flow and flow through a second passive channel or leg.

The amplifier has two output receivers arranged so that the left receiver O1 admits essentially all the flow in the absence of a control signal. This is due to the deflexion of the main stream by the passive leg flow. When a control signal is applied, the momentum flux is increased in the main stream thereby deflecting the output jet to the right. The action is proportional and amplification results due to the sensitivity of the separation mechanism to control flow signals.

The device is essentially a flow modulator, giving very high no load flow gains in the order of 200. Fig. 1.38 shows typical mass flow transfer

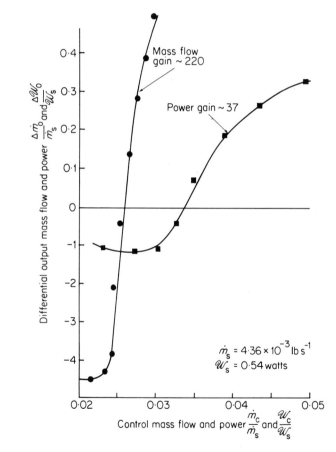

Fig. 1.38 Flow gain characteristics of a double leg amplifier (*N.A.S.A.*, CR–245, 4–18, Fig. 4.2.5-3)

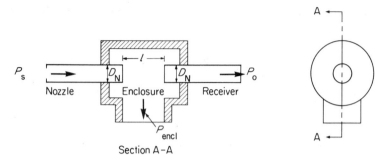

Fig. 1.39 Confined jet amplifier after Mayer (*Ref. 18*, Fig. 1)

characteristics. However, pressure gains are correspondingly poor, making it difficult to match in a cascade of similar devices. The dynamic response is also slow and noise is reported to be a problem. These factors have contributed to the lack of widespread usage of this device in fluidic circuits.

1.10 Modulated Pressure Field Analogue Devices

1.10.1 *Confined Jet Amplifier*

The confined jet amplifier, which was developed by Mayer[18], consists of a supply nozzle, an axially aligned receiver and an enclosure around

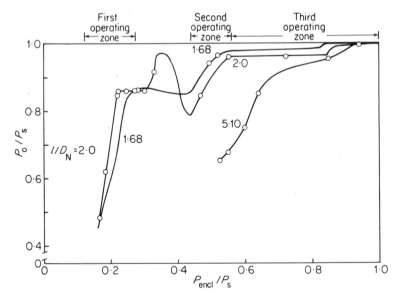

Fig. 1.40 Effect of gap length on performance ($l/D_N > 1$) (Mayer, *Ref. 18*, Fig. 5)

both nozzle and receiver. Fig. 1.39 shows the basic arrangement. The input signal to the device is the pressure within the enclosure, which modulates the output to provide pressure amplification. The confined jet amplifier can operate efficiently over a wide range of gas pressures (up to 1000 lb in^{-2}) with output pressure recovery usually well in excess of 80 per cent. It is thought that the device is strongly dependent on the position of a shock wave relative to the receiver which in turn is controlled by the pressure field. For the limited subsonic operating range a rather different effect takes place with the receiver pressure recovery being governed by the spread characteristics of the confined jet.

Mayer gives typical input-output non-dimentionalized pressure characteristics, under blocked outlet load conditions, for the geometry in which the diameters of the supply nozzle and receiver are equal, as shown in Fig. 1.40. Three distinct zones of operation are observed in which the pressure recovery successively increases at the expense of reduced pressure gain as the gap between nozzle and receiver is increased. Generally, in zone 1 the best operating conditions occur with pressure recoveries in the region of 86 per cent and corresponding gains of about 7. In zone 2 the balance alters to 96 per cent and 2 respectively while in zone 3 the pressure recovery has increased a little to 97·5 per cent but with a significant fall in gain to 0·3.

In most of the applications of the confined jet amplifier a vortex amplifier is used for the control of the enclosure pressure. With suitable design, a vortex amplifier can act as a pressure regulator so that, when controlling a confined jet amplifier, any load flow variations in the jet receiver has a reduced effect on the output pressure. Very high pressure gains are possible from combined confined jet and vortex amplifier combinations which are useful in high pressure fluidic power control circuits.

Otsap[19] has described an asymmetric confined jet amplifier which is controlled directly by a vortex chamber adjacent to one side of the supply jet. This gives the device the interesting characteristic that the vent pressure of the chamber may be held constant over a wide range of operating supply pressures and may be adjusted by varying the supply and vent restriction areas.

1.11 Position Sensors

One of the attractions of fluidics is the simplicity, low cost and ruggedness of fluid sensors compatible with both analogue and digital devices. The commonest types of pure fluid position sensor exploit some characteristic of a fluid jet which usually modulates the pressure with very small

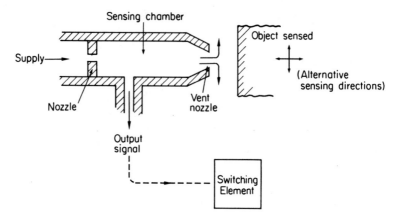

Fig. 1.41 Back pressure sensor

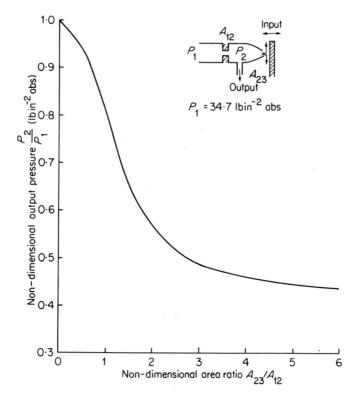

Fig. 1.42 Back pressure sensor characteristic

[1.11] POSITION SENSORS

flow changes. One type, known as the back pressure sensor or amplifier has been used for many years in both pneumatic and hydraulic systems. With the increase of interest in fluidics these sensors have been used more widely for such applications as encoders (see Chapter 13) and A.C. signal generators and have been augmented by more sophisticated jet configurations.

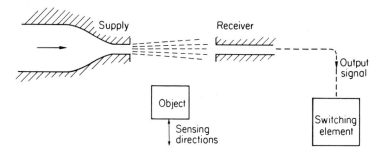

Fig. 1.43 Simple interruptible jet sensing

1.11.1 Back Pressure Type

Fig. 1.41 shows the essential features of a back pressure sensor, which consists of a constant pressure supply separated from a back pressure chamber by a nozzle restriction. A nozzle at the outlet from the chamber allows a small flow to vent through the sensor. An object placed close to the venting jet flow causes the pressure in the chamber to increase and, in the limiting case, when complete blocking of the vent occurs the chamber pressure becomes identical to the supply pressure. The non-dimensionalized chamber pressure variation with position of the object from the nozzle is shown in Fig. 1.42 for a supply pressure of 34·7 lb in^{-2} abs. using air as the operating medium. Area A_{23} is based on the circumferential area restriction caused by the moving surface. In analogue system applications the middle portion of the characteristic is used as it is approximately linear. For digital applications the extreme excursions of the curve are the only points of interest so that a pressure is achieved above and below the switching pressure of the fluidic device downstream from the sensor. The signal generated by the sensor usually activates the fluidic device directly so the device control nozzle must be sufficiently small so that no appreciable flow is taken from the sensing element. Although this sensor gives good signal changes, generally speaking the working gaps are very small, 0·003 to 0·005 in being typical, which may be a practical limitation in some applications. Because the resistance of the restricter and the capacitance of the chamber volume dominate the sensor dynamics, a simple time constant representation is possible based on these two lumped parameters. Moses et al.[20] give experimental data supporting this approach for a pneumatic back pressure sensor with a laminar supply nozzle and conclude that the model is satisfactory up to 500 Hz. An analysis of the back pressure device is given in Chapter 2.

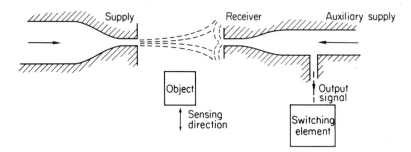

Fig. 1.44 Interruptible jet with pressurized receiver

1.11.2 Interruptible Jet Type

The simplest form of interruptible jet configuration is shown in Fig. 1.43 and consists of a supply nozzle and receiver axially aligned but separated by a gap. Any object passing through this gap interrupts the jet stream and reduces the receiver pressure nominally to ambient pressure. This form of sensor has been widely used for the high speed detection of the presence of objects in industrial sequencing circuits and also for pneumatic encoders. It has the important advantages that the gap setting is not critical, the texture of the objects is unimportant and the dynamic response of the sensor is rather better than the equivalent sized back pressure device. One disadvantage is that the receiver entrains ambient air with the jet so that the downstream circuit may become contaminated, should the environment be unclean.

This problem may be overcome by slightly pressurizing the receiver, in addition to the supply nozzle, as shown in Fig. 1.44 so that air is always vented from the receiver. When the gap is open the main jet impinges on the lower pressure jet from the receiver thereby constricting the vent flow which in turn raises the pressure at the outlet from the sensor. When an object interrupts the main jet, the receiver vent flow increases and the outlet pressure is correspondingly reduced. Both these types of jet sensor may be used with gaps typically up to 1 inch with nozzle diameters in the region of 0·020 in.

By exploiting the sensitive laminar jet structure of the Turbulence Amplifier much larger gaps may be used with multiple jet configurations.

[1.11] POSITION SENSORS 43

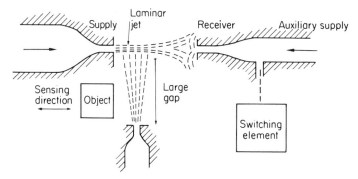

Fig. 1.45 Complex interruptible jet configuration for large gaps

Fig. 1.45 illustrates the use of a supplementary jet to trigger turbulence in the main jet, the gap now being between the supplementary jet and the main jet. When an object interrupts the supplementary jet gap the main jet is not disturbed and causes vent flow to fall thereby increasing the sensor output pressure. The reverse effect occurs when the object is

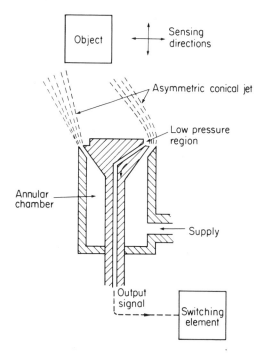

Fig. 1.46 Conical jet sensor after Auger (*Ref. 21*)

removed. This has the advantage that the new gap may be increased by at least an order of magnitude over the previous configuration provided the greater power consumption associated with the higher supply pressure required is tolerable.

1.11.3 Conical Jet Sensor

Auger[21] has described a novel position sensor based on the structure of a conical jet issuing from an annular nozzle as shown in Fig. 1.46. If two free jets are close to each other the fluid between them will try to entrain into both jets thereby locally reducing the pressure. When the configuration is modified to make a conical jet the pressure inside the volume contained by the jet will be significantly reduced as entrainment flow can no longer be made up by the ambient fluid.

By a suitable choice of cone angle the conical jet can be made very sensitive to the presence of an object up to several annular diameters downstream. This has the effect of deflecting the jet into an asymmetric configuration so that an output sensing nozzle located close to the most deflected portion of the jet would record a relative reduction in pressure. The mechanism of the deflexion of the jet cone is as yet not well understood but Auger attributes it to the relative strength of vortices shed from the inner surfaces of the cone.

Although signal changes may be as large as 10 per cent of the supply pressure the dynamic response of the sensor tends to be slower than types previously described, probably due to the difficulties of miniaturization and the finite time required to re-establish the complex flow field. Due to the large effective nozzle areas used, air consumption is likely to be significant.

References

1. Steptoe, B. J. Steady state and dynamic characteristics in digital wall attachment devices, *C.F.C.* (2), Paper B3.
2. Foster, K. and Retallick, D. A. Some experiments on a free foil switching device, *C.F.C.* (2), Paper D1.
3. Parker, G. A. and Jones, B. Experiments with AND and EXCLUSIVE–OR passive elements, *C.F.C.* (2), Paper C4.
4. Greenwood, J. R. The design and development of a fluid logic element, *B.Sc. Thesis* (1960), M.I.T.
5. Greenwood, J. R. and Ezekiel, F. D. Hydraulics half-add binary numbers, *Control Eng.*, **8.2,** 145 (1961).
6. Gray, W. E. and Stern, H. Fluid amplifiers, capabilities and applications, *Control Eng.*, **11**, 57 (1964).
7. Mitchell, A. E. Calculating with jets, *New Scientist*, **17**, No. 329 (1963).
8. Glaettli, H. H. Circuits using fluid dynamic components, *C.F.C.* (1), Paper D4.

REFERENCES

9. Barker, R. A. Development of an 'equality' fluid logic element for use in a control system comparator, *M.Sc. Thesis* (1965), Birmingham University.
10. Parker, G. A. and Jones, B. A fluidic subtractor for digital closed loop control systems, *C.F.C.* (2), Paper H1.
11. Hellbaum, R. F. Wall attachment crossover AND gate, *A.F.*, 187.
12. Auger, R. N. The turbulence amplifier in control systems, *H.D.L.* (2), **II**, 261.
13. Verhelst, H. A. M. On the design, characteristics and production of turbulence amplifiers, *C.F.C.* (2), Paper F2.
14. Swartz, E. L. A flueric induction AND gate, *H.D.L.* (3), **IV**, 165.
15. König, G. R. Design of inputs and outputs of digital pneumatic jet components for adaptation to the standardised pressure range, *C.F.C.* (2), Paper B5.
16. Walker, M. L. and Trask, R. P. Feasibility study of a laminar NOR unit, *A.F.*, 162.
17. Zisfein, M. B. and Curtiss, H. A. A high gain proportional fluid state flow amplifier, *H.D.L.* (2), **I**, 375.
18. Mayer, E. A. Confined–jet amplifier, *J. Basic Eng.*, **90**, 103 (1968).
19. Otsap, B. A. Experimental study of a proportional vortex fluid amplifier, *H.D.L.* (2), **II**, 85.
20. Moses, H. L., Small, D. A. and Cotta, G. A. Response of a fluidic air gauge, *A.S.M.E. Paper* 68–WA/FE–16.
21. Auger, R. N. A fluidic proximity detector, *C.F.C.* (3), Paper E7.

2
Lumped and Distributed Parameter Fluid System Components

2.1 General

The design of fluidic control systems requires an adequate knowledge of the control device characteristics, interface elements and also the fluid components used for interconnexions. This chapter is concerned with the steady state and dynamic characteristics of fluid components occurring in fluidic networks. Generally the dynamic signals transmitted in a fluidic analogue circuit are sinusoidal and may be analysed using existing linearized theory. In digital systems the transient or dynamic changes are associated with step and pulse signal waveforms, the analysis of which is also well known for linear systems. It will be shown that fluid components are not directly amenable to linear analysis, except in special cases, although satisfactory linearizing techniques may be adopted at the expense of increased complexity.

Two techniques are available for linear analysis of circuits, depending on the time scale of the signal changes. When the time scale is relatively long, in other words for low frequencies, all the system components may be considered to have discrete properties or lumped parameters. At higher frequencies or shorter time scales the component properties become dependent on the physical dimensions of the system so that they are considered to have distributed parameters.

In order to establish a compatible system of units for fluid systems, it is advantageous to consider the equivalent electrical units, enabling the parameters used to be compared to control system analysis based on electrical networks. The parameters of an electrical system most easily measured are potential difference and current, which may be used to describe resistance, inductance and capacitance. The product of potential difference and current gives power in consistent units based on unit charge. However, in a fluid medium the two equivalent commonest measurements are pressure difference and volume flow which, while leading to a consistent set of units, must be treated with caution. In electrical circuits, by Kirchoff's Law, current is conserved in a network but the fluid equivalent of volume flow is only conserved under incompressible flow conditions. As there is a loss of generality by excluding compressible flow, it is more satisfactory to consider either mass or weight flow to be equivalent to current. By so doing the power product does not lead to consistent units. Both Kirshner[1] and Taplin[2] have suggested modifications to the pressure difference parameter to retain consistent units when using weight flow as the other parameter. Kirshner has proposed on a mechanical potential from fluid mechanic considerations and Taplin from a fundamental thermodynamic approach uses **R**T.ln (pressure ratio) to replace pressure difference. However, because the simpler pressure difference and weight flow units are more easily understood they will be adopted throughout the text.

The distinction between weight and volume flow arises because of compressibility considerations within the fluid caused by density variations at different points in the system. Generally speaking compressibility effects are not important in liquids except when considering the dynamic response of servo-mechanisms or very large volumes of liquids. However, for gases compressibility is far more significant as large density changes can occur over a wide range of flow conditions. As a rough guideline, for gas velocities exceeding about 20 per cent of the local acoustic velocity the flow should be considered as compressible. The acoustic velocity a is given by

$$a^2 = \gamma \mathbf{R} T \tag{2.1}$$

2.2 Basic System Parameters

2.2.1 Impedance

When basing the analysis of component performance on the sinusoidal frequency response, it is found that some components, such as capacitances, have values which are frequency dependent. In linear electrical circuit theory it has been found convenient to treat the ratio of the voltage to current as a system parameter, known as the impedance Z. In general, the impedance of a circuit element may also be frequency dependent so that the amplitude and phase of the voltage signal changes relative to that of the current signal as the frequency changes. In control theory it is common to express the impedance under such conditions as a complex number. That is

$$Z(j\omega) = A + jB \tag{2.2}$$

where A contains all resistive terms and B contains all frequency dependent terms which are known as reactances.

Adopting pressure difference ΔP and weight flow, W, as the basic system parameters in a fluid circuit, we may define the fluid impedance of an element at a frequency ω as

$$Z(j\omega) = \frac{\Delta p}{w} \tag{2.3}$$

where Δp and w are the respective small signal changes in pressure difference and weight flow.

With steady state conditions, the reactance term B of (2.2) is not present and the impedance becomes a constant, termed the static impedance.

Resistive, capacitive and inertia circuit elements may all be identified in fluid circuits and described in terms of their impedances. For linear circuits this allows the impedances of individual elements within the system to be manipulated by the ordinary rules of algebra.

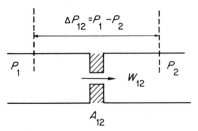

Fig. 2.1 Fluid resistance notation

2.2.2 Fluid Resistance

All fluidic devices, because of viscous forces in the fluid, exhibit resistance to the fluid flow which results in a loss of energy associated with a pressure drop across the element. The steady state fluid resistance R of a fluid element is defined as the pressure drop across it divided by the flow. Hence the resistance is:

$$R_{12} = \frac{\Delta P_{12}}{W_{12}} \tag{2.4}$$

with units of TL^{-2} and may be conveniently represented as shown in Fig. 2.1. Note that suffix 1 is used to denote upstream conditions and

[2.2] BASIC SYSTEM PARAMETERS

suffix 2 for downstream conditions while parameters, such as weight flow, which are transferred between these states are notated as suffix 12. This convention is used throughout the chapter.

The resistance of fluid elements is dependent on the fluid flow regime and also the geometry of the element. Several common cases will now be considered.

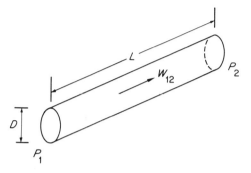

Fig. 2.2 Tubular resistance

2.2.2.1 Tube resistance. Provided the Reynolds Number, $Re = \rho \bar{u} D/\mu < 2000$ the flow is considered to be laminar so that the Hagen-Poiseuille Law applies for the relationship between pressure drop and flow. Using the notation shown in Fig. 2.2 the equation is

$$(P_1 - P_2) = \Delta P_{12} = \frac{128 \mu L}{\pi D^4 \mathscr{S}} W_{12}$$

or

$$R_{12} = \frac{128 \mu L}{\pi D^4 \mathscr{S}} \qquad (2.5)$$

The important property of the laminar resistor is that it is essentially linear as its resistance is independent of ΔP and W. Equation (2.5) is not precisely true as entrance and exit flow conditions introduce corrections, but these may be minimized with careful design. A value of 140 instead of 128 is sometimes preferred as giving a closer empirical relationship. In practice it also is difficult to obtain a calculated value for R with tubular resistances due to the strong dependence on the tube diameter. Slight geometric changes of diameter are difficult to measure and eliminate.

For air at 68°F and atmospheric pressure the resistance equation becomes

$$R = 2 \cdot 44 \times 10^{-3} \frac{L}{D^4} \sec \text{in}^{-2} \qquad (2.6)$$

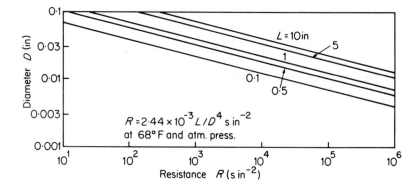

Fig. 2.3 Laminar flow resistance of air filled tubes (N.A.S.A., CR-245, 2–25, Fig. 2.2.2–1)

Fig. 2.3 illustrates the laminar flow resistance of tubes filled with air at ambient conditions.

The corresponding resistive impedance expression for laminar flow between parallel plates, see Fig. 2.4, is:

$$R_{12} = \frac{3\mu L}{4bh^3 \mathscr{S}} \quad \text{for } b \gg h \tag{2.7}$$

and has a similar form to (2.5) for the tube configuration. A more exact analysis of steady laminar flow in rectangular channels of low aspect ratio (b/h) results in the following relationship:

$$R_{12} = \frac{3\mu L}{4bh^3 \mathscr{S}} \left[1 - \frac{h}{b} f\left(\frac{b}{h}\right)\right]^{-1} \tag{2.8}$$

where

$$f\left(\frac{b}{h}\right) = \frac{192}{\pi^5} \sum_{n=1}^{\infty} \frac{1}{n^5} \tanh\left(\frac{n\pi b}{h}\right)$$

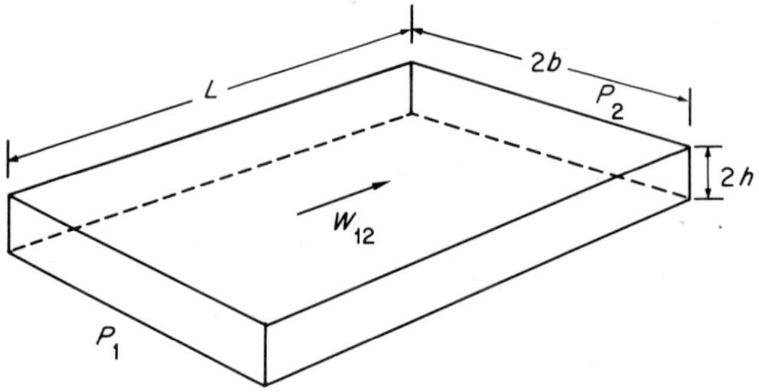

Fig. 2.4 Rectangular channel resistance

[2.2] BASIC SYSTEM PARAMETERS

The resistance variation at low aspect ratios for a particular linear capillary is shown in Fig. 2.5. In practice this equation should include a non-linear term to account for entrance loss effects but for capillaries of appreciable length to width ratios (> 100) this is small and is usually ignored.

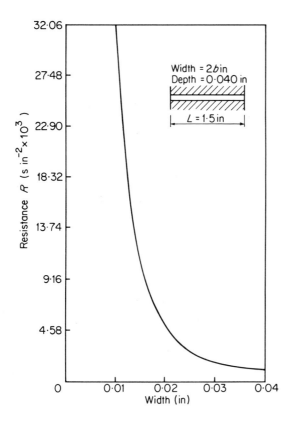

Fig. 2.5 Resistance variation of a rectangular pneumatic capillary at low aspect ratios (Parker and Jones, C.F.C.(3), Paper K1, Fig. 5)

For dynamic analysis of laminar resistances, R may be identically replaced by the resistive impedance Z_r as the resistance value is linear. This is not the case with the remaining types of fluid resistance discussed in this chapter, which must be linearized, as shown in Section 2.3, before the resistive impedance is found.

For Reynolds Numbers well in excess of 2000 the flow in a tube is turbulent and becomes more complex to describe. In terms of the resistance it may be stated as:

$$R_{12} = \left(\frac{4}{\pi}\right)^{\frac{7}{4}} \left(\frac{0.316L}{\mathscr{G}}\right) \left(\frac{\mu}{16g^3 D^{19}}\right)^{\frac{1}{4}} W_{12}^{\frac{3}{4}} \qquad (2.9)$$

It will be noted that the resistance is now a function of the weight flow and hence is non-linear. Fig. 2.6 shows the limits of linear resistance for incompressible air flow through a 1 in length of tubing of various diameters.

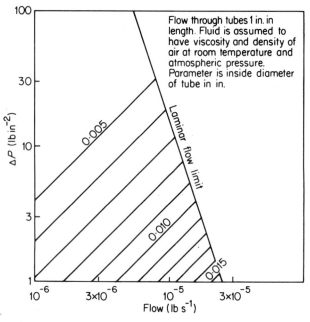

Fig. 2.6 The flow regimes for air tubes (*N.A.S.A.*, CR-245, 2–34, Fig. 2.2.3–4)

2.2.2.2 Orifice and nozzle resistance. When the cross-sectional area of a channel is abruptly reduced to form a nozzle or orifice, the resistance characteristics also change significantly from the channel flow case. Using the notation of Fig. 2.1 and assuming incompressible flow, Bernoulli's equation together with the equation of continuity gives the weight flow as:

$$W_{12} = C_D A_{12} \sqrt{2g\rho_2(P_1 - P_2)} \qquad (2.10)$$

if the upstream velocity is negligible. If this latter assumption cannot be

[2.2] BASIC SYSTEM PARAMETERS

made then P_1 is the total rather than the static pressure. Putting $(P_1 - P_2) = \Delta P_{12}$, then the resistance may be written as

$$R_{12} = \frac{\Delta P_{12}}{W_{12}} = \frac{\sqrt{\Delta P_{12}}}{C_D A_{12} \sqrt{2g\rho_2}} \quad (2.11)$$

which is a very common form of non-linear resistance.

When the flow is compressible, the orifice resistance for a gas becomes complex. As the pressure drop across the orifice increases the particle velocity increases until, when the velocity is just sonic, the flow becomes choked. At supersonic velocities the form of the equation changes and the weight flow ceases to be affected by the downstream conditions. For isentropic conditions the weight flow through an orifice is given by:

$$W_{12} = \frac{A_{12} P_1}{\sqrt{T_1}} \left\{ \frac{2\gamma g}{(\gamma - 1)R} \left[\left(\frac{P_2}{P_1}\right)^{\frac{2}{\gamma}} - \left(\frac{P_2}{P_1}\right)^{\frac{\gamma+1}{\gamma}} \right] \right\}^{\frac{1}{2}} \quad (2.12)$$

The flow becomes choked at a critical pressure ratio P_2/P_1, which for air is 0·528. The corresponding critical weight flow W_{crit} may be written as

$$W_{\text{crit}} = \frac{A_{12} P_1}{\sqrt{T_1}} \left[\frac{\gamma g}{R} \left(\frac{2}{\gamma + 1}\right)^{\frac{\gamma+1}{\gamma-1}} \right]^{\frac{1}{2}} \quad (2.13)$$

Following the work of Anderson[3], from which source much of the following analysis is taken, a non-dimensional parameter N_{12} may be defined as the ratio of actual to critical weight flow so that (2.12) reduces to:

$$W_{12} = \frac{K P_1 A_{12} N_{12}}{\sqrt{T_1}} \quad (2.14)$$

where

$$N_{12} = \frac{W_{12}}{W_{\text{crit}}} = \left[\frac{\left(\frac{P_2}{P_1}\right)^{\frac{2}{\gamma}} - \left(\frac{P_2}{P_1}\right)^{\frac{\gamma+1}{\gamma}}}{\left(\frac{\gamma - 1}{2}\right)\left(\frac{2}{\gamma + 1}\right)^{\frac{\gamma+1}{\gamma-1}}} \right]^{\frac{1}{2}} \quad (2.15)$$

$$K = \left[\frac{\gamma g}{R} \left(\frac{2}{\gamma + 1}\right)^{\frac{\gamma+1}{\gamma-1}} \right]^{\frac{1}{2}} \quad (2.16)$$

Equation (2.14) is in a suitable form for computation as K is a constant for any gas (0·532 $\sqrt{°R}$ s^{-1} for air), the upstream conditions of pressure and temperature are easily found and the only complexity is in N_{12}. For any gas, N_{12} is a function of the pressure ratio P_1/P_2 which may be tabulated or presented in graphical form. Fig. 2.7 shows the variation of N_{12} with

P_1/P_2 for air up to critical conditions. In the general case, the resistive impedance cannot be directly found as the weight flow may not be dependent on the pressure difference. From known conditions the pressure difference and weight flow can be calculated and the ratio taken to give the resistance.

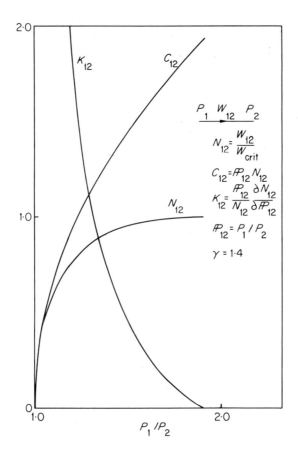

Fig. 2.7 Compressible flow nozzle equation coefficients

An approximate compressible flow orifice equation may be derived which is useful in that it has a similar form to the incompressible flow case given by (2.10). The important restriction to its usage is that pressure differences are small compared with the system pressure which implies that density changes are also small.

[2.2] BASIC SYSTEM PARAMETERS

Let $P_2 = P_1 - \Delta P$ with ΔP small. Then, neglecting squared terms and higher

$$\left(\frac{P_2}{P_1}\right)^{\frac{2}{\gamma}} \simeq 1 - \frac{2}{\gamma}\frac{\Delta P}{P_1}$$

$$\left(\frac{P_2}{P_1}\right)^{\frac{\gamma+1}{\gamma}} \simeq 1 - \left(\frac{\gamma+1}{\gamma}\right)\frac{\Delta P}{P_1}$$

So that N_{12} becomes

$$N_{12} = \left[\frac{2\Delta P\left(1 - \frac{3\Delta P}{2\gamma P_1}\right)^{\frac{1}{2}}}{\gamma P_1\left(\frac{2}{\gamma+1}\right)^{\frac{\gamma+1}{\gamma-1}}}\right] \quad (2.17)$$

Also

$$\frac{P_1}{RT_1} = \rho_1 = \frac{\rho_2}{\left(1 - \frac{\Delta P}{P_1}\right)^{\frac{1}{\gamma}}} \simeq \frac{\rho_2}{1 - \frac{1}{\gamma}\frac{\Delta P}{P_1}} \quad (2.18)$$

Substituting (2.17) and (2.18) in the orifice equation (2.14) the approximate formula is obtained

$$W_{12} = \left[\frac{1 - \frac{3}{2\gamma}\frac{\Delta P}{P_1}}{1 - \frac{1}{\gamma}\frac{\Delta P}{P_1}}\right]^{\frac{1}{2}} A_{12}\sqrt{2g\rho_2 \Delta P} \quad (2.19)$$

The square bracket term expresses the deviation from the incompressible case which for $\Delta P/P_1 = 0.01$ gives an error in using the incompressible equation of 0.2 per cent. However, for $\Delta P/P_1 = 0.1$ the corresponding error increases to 2.5 per cent.

Using the approximate compressible flow equation, the resistance may be conveniently written as:

$$R_{12} = \frac{\Delta P}{W_{12}} = \left[\frac{1 - \frac{1}{\gamma}\frac{\Delta P}{P_1}}{1 - \frac{3}{2\gamma}\frac{\Delta P}{P_1}}\right]^{\frac{1}{2}} \frac{\sqrt{\Delta P}}{A_{12}\sqrt{2g\rho_2}} \quad (2.20)$$

2.2.3 Fluid Capacitance

In a system containing a compressible fluid, such as a gas, any volume has a fluid capacity associated with it so that a change of pressure within the volume produces a change in weight of fluid contained in it. Defining the fluid capacity C of a volume V as the ratio of the change in weight contained within the volume divided by the associated pressure difference change, then if M is the weight of fluid in volume V:

$$C = \frac{\int w\,dt}{\Delta p} = \left.\frac{\partial M}{\partial P}\right|_{V=\text{const}} \qquad (2.21)$$

with units of L^2. More conveniently, the ratio may be rewritten, using the La Place operator s, in impedance form as:

$$Z_C = \frac{\Delta p}{w} = \frac{1}{sC} \qquad (2.22)$$

where Z_C is the capacitive impedance.

For an ideal gas:

$$PV = M\mathbf{R}T$$
$$\therefore\ \partial P V + \partial V P = \partial M \mathbf{R} T + M \mathbf{R}\,\partial T$$

For a fixed volume $\partial V = 0$ so that, in general, the capacitance may be expressed as:

$$C = \left.\frac{\partial M}{\partial P}\right|_{V=\text{const.}} = \frac{V}{\mathbf{R}T} - \frac{M}{T}\frac{\partial T}{\partial P} \qquad (2.23)$$

When the pressure changes very slowly in the volume, isothermal conditions occur within the gas so that $\partial T = 0$. In this case

$$C = \frac{V}{\mathbf{R}T} \qquad (2.24)$$

More commonly the pressure fluctuations are sufficiently rapid to prevent temperature equalization and the polytropic relationship must be used. One form is:

$$\frac{P^{\frac{n-1}{n}}}{T} = \text{const.} \qquad (2.25)$$

where n is the polytropic constant which may be written in partial differential form (see Section 2.3) as:

$$\frac{\partial T}{\partial P} = \left(\frac{n-1}{n}\right)\frac{T}{P} \qquad (2.26)$$

[2.2] BASIC SYSTEM PARAMETERS

Substituting in the general equation (2.23) gives the polytropic capacitance as:

$$C = \frac{V}{n\mathbf{R}T} \qquad (2.27)$$

from which it is seen that the isothermal value is reduced by the value of the polytropic constant. For air at 68°F these relationships for the isothermal and adiabatic cases reduce to:

$$\begin{aligned} C_{\text{isothermal}} &= 2{\cdot}95 \times 10^{-6} V \text{ in}^2 \\ C_{\text{adiabatic}} &= 2{\cdot}11 \times 10^{-6} V \text{ in}^2 \end{aligned} \qquad (2.28)$$

which are illustrated in Fig. 2.8.

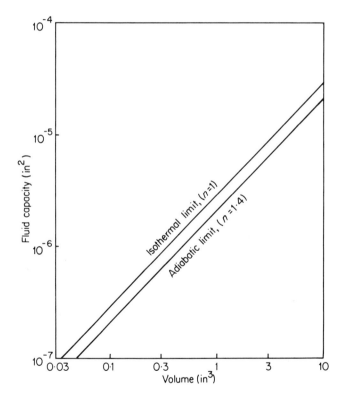

Fig. 2.8 Fluid capacity of a volume (*N.A.S.A.*, CR-245, 2–48, Fig. 2.2.6–1)

It has already been shown in (2.22) that the capacitance involves a time rate effect. If the impedance is expressed in frequency form then putting $s = j\omega$

$$Z_C(j\omega) = -j\frac{1}{\omega C} \tag{2.29}$$

which is known as the capacitive reactance. This is plotted for air at 68°F in Fig. 2.9 using the isothermal capacitance values.

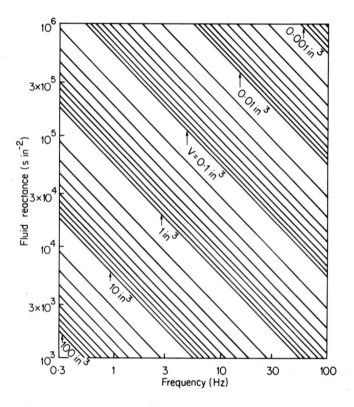

Fig. 2.9 Isothermal capacitive reactance (*N.A.S.A.*, CR-245, 2–50, Fig. 2.2.6–2)

2.2.4 *Fluid Inertia or Inductance*

In fluid circuitry, wherever high velocity transient flows occur a mass of fluid is accelerated or decelerated causing a significant pressure change due to the fluid inertia. Defining the fluid inertia or inductance H as the

[2.2] BASIC SYSTEM PARAMETERS

ratio of the pressure difference change across the element caused by the rate of fluid weight flow then

$$H = \frac{\Delta p}{\dfrac{dw}{dt}} \qquad (2.30)$$

with units of T^2L^{-2}. In impedance form

$$Z_H = \frac{\Delta p}{w} = sH \qquad (2.31)$$

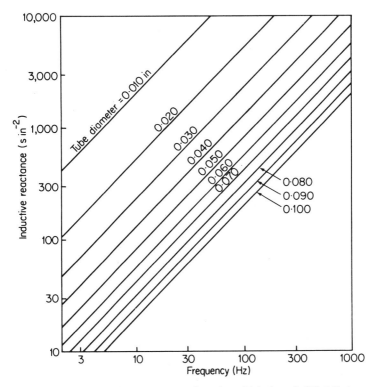

Fig. 2.10 Inductive reactance of a tube of 1 in length (*N.A.S.A*, CR-245, 2-56, Fig. 2.2.7-1)

For incompressible conditions, the fluid inductance associated with flow in a tube may be found by comparing (2.30) with the equation of motion for an element of fluid. If $\bar{u}$ represents the sectional mean velocity

$$w = \rho q = \rho A \bar{u}$$

so that

$$\Delta p = H\rho A \frac{d\bar{u}}{dt} \quad (2.32)$$

The fluid acceleration $d\bar{u}/dt$ exerts a force causing a pressure difference giving a further relationship

$$\Delta p A = \frac{\rho A L}{g} \frac{d\bar{u}}{dt} \quad (2.33)$$

Comparing (2.32) and (2.33) it is seen that:

$$H = \frac{L}{Ag} = \frac{4L}{\pi D^2 g} \quad \text{for round tubing} \quad (2.34)$$

It is interesting to note that the fluid inductance of tubing is a function only of the element dimensions, unlike fluid capacitance which is a function both of geometry and fluid properties.

The inductive reactance is found by substituting $s = j\omega$ in (2.31) giving:

$$Z_H(j\omega) = j\omega H \quad (2.35)$$

This is plotted in Fig. 2.10 for a tube of 1 in length.

Generally speaking fluid inductance is associated with high fluid resistance values in conjunction with fast transient conditions. Almost invariably the inductive effect is at least one order of magnitude smaller than both resistive and capacitive effects and in many dynamic analyses of pneumatic circuits it can be neglected. One important exception to this is when a system is in or near resonance as the inherent transfer of energy between the fluid inductance and capacitance causes the characteristic resonant frequency. The well known Helmholtz resonator is basically a fluid system in this condition. Also in high frequency fluidic circuits inductive effects may be significant.

2.3 Linearization of System Parameters

2.3.1 General

It has been shown that some of the system parameters, such as incompressible flow resistance, give rise to non-linear relationships. In general, dynamic circuit analysis is considerably simplified if algebraic manipulations of linear functions can be used. By using linearizing approximations for non-linear quantities, considerable insight into more complex systems may be achieved although caution must be shown in not

[2.3] LINEARIZATION OF SYSTEM PARAMETERS

destroying the essential features of the non-linear phenomenon. Fig. 2.11 illustrates the straight line approximation for the non-linear function at an operating point O.

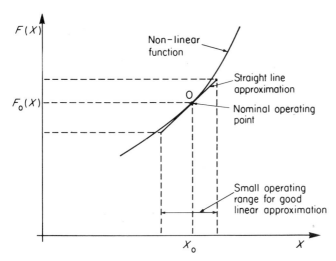

Fig. 2.11 Straight line approximation of a non-linear function

The incremented variation ΔY for a function $Y = F(X_1, X_2, \ldots, X_n)$ of n independent variables is given by:

$$\Delta Y = \left.\frac{\partial Y}{\partial X_1}\right|_0 \Delta X_1 + \left.\frac{\partial Y}{\partial X_2}\right|_0 \Delta X_2 + \cdots + \left.\frac{\partial Y}{\partial X_n}\right|_0 \Delta X_n$$

where 0 is a reference condition. Using lower case letters to represent variations from the reference conditions, then:

$$Y - Y_0 = \Delta Y = y$$
$$X_1 - X_0 = \Delta X_1 = x_1$$
$$\text{etc.}$$

and

$$y = C_1 x_1 + C_2 x_2 + C_3 x_3 + \cdots + C_n x_n \qquad (2.36)$$

where

$$C_1 = \frac{\partial Y}{\partial X_1}, \quad C_2 = \frac{\partial Y}{\partial X_2}, \ldots,$$

alternatively, if:

$$Y \propto X_1^\alpha X_2^\beta X_3^\gamma \ldots,$$

then

$$\frac{y}{Y} = \alpha \frac{x_1}{X_1} + \beta \frac{x_2}{X_2} + \gamma \frac{x_3}{X_3} + \cdots, \qquad (2.37)$$

which is a useful form sometimes preferred to (2.36).

Because of the restriction on the operating range introduced by linearization, the use of this method for the analysis of system dynamics is confined to time variant signals of small amplitude, often referred to a small signal analysis. Frequently, particularly with complex non-linear functions, the steady state functional relationships are assumed to apply to the dynamic case when linearizing; that is, a quasi-steady state analysis is adopted.

Throughout the chapter, lower case letters are used to represent changes of the variables and system parameters from reference conditions.

2.3.2 Incompressible Flow Orifice Impedance

It has been stated in (2.10) that the weight flow W_{12} for an orifice of cross-sectional area A_{12} is:

$$W_{12} = C_D A_{12} \sqrt{2g\rho_2(P_1 - P_2)} \qquad (2.38)$$

For incompressible flow, assuming that $W_{12} = F(A_{12}, \Delta P_{12})$ only, then linearizing this equation using (2.36):

$$w_{12} = C_1 a_{12} + C_2 \Delta p \qquad (2.39)$$

where

$$C_1 = \left.\frac{\partial W_{12}}{\partial A_{12}}\right|_0 = C_D \sqrt{2g\rho_2 \, \Delta P_{12}}$$

$$C_2 = \left.\frac{\partial W_{12}}{\partial P_{12}}\right|_0 = \frac{C_D A_{12} \sqrt{2g\rho_2}}{2\sqrt{\Delta P_{12}}}$$

Now the linearized incremental resistance r for a fixed area orifice ($a_{12} = 0$) may be written as:

$$r_{12} = \left.\frac{\partial \Delta P_{12}}{\partial W_{12}}\right|_{A_{12}=\text{const.}} = \frac{1}{C_2} = \frac{2\sqrt{\Delta P_{12}}}{C_D A_{12} \sqrt{2g\rho_2}} \qquad (2.40)$$

Comparing the incremental resistance with the steady state fluid resistance, given by (2.11) for incompressible flow, it is seen that:

$$r = 2R$$

It should be noted that the incremental resistance assumes that non-

[2.3] LINEARIZATION OF SYSTEM PARAMETERS

steady state conditions in the resistance may be characterized by the steady state relationships.

2.3.3 General Nozzle Flow Equation

Equation (2.14) gave a useful form of the general nozzle flow equation for computation which may be restated as:

$$W_{12} = \frac{KP_1 A_{12} N_{12}}{\sqrt{T_1}} \quad (2.41)$$

Linearizing

$$w_{12} = C_1 p_1 + C_2 a_{12} + C_3 n_{12} + C_4 t_1 \quad (2.42)$$

where

$$C_1 = \frac{\partial W_{12}}{\partial P_1} = \frac{KA_{12}N_{12}}{\sqrt{T_1}}$$

$$C_2 = \frac{\partial W_{12}}{\partial A_{12}} = \frac{KP_1 N_{12}}{\sqrt{T_1}}$$

$$C_3 = \frac{\partial W_{12}}{\partial N_{12}} = \frac{KP_1 A_{12}}{\sqrt{T_1}}$$

$$C_4 = \frac{\partial W_{12}}{\partial T_1} = -\frac{1}{2}\frac{KP_1 A_{12} N_{12}}{T^{\frac{3}{2}}}$$

or, alternatively

$$\frac{w_{12}}{W_{12}} = \frac{p_1}{P_1} + \frac{a_{12}}{A_{12}} + \frac{n_{12}}{N_{12}} - \frac{1}{2}\frac{t_1}{T_1} \quad (2.43)$$

clearly, from (2.15) $N_{12} = F(P_1, P_2)$ so that

$$n_{12} = C_5 p_1 + C_6 p_2 \quad (2.44)$$

where

$$C_5 = \frac{\partial N_{12}}{\partial P_1}$$

$$C_6 = \frac{\partial N_{12}}{\partial P_2} = -C_5$$

Evaluation of C_5 and C_6 from (2.15) gives

$$C_5 = -C_6 = K_{12}\frac{N_{12}}{P_1} \quad (2.45)$$

where

$$K_{12} = \frac{\frac{n-1}{2n}}{\left(\frac{P_1}{P_2}\right)^{\frac{n-1}{n}} - 1} - \frac{1}{n} \qquad (2.46)$$

which is shown in Fig. 2.7 for gases with $n = 1\cdot 4$ over a range of pressure ratios up to sonic conditions. Substituting (2.44) and (2.45) in (2.43) gives the general linearized form for nozzle flow:

$$\frac{w_{12}}{W_{12}} = (1 + K_{12})\frac{p_1}{P_1} - K_{12}\frac{p_2}{P_2} + \frac{a_{12}}{A_{12}} - \frac{1}{2}\frac{t_1}{T_1} \qquad (2.47)$$

Usually for steady upstream conditions $t_1 = 0$ but this is not true for flow fluctuations within the system. In this case, for ideal gas flow

$$\frac{P^{\frac{n-1}{n}}}{T} = \text{const}$$

so that

$$\frac{t}{T} = \frac{n-1}{n}\frac{p}{P} \qquad (2.48)$$

which may be used to simplify the general equation to

$$\frac{w_{12}}{W_{12}} = \left[\left(\frac{n+1}{2n}\right) + K_{12}\right]\frac{p_1}{P_1} - K_{12}\frac{p_2}{P_2} + \frac{a_{12}}{A_{12}} \qquad (2.49)$$

although this does not give a value for incremental resistance directly. For a fixed nozzle ($a_{12} = 0$) and constant upstream conditions ($p_1 = 0$) the equation reduces to:

$$r = \frac{-p_2}{w_{12}} = \frac{P_2}{W_{12}K_{12}} \qquad (2.50)$$

The corresponding incremental resistance for constant downstream conditions ($p_2 = 0$) is:

$$r = \frac{p_1}{w_{12}} = \frac{P_1}{W_{12}\left[\left(\frac{n+1}{2n}\right) + K_{12}\right]} \qquad (2.51)$$

2.4 Lumped Parameter Fluid Networks

2.4.1 General

Fluid inductance commonly occurs with significant fluid resistance so that fluid networks consisting of lumped parameter inductive terms alone

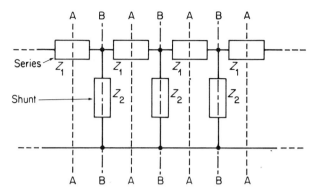

Fig. 2.12 Two terminal ladder network (Jackson, *Wave filters* (Methuen Monograph))

are difficult to achieve. On the other hand capacitive networks with and without resistance added are quite common. The lumped parameter model of a network will be largely dependent upon the range of frequencies under consideration. It has already been shown that, as with electronic components, the capacitive effect reduces and the inductive effect increases as the frequency is raised. Often low frequency network representations include only resistance and capacitance, but include an inductive term at high frequencies. This is discussed further in Section 2.4.4.

Generally linear system networks may be represented by combinations of impedances with two terminals, as shown in Fig. 2.12, which may be algebraically summated. In the figure Z_1 represents the series impedance while Z_2 represents the shunt impedance. The 'ladder' network illustrated is built up by the continuous repetition of identical sections obtained by cutting the circuit along lines A–A. Hence the network may be redrawn as a number of discrete sections with the series impedances Z_1 split in two as shown in Fig. 2.13.

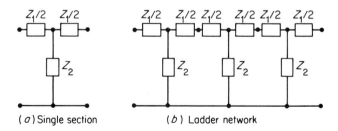

Fig. 2.13 Modified form of network (Jackson, *Wave filters* (Methuen Monograph))

Consider one filter section with series impedances $Z_A = Z_1/2$ and shunt impedance $Z_B = Z_2$, as shown in Fig. 2.14, loaded at the output with impedance Z_R. The impedance between the terminals A and B, Z_{AB}, is given by

$$Z_{AB} = Z_A + \frac{Z_B(Z_A + Z_R)}{Z_A + Z_B + Z_R} \tag{2.52}$$

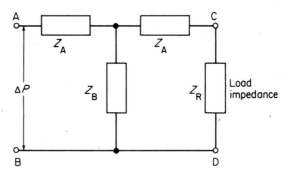

Fig. 2.14 Loaded network section

If the load impedance Z_R is the next section of the ladder network then, because of symmetry,

$$Z_{AB} = Z_R$$

must be true. Substituting in equation (2.52) and re-arranging

$$Z_{AB}^2 = Z_A(Z_A + 2Z_B) \tag{2.53}$$

But for a multistage filter

$$Z_A = \frac{Z_1}{2} \quad \text{and} \quad Z_B = Z_2$$

hence

$$Z_{AB}^2 = \frac{4Z_1Z_2 + Z_1^2}{4} \tag{2.54}$$

With these special conditions of network symmetry resulting in the input and output impedance of each network section being identical, the impedance Z_{AB} is termed the characteristic impedance, Z_{ch}. We may write the characteristic impedance from (2.54) as

$$Z_{ch} = \tfrac{1}{2}\sqrt{4Z_1Z_2 + Z_1^2} \tag{2.55}$$

[2.4] LUMPED PARAMETER FLUID NETWORKS

and is an important quantity in connexion with transmission lines as well as filters.

If, on the other hand, the load impedance Z_R is made equal to zero then the weight flow W_2 between terminals C and D does not involve any pressure drop, i.e. short circuit conditions. Hence from (2.52) the input impedance is

$$Z_{sc} = Z_A + \frac{Z_A Z_B}{Z_A + Z_B} \quad (2.56)$$

If the load impedance is infinite $Z_R = \infty$ and no flow across terminals C and D occurs, i.e. blocked load conditions. The input impedance is now

$$Z_{bl} = Z_A + Z_B \quad (2.57)$$

Combining (2.55), (2.56) and (2.57) we have

$$Z_{ch} = Z_{sc} Z_{bl} \quad (2.58)$$

which represents a useful way of determining Z_{ch} in practice.

2.4.2 Cascading Fluid Resistances

In the steady state the weight flows through each fluid resistance are equal and interconnexion capacitance may be neglected. Using the notation of Fig. 2.15, this implies that

$$W_{12} = W_{23} \quad (2.59)$$

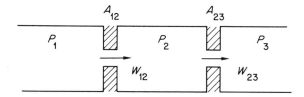

Fig. 2.15 Two nozzles in cascade

which for incompressible flow through orifices based on (2.10) gives:

$$C_{D_{12}} A_{12} \sqrt{2g\rho_2 (P_1 - P_2)} = C_{D_{23}} A_{23} \sqrt{2g\rho_3 (P_2 - P_3)}$$

assuming $\rho_2 = \rho_3$:

$$\frac{P_1 - P_2}{P_2 - P_3} = \left[\frac{C_{D_{23}} A_{23}}{C_{D_{12}} A_{12}} \right]^2 \quad (2.60)$$

In the general case of compressible nozzle flow using (2.14), the weight flow equation becomes:

$$\frac{KP_1 A_{12} N_{12}}{\sqrt{T_1}} = \frac{KP_2 A_{23} N_{23}}{\sqrt{T_2}} \qquad (2.61)$$

Unlike the incompressible case, the independent variables of pressure and area ratio are not easily separated allowing a rapid evaluation of the steady state conditions. This is because N is dependent on the pressure ratio and changes in flow conditions when the nozzles become choked. Again following the notation of Anderson[3] and assuming that $T_1 = T_2$, (2.61) may be re-written as:

$$\frac{P_1}{P_2} N_{12} = \frac{A_{23}}{A_{12}} N_{23} = C_{12} \qquad (2.62)$$

where the new parameter C_{12} is still a function of pressure ratio P_1/P_2 and γ. The variation of C_{12} with pressure ratio for air was shown in Fig. 2.7.

Four cases exist for a two nozzle configuration. These are:

A. Sonic flow in both nozzles ($N_{12} = N_{23} = 1$).
B. Subsonic flow in A_{12}, sonic in A_{23} ($N_{23} = 1$).
C. Sonic flow in A_{12}, subsonic in A_{23} ($N_{12} = 1$).
D. Subsonic flow in both nozzles.

The four regions are shown for air flow through the nozzles in Fig. 2.16 with lines of constant area ratio superimposed. Case D is the only one to present any computational difficulty as the full form of (2.62) must be used, requiring an iterative solution.

Clearly, due to inherent non-linearities, the incompressible and compressible flow conditions in fluid nozzle cascades do not give a pressure dividing effect analogous to voltage dividing in linear electronic circuits. The only fluid resistance network giving this relationship would be based on laminar capillary resistances given by (2.5) and (2.7). In this case

$$\frac{P_2 - P_3}{P_1 - P_3} = \frac{R_{23}}{R_{12} + R_{23}} \qquad (2.63)$$

2.4.3 Fluid Resistance and Capacitance Networks

Fluid capacitance has been considered in Section 2.2.3 associated with the properties of a fixed volume of fluid. During transient flow changes within the volume the conditions of the fluid, usually described by the Perfect Gas Law for a gas, change relative to ambient conditions. If the linearized fluid components representations are used for transient con-

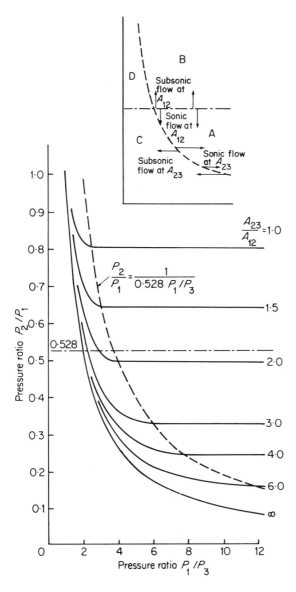

Fig. 2.16 Two nozzles in cascade—equilibrium conditions (Raven, F. H., *Automatic Control Engineering*, McGraw–Hill, 256, Fig. 12.14)

ditions, analogies may be drawn with equivalent electrical quantities. Thus in the case of a fluid capacitance, its electrical equivalent is a capacitance to earth, i.e. has values relative to ambient conditions. There is no non-moving part fluid equivalent of an electrical series capacitance.

A very common case of a fluid resistance and capacitance network is the two cascaded restrictions venting to atmosphere separated by a volume. The network is illustrated in Fig. 2.17, and includes the equivalent electrical circuit. Each of the fluid resistive impedances, denoted by Z_{12}

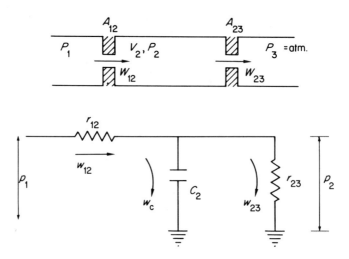

Fig. 2.17 Equivalent network for dynamic small signal analysis of two cascaded nozzles separated by a volume venting to atmosphere

and Z_{23} can be a laminar flow capillary, an incompressible flow nozzle or a compressible flow nozzle provided linearized forms are used in the latter two cases. From weight flow continuity:

$$w_{12} = w_C + w_{23} \tag{2.64}$$

Using equations (2.4) and (2.22), the transfer function of the circuit $G(s)$ is given by:

$$G(s) = \frac{p_2}{p_1} = \frac{r_{23}}{r_{12} + r_{23}} \frac{1}{1 + sr_p C} \tag{2.65}$$

where r_p is the parallel combination of r_{12} and r_{23}. The transfer function shows that such circuits are characterized by a first order lag with a time

[2.4] LUMPED PARAMETER FLUID NETWORKS

constant $r_p C$. The frequency response gives a bandwidth from steady state conditions up to the break frequency ω_b where:

$$\omega_b = \frac{1}{r_p C} \quad (2.66)$$

thus acting as a low pass filter.

The well known back pressure sensor or flapper amplifier discussed in Chapter 1 is a modification of this circuit. A flapper arm moving close to the downstream nozzles modifies its resistance thereby modulating the pressure P_2 in the interconnecting volume V_2, as shown in Fig. 2.18.

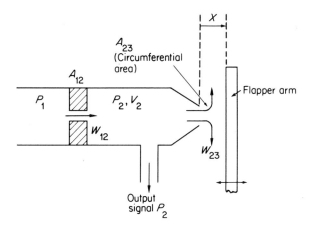

Fig. 2.18 Back pressure amplifier

Using the general nozzle flow equations for a gas in linearized form, the input flow is:

$$w_{12} = -C_1 p_2 \quad (2.67)$$

where

$$C_1 = \frac{W_{12} K_{12}}{P_2} \quad \text{from equation (2.50)} \quad (2.68)$$

The vented flow is dependent on flapper position X as well as the upstream pressure P_2, so that the linearized equation in this case is:

$$w_{23} = C_2 x + C_3 p_2 \quad (2.69)$$

where

$$C_2 = \frac{W_{23}}{X} \quad (2.70)$$

$$C_3 = \frac{W_{23}}{P_2}\left[\left(\frac{n+1}{2n}\right) + K_{23}\right] \quad \text{from equation (2.49)}$$

The change in weight flow within the volume V_2 is:

$$w_{\text{vol}} = C_4 s p_2 \tag{2.71}$$

where

$$C_4 = \frac{V_2}{n\mathbf{R}T_2} \tag{2.72}$$

Combining (2.67), (2.69) and (2.71) together using the weight continuity equation gives the transfer function as:

$$G(s) = \frac{p_2}{x} = -\frac{K}{1+s\tau} \tag{2.73}$$

where

$$\tau = \frac{C_4}{C_1 + C_3} \tag{2.74}$$

$$K = \frac{C_2}{C_1 + C_3}$$

Using the values for the constants C_1 to C_4 and remembering that in the steady state $W_{12} = W_{23}$ then

$$\tau = \frac{M_2}{nW_{12}}\left[\frac{1}{K_{12} + K_{23} + \left(\frac{n+1}{2n}\right)}\right] \tag{2.75}$$

$$K = \frac{P_2}{X}\left[\frac{1}{K_{12} + K_{23} + \left(\frac{n+1}{2n}\right)}\right]$$

Generally the gain of the flapper amplifier is high although for practical reasons, such as air economy, this is usually achieved with very small movements (typically 0·005 in). The time constant can be made small by suitable sizing of the chamber so that the frequency response is compatible with fluidic systems[4]. Essentially the amplifier is a high impedance device so that only very small load flows may be taken from the chamber without appreciably altering the gain. In digital systems, the sensor is operated non-critically between significantly different pressure levels so that the output flow limitation is less important.

The choice of geometry from the steady state conditions is important in determining the operating conditions of the flapper amplifier so that high gain is achieved. In the steady state, the system reduces to a cascade

[2.4] LUMPED PARAMETER FLUID NETWORKS

of two fluid resistances, one of which is variable dependent on the position of the flapper arm. For the general compressible flow case, equation (2.62) gives:

$$\frac{P_1}{P_2} N_{12} = \frac{A_{23}}{A_{12}} N_{23} = C_{12} \qquad (2.76)$$

where

$$A_{12} = C_{D_{12}} \frac{\pi D_{12}^2}{4}$$
$$A_{23} = C_{D_{23}} \pi D_{23} X \qquad (2.77)$$

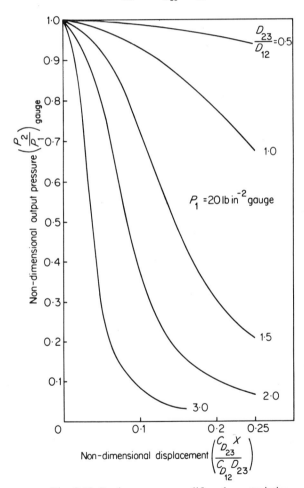

Fig. 2.19 Back pressure amplifier characteristics

The corresponding expression for incompressible flow given by (2.60) is:

$$\frac{P_1 - P_2}{P_2 - P_3} = \left[\frac{C_{D_{23}} A_{23}}{C_{D_{12}} A_{12}}\right]^2 \qquad (2.78)$$

If pressures are given as gauge values $P_3 = 0$ then:

$$\frac{P_1}{P_2} = \left[1 + \left(\frac{C_{D_{23}} A_{23}}{C_{D_{12}} A_{12}}\right)^2\right]^{-1} \qquad (2.79)$$

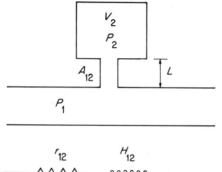

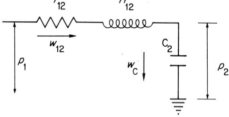

Fig. 2.20 Helmholtz resonator and equivalent circuit

Equations (2.62) and (2.77) may be combined to give a non-dimensional plot of pressure ratio against flapper displacement. Fig. 2.19 illustrates curves of constant diameter ratio D_{23}/D_{12} for a pneumatic flapper amplifier operating at 20 lb in^{-2} gauge. Usually the ratio of the gauge pressures P_1/P_2 is chosen in the region of 0·5 and the ratio of nozzle diameters selected for the desired gain and displacement. Commonly D_{23}/D_{12} is in the range 1·5 to 3·0.

2.4.4 Fluid Resistance, Capacitance and Inductance Networks

Any fluid network containing transient energy storing elements of capacitance and inductance acting in opposition will resonate if the forcing frequency approaches a critical value. The simplest of such fluid circuits is shown in Fig. 2.20 and is known as a Helmholtz resonator. Due to the high velocities occurring in the neck of the valve both fluid resistive and inductive components are present.

The overall impedance may be expressed as:

$$Z(j\omega) = r_{12} + j\left(\omega H_{12} - \frac{1}{\omega C_2}\right) \qquad (2.80)$$

At resonance the reactance is zero and $\omega = \omega_r$ where:

$$\omega_r = \frac{1}{\sqrt{C_2 H_{12}}} = \sqrt{\frac{nA_{12}g\mathbf{R}T_2}{V_2 L}} \quad \text{for a gas} \qquad (2.81)$$

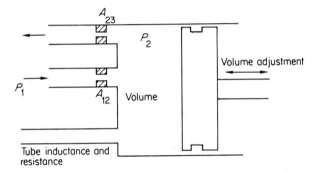

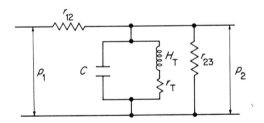

Fig. 2.21 Adjustable parallel Helmholtz resonator and equivalent circuit (*Ref. 5*)

A more useful Helmholtz resonator[5] has been used in adjustable form for carrier frequency modulation techniques, which are discussed in Chapter 5. It consists of an inductance and capacitance in parallel, rather than in series, as shown in Fig. 2.21 which has a maximum impedance at resonance thereby allowing an optimum load matching between the frequency source and the resonator. The parallel resonator has a transfer function of the form

$$\frac{p_2}{p_1} = \frac{K(1 + \tau s)}{\left(\dfrac{s^2}{\omega_r^2} + 2\zeta \dfrac{s}{\omega_r} + 1\right)} \qquad (2.82)$$

where for

$$r_{12} \gg r_T$$

$$r_{23} \gg r_T$$

$$\omega_r = \frac{1}{\sqrt{CH_T}}$$

$$\zeta = \frac{1}{2\omega_r}\left[r_T C + H_T\left(\frac{r_{12} + r_{23}}{r_{12} r_{23}}\right)\right]$$

$$K = \frac{r_T}{r_{12}}$$

$$\tau = \frac{H_T}{r_T}$$

If τ is much greater than $1/\omega_r$, the Helmholtz resonator alone has little phase lag at the resonant frequency.

2.5 Fluid Transmission Lines

NOTATION

A	= tube cross-sectional area		γ	= ratio of specific heats
a	= tube radius		θ	= temperature
c	= propagation velocity		ρ	= density
C_0	= adiabatic capacitance		ρ'	= frequency dependent density coefficient
k	= attenuation constant			
K, K'	= resistance coefficients		μ	= absolute viscosity
L	= transmission line length		ω	= angular frequency
p	= pressure		ν	= μ/ρ = kinematic viscosity
r	= radial distance		ϕ	= phase angle
t	= time		σ^2	= Prandtl Number
v	= axial particle velocity			
$\bar{v}$	= sectional mean velocity		Subscript	
α	= propagation constant		0	= reference conditions
β	= liquid bulk modulus			

2.5.1 General

It is often important to know the dynamic behaviour of the channels and tubes interconnecting fluidic devices in a circuit. A linear analysis for sinusoidal signals can be made with good accuracy at low frequencies by representing the channels as lumped parameter networks using the analysis of the previous Section. However, in typical fluidic systems the signal frequency range of interest is usually 1 to 1000 Hz which is transmitted through tubes of diameter in the range 0·1 in to 0·04 in over lengths from a few inches to 10 ft. This range of configurations requires a complex theoretical analysis for prediction of signal conditions. This is because the system parameters have a spatial, as well as time, distribution when the signal wavelength is short compared with the length of the transmission line. Under such conditions the signal amplitude changes appreciably in a short distance thereby requiring the use of a distributed parameter analysis. Details of transient and frequency response analysis of fluid transmission lines can be found in the literature[6-10].

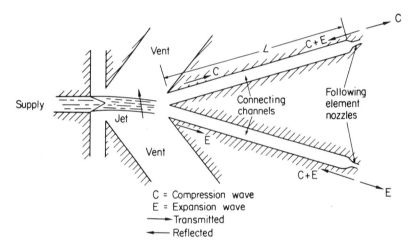

Fig. 2.22 Wave effects of supply jet deflexion in a beam deflexion amplifier

When a beam deflexion fluid amplifier output state changes appreciably, the supply jet is deflected out of one output channel into the other. Under dynamic conditions the new signals are propagated as waves as shown in Fig. 2.22. A compression wave (pressure increase) is propagated through the channel into which the jet is deflected. If each channel, of length L, is closed-ended, the incident wave is reflected back in the same sense, i.e. compression as compression and expansion as expansion. However, for a completely open channel the opposite occurs, i.e. a compression wave is reflected as an expansion wave and vice versa. This maintains the ambient condition constant. The prediction of individual waves within the channels is only important when the frequency is high enough so that the channel lengths L are comparable to the wavelength.

In practice, the connecting channel is terminated by the next element in the network so that the load on the amplifier is the input nozzle of this element. Under these conditions the channel is partially open so that an incident wave will be partly reflected as a compression wave and partly as an expansion wave. There exists an opening for the nozzle such that these reflexions will cancel each other and it is in this condition that we say the channel is dynamically matched to the load. Often the terminating load can only be matched to the pipe at one particular frequency.

In general, a transmission line may be considered as an infinite number of sections of a two terminal ladder network as defined in Fig. 2.12. If each section is of length dx and the series impedance per unit length is z_1, then the series impedance of a section is

$$Z_1 = z_1 \, dx$$

Usually the inverse of the shunt impedance, called the shunt admittance Y_2, is used: that is

$$Y_2 = \frac{1}{Z_2}$$

and the shunt admittance per unit length y_2 is given by

$$Y_2 = y_2 \, dx$$

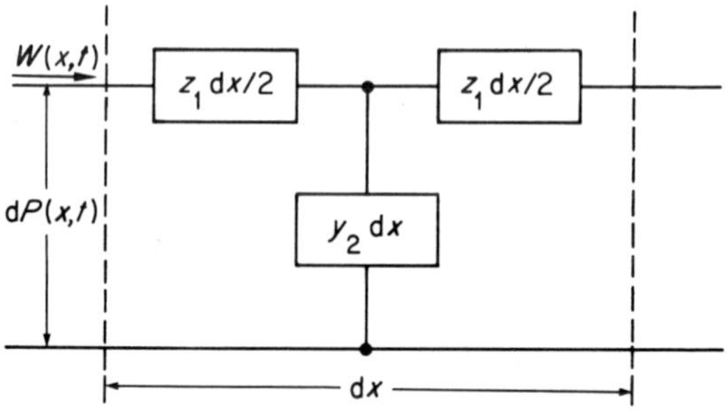

Fig. 2.23 Section of a transmission pipe

Fig. 2.23 illustrates a section of such a transmission line. For a distributed parameter system, we may define the impedance per unit length as

$$z = -\frac{\frac{\partial p}{\partial x}}{W} \qquad (2.83)$$

and the admittance per unit length as

$$y = \frac{-\frac{\partial W}{\partial x}}{p} \quad (2.84)$$

remembering that the flow and pressure decrease as distance x increases. The following analysis for liquid and pneumatic lines assumes small amplitude signals which implies that the velocity of the fluid particles is not large compared with the acoustic propagation velocity. Large amplitude waves will be considered later in Section 2.5.4.

2.5.2 Liquid Transmission Lines

The application of fluidics to liquid systems is limited at the present time so that the analysis of liquid transmission lines only has restricted usefulness. However, it gives considerable insight into the more complex case of a gas filled line and is more easily used. The main simplification in the analysis of liquid lines occurs because it is possible to neglect internal heat exchange processes produced by the transmission of a signal fluctuation through the liquid. This is not possible, particularly at low frequencies, in a pneumatic line because the equation of state for a gas shows that the pressure change must be expressed as

$$dp = p_0 \cdot \frac{d\rho}{\rho_0} + p_0 \cdot \frac{d\theta}{\theta_0} \quad (2.85)$$

which, because of heat transfer, means that the second term is not negligible. In contrast, the corresponding expression for a liquid line is

$$dp = \beta \frac{d\rho}{\rho_0} \quad (2.86)$$

Equations (2.85) and (2.86) give a measure of the compressibility of the fluid concerned and are associated with the distributed capacitance of the line.

In pipes of uniform cross-section it is often more convenient to use pressure, p, and particle velocity, v, as the two variables to describe the system. Conversion to weight flow is easily achieved. It can be shown[11] that from continuity and momentum considerations the pressure and velocity gradients respectively are

$$\left. \begin{array}{l} -\dfrac{\partial p}{\partial x} = \rho \dfrac{\partial v}{\partial t} - \dfrac{\mu}{r} \dfrac{\partial}{\partial r}\left(r \dfrac{\partial v}{\partial r}\right) \\ \\ -\dfrac{\partial v}{\partial x} = \dfrac{1}{\beta} \dfrac{\partial p}{\partial t} \end{array} \right\} \quad (2.87)$$

on the assumption that:
- (a) The signal amplitudes are small.
- (b) Only longitudinal waves are present ($\omega v/c^2 \ll 1$).
- (c) Uniform pressure exists across a pipe cross-section ($v/ac \ll 1$).
- (d) Laminar flow.
- (e) Rigid walls.
- (f) Isothermal conditions at the walls.

Defining a mean velocity $\bar{v}$ for each pipe section and allowing only sinusoidal fluctuations of angular frequency ω (2.87) becomes

$$\left. \begin{aligned} -\frac{\partial p}{\partial x} &= \rho' \frac{\partial \bar{v}}{\partial t} + K'\bar{v} \\ -\frac{\partial \bar{v}}{\partial x} &= \frac{1}{\beta} \frac{\partial p}{\partial t} \end{aligned} \right\} \quad (2.88)$$

where ρ' and K' are frequency dependent density and resistance coefficient terms respectively. As $\omega \to 0$, $\rho' \to 4\rho/3$ and $K' \to K$ where

$$K = \frac{8}{a^2}$$

An extremely useful non-dimensional frequency number, which occurs commonly in fluid transmission line theory is

$$a\sqrt{\frac{\omega}{v}}$$

which can be used to define the different dynamic flow regimes. Fig. 2.24

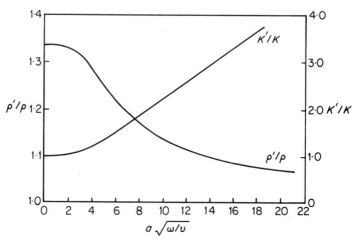

Fig. 2.24 Effect of frequency on the inertia and resistance of a liquid transmission line (Foster and Parker, *Ref. 11*, 599)

[2.5] FLUID TRANSMISSION LINES

shows the effect of non-dimensional frequency on both the density and resistance coefficients. It can be clearly seen that with increasing frequency the resistance coefficient increases appreciably.

For sinusoidal signals we may substitute $d/dt = D = j\omega$ in (2.88) and obtain expressions for the series impedance, z_1, and shunt admittance, y_2. Thus

$$z_1 = \frac{-\frac{\partial p}{\partial x}}{W} = \frac{1}{\rho A}[K' + j\omega \rho']$$

$$y_2 = \frac{-\frac{\partial W}{\partial x}}{p} = j\omega \frac{\rho A}{\beta}$$

(2.89)

which shows that the series impedance is dependent only on resistance and inertia effects while the shunt admittance is a capacitive effect. This means that the partial differential equation (2.88) representing the liquid line is

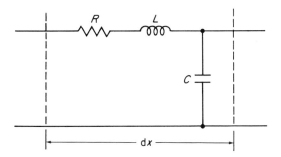

Fig. 2.25 Electrical analogue of a liquid transmission line

analogous to the well-known electrical telegraph line equations without the term for conductance. Fig. 2.25 shows the electrical analogue of the liquid line. The electrical and fluid equivalents are:

$$R \equiv \frac{K'}{\rho A} \qquad L \equiv \frac{\rho'}{\rho A} \qquad C \equiv \frac{\rho A}{\beta}$$

We may solve the partial differential (2.86) for sinusoidal forced oscillation of the line by writing:

$$\bar{v} = V e^{j\omega t} \qquad p = P e^{j(\omega t + \phi)}$$

so that separating the variables:

$$\frac{d^2\bar{v}}{dx^2} - \alpha^2\bar{v} = 0 \qquad \frac{d^2p}{dx^2} - \alpha^2 p = 0 \qquad (2.90)$$

where α is known as the propagation constant with:

$$\alpha = \frac{1}{c}\sqrt{j\omega(j\omega + k)} = \sqrt{z_1 y_2}$$

$$c = \sqrt{\frac{\beta}{\rho'}} \quad \text{is the propagation velocity}$$

$$k = \frac{K'}{\rho'} \quad \text{is the attentuation constant}$$

The generalized solutions to (2.90) are:

$$\left.\begin{array}{l}\bar{v} = C_1 \cosh \alpha x + C_2 \sinh \alpha x \\ p = C_3 \cosh \alpha x + C_4 \sinh \alpha x\end{array}\right\} \qquad (2.91)$$

Where C_1, C_2, C_3, C_4 are constants.

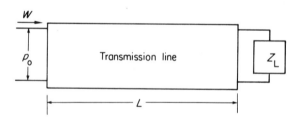

Fig. 2.26 Load termination of a transmission line

Considering a transmission line of length L as a two terminal network with a general load impedance Z_L attached to it, as shown in Fig. 2.26, then the input impedance, Z_0, is given by:

$$Z_0 = \left.\frac{p}{\rho A \bar{v}}\right|_{x=0} \qquad (2.92)$$

and the load impedance,

$$Z_L = \left.\frac{p}{\rho A \bar{v}}\right|_{x=L} \qquad (2.93)$$

[2.5] FLUID TRANSMISSION LINES

using (2.92) and (2.93) in (2.91) gives the relationship:

$$\frac{Z_0}{Z_L} = \frac{\tanh \alpha L + \dfrac{Z_L}{Z_{ch}}}{1 + \tanh \alpha L \dfrac{Z_L}{Z_{ch}}} \qquad (2.94)$$

where the characteristic impedance

$$Z_{ch} = \frac{1}{\rho A}\sqrt{\frac{\beta\rho'(j\omega + k)}{j\omega}} = \sqrt{\frac{z_1}{y_2}} \qquad (2.95)$$

Inspection of (2.94) shows that when $Z_L = Z_{ch}$ then also $Z_0 = Z_{ch}$ and the impedance remains unaltered along the length of the pipe. Thus the pipe behaves as though it were infinitely long and is a very important condition for matching. This is because a signal wave travelling along the pipe will not be reflected to cause interference with the next signal and so can be thought of as perfect matching. Equation (2.95) representing the characteristic impedance is a complex quantity, i.e. it has magnitude and phase, which will change with frequency. Brown[9] has computed the characteristic values for both liquid and air lines which are shown in Fig. 2.27. The impedance marked Z_{co} has a value of

$$\left. \begin{aligned} Z_{co} &= \frac{1}{A}\sqrt{\frac{\gamma P_0}{\rho_0}} \quad \text{for a gas} \\ Z_{co} &= \frac{1}{A}\sqrt{\frac{\beta}{\rho_0}} \quad \text{for a liquid} \end{aligned} \right\} \qquad (2.96)$$

Comparing the characteristic impedance for a line with that of a lumped parameter network obtained in Section 2.4.1, it is seen that a line does not have a cut-off frequency nor does the characteristic impedance become zero at any frequency. Instead the characteristic impedance reduces to approximately a constant value, when the non-dimensional frequency

$$a\sqrt{\frac{\omega}{v}} > 30$$

Sometimes it is not possible to make the load impedance have the characteristic value. In this case it is still possible to prevent reflexions in the transmission pipe by the use of small branch pipes or stubs connected to the pipe close to the load. Usually a closed ended tube is chosen which has a length designed to provide a reflexion which will cancel with the load reflexion wave at a particular frequency although of necessity this does not provide matching over a large range of operating conditions.

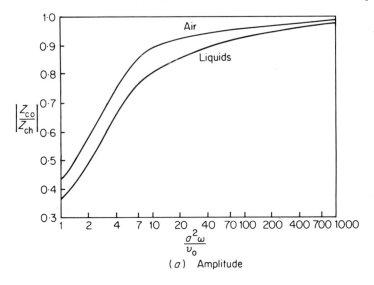

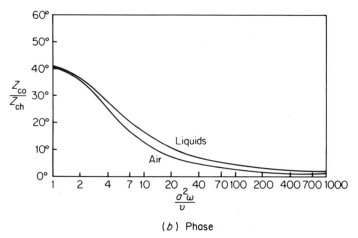

Fig. 2.27 Fluid characteristic impedance variations with frequency (Brown, *Ref. 9*)

2.5.3 Pneumatic Transmission Lines

Nichols[10] has given a linear theory for pneumatic transmission lines which includes instantaneous heat transfer effects. Some of his approximations for low frequency predictions have been improved by Krishnaiyer and Lechner[12] from which source the following analysis is taken.

Nichols uses two characteristic frequencies, similar to the one defined for liquid lines, which are:

$$\omega_v = \frac{8\pi v}{A} \qquad \omega_T = \frac{\omega_v}{\sigma^2} \tag{2.97}$$

where σ^2 is the Prandtl number, a dimensionless property of the fluid with a value of 0·708 for air. The forcing frequency ω is written non-dimensionally in each case as:

$$h_v = 2\sqrt{\frac{\omega}{\omega_v}} \qquad h_T = 2\sqrt{\frac{\omega}{\omega_T}} \tag{2.98}$$

Nichols' analysis based on volume rather than weight flow results in the following equations

$$z_1 = \frac{8\pi\mu}{A^2}\frac{jh_v^2}{4}\left[\frac{1}{1 - J(h_v\sqrt{2})} - 1\right] + j\omega\frac{\rho}{A} \tag{2.99}$$

$$y_2 = j\omega\frac{A}{\gamma p}[1 + (\gamma - 1)J(h_T\sqrt{2})] \tag{2.100}$$

By curve fitting for the Bessel functions, the series impedance and shunt admittance may be closely approximated to:

$$z_1 \simeq \frac{8\pi\mu}{A^2}C_1 + j\left[\frac{\omega\rho}{A} + \frac{8\pi\mu}{A^2}C_2\right] \tag{2.101}$$

$$y_2 \simeq \frac{\omega(\gamma - 1)AC_4}{\gamma p(C_3^2 + C_4^2)} + j\omega\left[\frac{A}{\gamma p} + \frac{(\gamma - 1)AC_3}{\gamma p(C_3^2 + C_4^2)}\right] \tag{2.102}$$

where

$$C_1 = \frac{3}{8} + \frac{h_v}{4} + \frac{3}{8h_v} \tag{2.103}$$

$$C_2 = \frac{1}{4h_v} - \frac{15}{64h_v} \tag{2.104}$$

$$C_3 = \frac{1}{4} + \frac{h_T}{2} + \frac{1}{4h_T} \tag{2.105}$$

$$C_4 = \frac{h_T}{2} - \frac{1}{4h_T} \tag{2.106}$$

These equations are in a form suitable for computation in the frequency range $0.1\omega_v < \omega < \infty$.

When the input pressure p_0 is known, which is usually the case in fluidic transmission lies, (2.94) can be expressed in alternative form to give the ratio of load to input pressure as:

$$\frac{p_L}{p_0} = \frac{2Z_{ch}Z_L \exp(-\alpha L)}{Z_{ch}(Z_L + Z_{ch}) + Z_{ch}(Z_L - Z_{ch}) \exp(-2\alpha L)} \qquad (2.107)$$

where the propagation constant

$$\alpha = \sqrt{z_1 y_2}$$

and characteristic impedance

$$Z_{ch} = \sqrt{\frac{z_1}{y_2}}$$

Although the basic Nichols theory explicitly excludes any mean flow through the pipe, Krishnaiyer and Lechner have found in practical tests that good correlation is still obtained with mean flow present. However, no upper mean flow limits are indicated by the authors.

Recently, the analysis of fluid transmission lines has been extended by Schaedel[13] to include lines of rectangular cross-section, although no experimental information is offered to confirm the theory by the author. Franke, Karam and others[14,15,16] using computer solutions have shown that the Nichols theory and its extension by Schaedel give good agreement for a wide range of experimental conditions. Fig. 2.28 shows typical response data for volume terminated transmission lines of circular and rectangular section given by Franke *et al*.

Volume terminated circular and rectangular lines show particularly good agreement between theory and experiment and this has been extended to include laminar and turbulent mean flow through the line. The theory can also be used to predict the performance of several series connected lines of different length and diameter. However, the most useful case for fluidic systems analysis, namely an orifice terminated line with mean flow present, has yet to be satisfactorily predicted.

Useful approximations to transmission line equations may be made for lines terminated by a volume operating at low dimensionless frequencies i.e. $a\sqrt{\omega/v} < 3$. Hougen, Martin and Walsh[17] replace the distributed parameter line by a second order system with good accuracy. The ratio of the load pressure p_L to the input pressure p_0 when the load is a volume V, is given by:

$$\frac{p_L}{p_0} \simeq \frac{1}{1 + \frac{2\xi j\omega}{\omega_n} - \frac{\omega^2}{\omega_n^2}} \qquad (2.108)$$

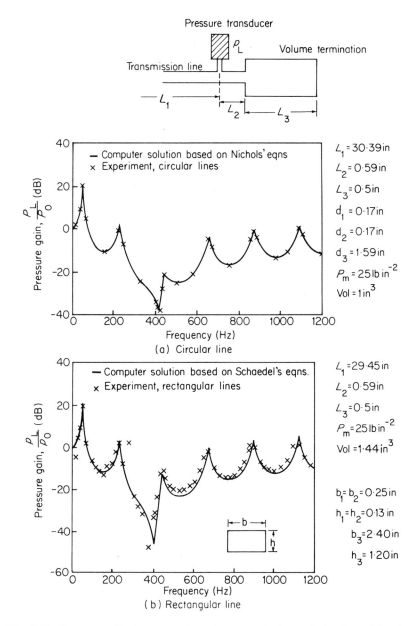

Fig. 2.28 Response of volume terminated pneumatic transmission lines (Franke et al., Ref. 16)

where

$$\omega_n = \frac{c'}{L\sqrt{\frac{1}{2} + \frac{V}{AL}}} \qquad \zeta = \frac{RL}{2\rho c'}\sqrt{\frac{1}{2} + \frac{V}{AL}} \qquad (2.109)$$

is the damping ratio

$$R = \text{resistance} = 32\frac{\mu}{d^2} \qquad (2.110)$$

$$c' = \text{effective acoustic velocity} = \sqrt{\frac{3}{4}\frac{\gamma p_0}{\rho_0}} \qquad (2.111)$$

If the damping ratio $\gg 1$ the system may be replaced by a first order system, in which case:

$$\frac{p_L}{p_0} \simeq \frac{e^{-j\omega\tau_d}}{1 + \frac{2\zeta j\omega}{\omega_n}} \qquad (2.112)$$

where the transport time, represented by the exponential term in (2.112), has a time delay given by

$$\tau_d = \frac{L}{c'} \qquad (2.113)$$

Samson[18] gives a similar approximate formula for the step input response of a line terminated by a volume. In this case,

$$\frac{p - p_0}{p_m - p_0} = 1 - e^{-\frac{t-\tau_d}{\tau_1}} \quad (t \geqslant \tau_d) \qquad (2.114)$$

where

$$p_0 = \text{initial pipe pressure}$$
$$p_m = \text{step input pressure}$$
$$\tau_d = \text{time delay} = \frac{L}{c'}$$
$$\tau_1 = \frac{RL^2}{\gamma p_0}\left(\frac{V}{AL} + \frac{1}{2}\right)$$

2.5.4 Transmission Lines—Large Amplitude Signals

If a digital device such as a bistable wall attachment element is switched from one output channel to the other the change in conditions is even more abrupt than changes in to a proportional amplifier, i.e. the signal will be in the form of a step change rather than a sinusoidal change. The

[2.5] FLUID TRANSMISSION LINES 89

connecting channel must then be analysed for its transient rather than frequency response characteristics. If the amplitude of the pressure signal is sufficiently large the small amplitude theory of Sections 2.5.2 and 2.5.3 breaks down. This is because the fluid particle velocity becomes significant compared with the acoustic velocity, causing distortion of the signal wave as it passes along the channel. Such conditions are non-linear and occur more readily in gases than liquids. Analytic techniques in this field rely on graphical analyses such as the method of characteristics[19,20].

The limits of the applicability of acoustic and finite wave theory may be determined by consideration of the signal propagation velocity and particle velocity in both cases. A basic assumption of acoustic theory is that the absolute propagation velocity, c, of the wave does not vary according to the strength of the signal pressure; that is, it is independent of the particle velocity.

The propagation velocity, a_0, of all points on a plane wave is given by:

$$a_0 = \sqrt{\frac{\gamma P_0}{\rho_0}} = \sqrt{\gamma R T_0} \qquad (2.115)$$

when moving through a gas whose absolute pressure, temperature and density are P_0, T_0 and ρ_0 respectively. Hence, the acoustic velocity is only dependent on temperature for a given gas, which leads to:

$$c = a_0 \qquad (2.116)$$

Although small, the particle velocity, v, is dependent on the gauge pressure p above the reference pressure P_0 according to the relation,

$$v = \frac{a_0 p}{\gamma P_0} = \frac{a_0}{\gamma}\left[\left(\frac{P}{P_0}\right) - 1\right] \qquad (2.117)$$

where P is the absolute pressure corresponding to gauge pressure p. In a finite amplitude wave the particle velocity changes appreciably and is dependent on the local conditions in the wave. The absolute propagation velocity in this case is the summation of the relative propagation velocity, a, and the particle velocity, v. i.e.

$$c = a + v \qquad (2.119)$$

The local values of pressure, density and temperature are P, ρ and T and since the compression or expansion process is adiabatic,

$$\frac{T}{T_0} = \left(\frac{P}{P_0}\right)^{\frac{\gamma-1}{\gamma}}$$

giving

$$\frac{a}{a_0} = \left(\frac{P}{P_0}\right)^{\frac{\gamma-1}{2\gamma}} \qquad (2.120)$$

Combining (2.119) and (2.120)

$$c = a_0 \left(\frac{P}{P_0}\right)^{\frac{\gamma-1}{2\gamma}} + v \qquad (2.121)$$

or for air with $\gamma = 1.4$

$$c = a_0 \left(\frac{P}{P_0}\right)^{\frac{1}{7}} + v \qquad (2.122)$$

It can also be shown that for a simple wave of finite amplitude the pressure–particle velocity relationship is:

$$v = \frac{2}{\gamma - 1} a_0 \left[\left(\frac{P}{P_0}\right)^{\frac{\gamma-1}{2\gamma}} - 1\right] \qquad (2.123)$$

Combining (2.121) and (2.123)

$$c = a_0 \left[6\left(\frac{P}{P_0}\right)^{\frac{1}{7}} - 5\right] \qquad (2.124)$$

$$v = 5a_0 \left[\left(\frac{P}{P_0}\right)^{\frac{1}{7}} - 1\right] \qquad (2.125)$$

Comparing (2.117) and (2.125) for the particle velocities, it is seen that in the acoustic case v has a linear relation with p, but in the finite wave case this is non-linear.

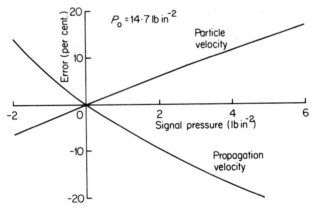

Fig. 2.29 Error in using small amplitude rather than finite wave theory

Fig. 2.29 shows the error introduced into the particle velocity and propagation velocity by the use of the acoustic theory rather than the more accurate finite wave theory. It is plotted for a range of signal gauge pressures, from atmospheric conditions, which are likely to be of interest in the majority of fluidic applications. The positive error indicates an acoustic wave prediction which is too large while the negative error a prediction which is too small. For a finite amplitude compression wave (positive pressures), the values of particle velocity indicated by the small wave theory are too high while for compression waves the opposite is true. Propagation velocities of points on waves of compression and expansion are respectively higher and lower than the acoustic velocity. The limit of applicability of small amplitude wave theory without introducing errors exceeding about 10 per cent shown by the figure is

$$0\cdot 90 < \frac{P}{P_0} < 1\cdot 14 \qquad (2.126)$$

or in terms of the gauge pressure amplitude p of the signal

$$1\cdot 5 \text{ lb in}^{-2} < p < +2 \text{ lb in}^{-2} \qquad (2.127)$$

If it is required to match a connecting channel to the input nozzle of the next fluidic element when the signals transmitted have a finite amplitude,

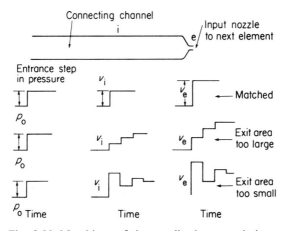

Fig. 2.30 Matching a finite amplitude transmission line to a nozzle (Kirshner, *Ref. 1*)

the nozzle area must be set to an optimum value, as shown in Fig. 2.30, so that no reflexions occur. As in the case of small amplitude waves, matching side branches may be connected to the connecting channel to aid this process[21].

References

1. Kirshner, J. M. *Fluid Amplifiers*, McGraw-Hill, New York, 1966.
2. Taplin, L. B. Small signal analysis of vortex amplifiers. Prepared for Agard Lecture Series, *Fluid Control—Components and Systems* (S. 1966).
3. Anderson, B. W. *The Analysis and Design of Pneumatic Systems*, Wiley, New York, 1967.
4. Moses, H. L., Small, D. A. and Cotta, G. A. Response of a fluidic air gauge. *A.S.M.E.* Paper 68-WA/FE-16.
5. Boothe, W. A., Ringwall, C. G. and Kelley, L. R. A fluid amplifier technique for speed governing using carrier frequency modulation, *I.F.A.C. Meeting* (London), Paper 31G (1966).
6. Iberall, A. S. Attenuation of oscillatory pressure in instrument lines, *J. Res. Natl. Bur. Stds.*, Paper RP 2115, **45** (July 1950).
7. Moise, J. C. Pneumatic transmission lines. *J. Instr. Soc. Am.*, **1**, 35 (Apr. 1954).
8. Schuder, C. B. and Binder, R. C. The response of pneumatic transmission lines to step inputs. *J. Basic Eng.*, **81**, 578 (1959).
9. Brown, F. T. The transient response of fluid lines. *J. Basic Eng.*, Paper 61WA-143.
10. Nichols, N. B. The linear properties of pneumatic transmission lines. *Trans. Instr. Soc. Am.*, **1**, No. 1, 5 (1962).
11. Foster, K. and Parker, G. A. Transmission of power by sinusoidal wave motion through hydraulic oil in a uniform pipe. *Proc. Inst. Mech. Engrs.*, **179**, Part 1 (1964).
12. Krishnaiyer, R. and Lechner, T. J. An experimental evaluation of fluidic transmission line theory. *A.F.*, **1967**, 367.
13. Schaedel, H. A theoretical investigation of fluidic transmission lines with rectangular cross-section. *C.F.C.* (3), Paper K3.
14. Karam, J. T. The frequency response of blocked pneumatic lines. *M.Sc. Thesis*, GAM/ME/66B-3, Air Force Inst Tech., Wright-Patterson AFB, Ohio, U.S.A. (1966).
15. Karam, J. T. and Franke, M. E. The frequency response of pneumatic lines. *J. Basic Eng.*, *A.S.M.E.*, Ser. D, **89**, 371–378 (1967).
16. Franke, M. E., Wilda, R. W., Miller, R. N. and Fada, C. V. The frequency response of volume terminated pneumatic lines with circular and rectangular cross-sections. *J.A.C.C.*, University of Colorado, Boulder, 410 (1969).
17. Hougen, Martin and Walsh. Dynamics of pneumatic transmission lines. *Control Eng.*, **10**, 114 (1963).
18. Samson, J. E. Dynamic characteristics of pneumatic transmission. *Trans. Soc. Instr. Tech.*, **10**, 117 (1958).
19. Shapiro, A. H. *The Dynamics and Thermodynamics of Compressible Fluid Flow*, Vols. I and II, Ronald, New York, 1953.
20. Rudinger, G. *Wave Diagram in Non-steady Flow in Ducts*, Van Nostrand, New York, 1955.
21. Walston, W. H. Transient response of a fluid line with and without bleeds. *H.D.L.* (3), **II**.

3
The Fluid Mechanics of Jets, Wall Attachment and Vortices

3.1 The Laminar Jet

If a jet emerges from a narrow slit or hole at a low enough velocity, the flow will remain laminar. Even under these conditions the jet carries with it some of the surrounding fluid because of the viscous shearing forces, and the resulting pattern of streamlines is approximately as indicated in Fig. 3.1. A concise account of the theoretical treatment of this situation is given by Schlichting[1] and is based on his own work and that of Bickley[2]. The assumption is made that the velocity profiles remain similar, as the problem does not possess any characteristic linear dimension providing

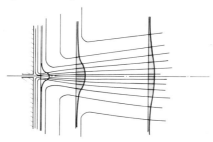

Fig. 3.1 An indication of the velocity profiles in a laminar jet. Fluid from the surroundings is accelerated by viscous forces, whilst that of the jet is slowed down. Thus the velocity profiles become 'flatter' with distances downstream and the effect is more marked as the Reynolds Number is reduced

the aperture from which the jet emerges is infinitesimally small. Taking coordinates x parallel to the axis of the jet measured from the nozzle and y perpendicular to the axis measured from the centre line, the velocity in the direction of the axis of the jet, u, is assumed to be a function of y/b, where b is the width of the jet, according to some convenient definition, and is itself proportional to x^q. It is shown in Ref. 1 that 'similar' solutions to the boundary layer equations exist if the velocity, and therefore the stream function, is proportional to a power of the distance from the stagnation point, and the following stream function is assumed

$$\psi \propto x^p f\left(\frac{y}{b}\right) = x^p f\left(\frac{y}{x^q}\right) \tag{3.1}$$

Two assumptions are now made

(a) Because the pressure throughout the flow field is almost uniform, the pressure gradient (dp/dx) is neglected, so that the total momentum in the x direction must be constant, i.e.

$$J = \rho \int_{-\infty}^{+\infty} u^2 \, dy = \text{a const.} \tag{3.2}$$

(b) Because the friction itself gives rise to the acceleration of the fluid which creates the jet 'spread', the friction and acceleration terms in the boundary layer equation must be of the same order of magnitude. This equation in two dimensions is:

$$u\frac{\partial u}{\partial x} + v\frac{\partial u}{\partial y} = U\frac{dU}{dx} + v\frac{d^2 u}{dy^2} \tag{3.3}$$

where the first term on the right-hand side may be neglected in this problem.

Using these two assumptions, two simultaneous equations in p and q are found which give $p = \frac{1}{3}$ and $q = \frac{2}{3}$. By making a suitable transformation of coordinates, a differential equation in ψ and a new co-ordinate ξ is obtained, which may be integrated. A similar approach is possible for the circular jet, but in this case $p = q = 1$. The results for the two cases are summarized in Table 3.1.

Notice that for the circular laminar jet, the volume flow is independent of the initial momentum of the jet. This means that a slower jet must spread much more than a fast jet in order to draw in the appropriate amount of fluid. A similar situation occurs in the two-dimensional laminar jet but not quite to the same extent, since the volume flow is proportional to $J^{\frac{1}{3}}$.

Beatty and Markland[3] make interesting observations on the matching

[3.1] THE LAMINAR JET   95

Table 3.1

	Two-dimensional jet	Circular jet
u	$0\cdot 4543\left(\dfrac{J^2}{v x \rho^2}\right)^{\frac{1}{3}}(1-\tanh^2\zeta)$	$\dfrac{3}{8\pi}\dfrac{J}{\rho v x}\dfrac{1}{(1+\frac{1}{4}\xi^2)^2}$
v	$0\cdot 5503\left(\dfrac{Jv}{x^2\rho}\right)^{\frac{1}{3}}\{2(1-\tanh^2\zeta)-\tanh\zeta\}$	$\dfrac{1}{4}\sqrt{\dfrac{3}{\pi}}\dfrac{\sqrt{J/\rho}}{x}\dfrac{(\xi-\frac{1}{4}\xi^3)}{(1+\frac{1}{4}\xi^2)^2}$
ζ	$0\cdot 2752\left(\dfrac{J}{\rho v^2}\right)^{\frac{1}{3}}\dfrac{y}{x^{\frac{2}{3}}}$	$\sqrt{\dfrac{3}{16\pi}}\dfrac{\sqrt{J/\rho}}{v}\dfrac{y}{x}$
Q	$3\cdot 3019\left(\dfrac{J}{\rho}vx\right)^{\frac{1}{3}}$	$8\pi v x$

of a real circular jet emerging from a hole of finite diameter with the theoretical jet emanating from a point source. The difficulty lies in the fact that the theoretical profile at the nozzle cannot be the correct one, so that the theoretical and actual values of the various important jet characteristics cannot all be identical. For the circular jet, they tabulate the ratios of various parameters for three types of matching given by the following alternative assumptions for the conditions at the nozzle exit:

(*a*) the radial component v_a of velocity at $y = a$ is zero, where a is the radius of the nozzle;
(*b*) the kinetic energy of the similarity solution is equal to that of the jet;
(*c*) the flow in the similarity solution is equal to that of the complete jet.

Table 3.2 compares the matching of the parameters for each assumption.

Whichever assumption is made, the kinematic momentum of the solution must be the same as that of the real jet at the plane of the nozzle. The similarity profile at this plane may be written for simplicity as

$$u = \frac{A\bar{u}_0}{\{1+B(y/a)^2\}^2} \tag{3.4}$$

where the constants A and B are given in the table, together with the other parameters, the most important of which is probably x_0/d Re which gives the position behind the nozzle plane of the virtual origin of the jet expressed as a 'similarity' solution.

3.2 The Transition from Laminar to Turbulence of a Jet

3.2.1 *Theoretical Ideas on Stability of a Laminar Jet*

If the velocity of a fluid jet is increased from a very low value, a critical Reynolds Number is reached where the flow becomes turbulent; the result-

Table 3.2

Inlet profile	Zero radial velocity	K.E. balance	Flow balance
	Uniform		
x_0/d Re	0·054	0·056	0·031
Re $v_a/\bar{u}_0$	0·000	0·148	−3·000
Ke_c/Ke_n	1·039	1·000	1·800
Q_c/Q_n	1·732	1·800	1·000
A	1·732	1·667	3·000
B	1·000	0·926	3·000

Inlet profile		Parabolic	
x_0/d Re	0·062	0·050	0·031
Re $v_a/\bar{u}_0$	0·000	−1·071	−3·840
Ke_c/Ke_n	0·800	1·000	1·600
Q_c/Q_n	2·000	1·600	1·000
A	2·000	2·500	4·000
B	1·000	1·563	4·000

$v_a/\bar{u}_0$ indicates the ratio of radial velocity component at $y = a$ in the similarity solution to the mean axial velocity in the nozzle;

Ke_c/Ke_n indicates the ratio of kinetic energy in the similarity solution to that in the nozzle;

Q_c/Q_n indicates the ratio of flow in the similarity solution to that in the nozzle;

A, B are constants defining the similarity solution velocity profile, see equation (3.4).

ing large increase in mixing between the high velocity fluid in the jet and the low velocity fluid surrounding it cause the jet to spread considerably and to draw in a large quantity of fluid from the surroundings. It is often assumed, probably because a distinct Critical Reynolds Number is quoted for a similar change in the flow conditions in a pipe, that such a specific critical value is applicable to the jet. This is not so, because the jet is a free stream flow and is therefore subject to external disturbances, and it appears that the frequency of these disturbances is important in determining the Reynolds Number at which the flow breaks into turbulence. It is further complicated by the fact that a laminar jet 'spreads' to a certain extent, as indicated by the results of Table 3.1, so that both the 'local' Reynolds Numbers and the local Strouhal Numbers increase with distance downstream from the nozzle; therefore, the onset of turbulence is initially localized and, as will become clear in the later discussion, becomes completely developed only at a sufficiently high Reynolds Number.

The behaviour is similar to that of the flow in a laminar boundary layer, and the considerable theoretical and experimental work that has been carried out on this problem is well summarized by Schlichting[1]. Briefly, the Navier–Stokes equations and the continuity equation for two-dimensional flow are developed for small perturbations by substituting

$$u = U + u', \quad v = v', \quad w = 0 \quad \text{and} \quad p = P + p'$$

This substitution implies that the mean velocity in the y direction is negligible compared with the mean velocity in the x direction, U, and that the perturbations in the two directions are important. The quadratic terms are neglected after the substitution, and it is also assumed that the mean flow itself satisfies the Navier–Stokes equations, resulting in the following three equations

$$\frac{\partial u'}{\partial t} + U \frac{\partial u'}{\partial x} + v' \frac{dU}{dy} + \frac{1}{\rho} \frac{\partial p'}{\partial x} = \nu \nabla^2 u' \tag{3.5}$$

$$\frac{\partial v'}{\partial t} + U \frac{\partial v'}{\partial x} + \frac{1}{\rho} \frac{\partial p'}{\partial y} = \nu \nabla^2 v' \tag{3.6}$$

$$\frac{\partial u'}{\partial x} + \frac{\partial v'}{\partial y} = 0 \tag{3.7}$$

The pressure p' may be eliminated from (3.5) and (3.6) and the continuity equation (3.7) is effectively replaced by assuming a stream function ψ for the disturbance. This stream function is assumed to be a function of y and also to behave in a manner similar to a wave, i.e. be a periodic function of both distance from the point of disturbance and of time, where the disturbances may exponentially increase or decrease. Thus the chosen function is

$$\psi(x, y, t) = \phi(y) \exp i(\alpha x - \beta t) \tag{3.8}$$

where α is real so that $2\pi/\alpha$ is the wavelength of a disturbance. $\beta = \beta_r + i\beta_i$ where β_r is the circular frequency of the oscillation and β_i

determines the degree of damping or amplification. It is also convenient to introduce $c = \beta/\alpha = c_r + ic_i$ where c_r denotes the velocity of wave propagation in the x-direction and c_i again denotes the degree of damping or amplification. This results in the Orr–Sommerfeld equation:

$$(U - c)(\phi'' - \alpha^2\phi) - U''\phi = -\frac{iv}{\alpha \, \text{Re}}(\phi'''' - 2\alpha^2\phi'' + \alpha^4\phi) \quad (3.9)$$

in which $\text{Re} = U_m b/v$, U_m being the maximum velocity of the mean flow and b is the local width of the jet suitably defined.

Equation (3.9) is a differential equation which is difficult to solve in full. Initially, solutions were restricted to the case of large Re, so that the right-hand side could be ignored. Later workers modified the results to include the effects of viscosity. The important point is that by substituting values of α and Re, one eigenvalue $\phi(y)$ and one value of $c = c_r + ic_i$ may be found for each case, so that a whole stability field may be found, in which a neutral stability curve given by $c_i = 0$ is plotted.

If the mean velocity distribution of a free two-dimensional jet in the axial direction is substituted for U, the resulting eigenvalues will be applicable to this particular situation. The first solution for the Orr–Sommerfeld equations for a two-dimensional jet was carried out by Savic[4] in 1941 for the case of infinite Reynolds Numbers and was extended by Lessen and Fox[5] in 1955, when they gave more detailed calculations of wave numbers and propagation velocity of various sets of amplification

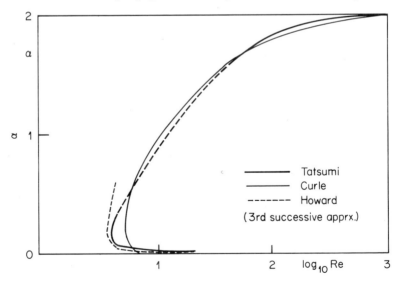

Fig. 3.2 Theoretical neutral stability curves for a laminar jet

rate. Theoretical calculations providing solutions for various Reynolds Numbers were carried out between 1957 and 1960 by Curle[6], Tatsumi and Katutani[7], Howard[8], Clenshaw and Elliot[9] and Sato[10]. Results due to Curle, Howard, and to Tatsumi and Katutani are given in Fig. 3.2, and it may be seen that the neutral stability curve extends down to very low Reynolds Numbers.

The minimum value of Re is that below which the flow is predicted to remain laminar whatever the disturbance, and it may be seen that this value (Re_{crit}) is predicted to be about 3·8 by two of the results and slightly higher by the other. However, the difference in actual values is small and depends upon the method of solution, the assumption made, and the numbers of terms from the right-hand side of (3.9) that are included. The finer points of the methods are still the subject of debate; for example, at low Reynolds Numbers the theoretical velocity profiles are apparently more accurate than at higher Reynolds numbers, but, as seen from Table 3.1, the jet spreads much more quickly so that the assumption of parallel flow (or pseudo-laminar jet) is a doubtful proposition. At higher Reynolds Numbers the jet profiles are shown by Sato[10] to differ significantly from those predicted by Bickley.

3.2.2 Experimental Investigations into the Stability of a Laminar Jet

Experimental evidence to qualitatively support the predictions was produced in 1939 by Andrade[11] from measurements on a two-dimensional water jet at low Reynolds Numbers. His results were insufficient to provide a quantitative discussion of the theory. In 1962, Chanaud and Powell[12] observed the stability of a two-dimensional water jet at Reynolds Numbers from about 20 up to 400 by applying a sinusoidal disturbance through two mechanically oscillated diaphragms, one in each side of the tank into which the jet issued. With this apparatus it was possible for them to vary both the frequency and amplitude of the disturbance. Their results, for a nozzle 0·250 in. wide and 6 in. deep with an inlet length of 18 in. to ensure a parabolic velocity distribution of the jet at the nozzle exit, are shown in Fig. 3.3. In this case, the ordinate is a Strouhal Number which is a dimensionless representation of frequency, i.e. $S = f(W/U_m)$ where f is the frequency, W is the slit width and U_m is the maximum velocity at the nozzle. The disturbance amplitude is also non-dimensionalized with respect to the slit width. The general shape of the neutral stability boundaries is similar to that predicted, but the amplitude of oscillation clearly has an effect. The difficulty in all such experiments is to define the neutral stability. This is because a disturbance in the unstable region close to the neutral boundary will take a long time to build up to a visible magnitude, and if only visual means are employed to detect the result (as was the case

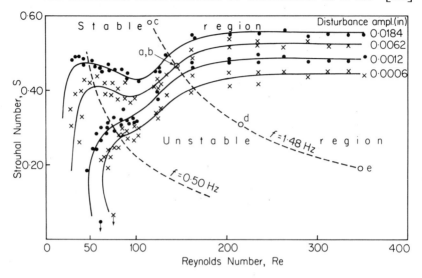

Fig. 3.3 Experimental neutral stability curves for a jet issuing from a nozzle 0·25 in. × 6 in. × 18 in. long for various amplitudes of excitation (Chanaud and Powell, *Ref. 12*, Fig. 3(a))

in Ref. 12) some inaccuracy will result. It will also be difficult to distinguish objectively between the effects of different disturbance amplitudes, and Chanaud and Powell try to clearly define their interpretation of 'neutral' stability.

This point is illustrated by a reproduction of their photographs of the water jet, made visible by the introduction of dye, which are shown in Fig. 3.4. The jet of Fig. 3.4(a) is neutrally stable whilst that shown in (b) has become unstable due to an increase in amplitude of the disturbance. The jet has been stabilized again in (c) by a decrease in the mean velocity of the flow and, therefore, of the Reynolds Number. By an increase in the mean velocity from that of (a), (d) results, and a further increase creates the flow situation shown in (e). The individual points representing these situations are plotted in Fig. 3.3, and it may be seen that the Reynolds Number for Fig. 3.4(e) is over 300. It is also clear from the photographs that it is not yet turbulent. However, from other available evidence it is known that the vortex pattern quickly changes into a fully turbulent jet a short distance downstream.

A further theoretical point to note is that the jet, according to Bickley's profile, spreads as the $\frac{2}{3}$ power of the distance from the nozzle and the centre line velocity decreases as the $\frac{1}{3}$ power, so that the Reynolds Number is proportional to the $\frac{1}{3}$ power. Howard[8] suggests that since the onset of

[3.2] THE TRANSITION FROM LAMINAR TO TURBULENCE OF A JET 101

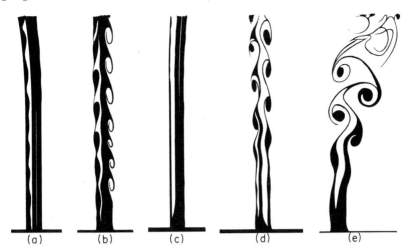

Fig. 3.4 (a) Neutrally stable jet; (b) Unstable jet due to increase in disturbance amplitude; (c) Stable jet due to decrease in mean velocity of (a); (d) Unstable jet due to increase in mean velocity of (a); (e) More unstable due to further increase of mean velocity of (a) (Chanaud and Powell, *Ref. 12*, Fig. 4)

instability depends on the local value of the Reynolds Number, the jet should be stable initially at low values and will later become unstable at a point well downstream from the nozzle exit. As the Reynolds Number increases, this point will gradually move upstream towards the nozzle. Although the photographs of Fig. 3.4 do not show this effect, there is sufficient evidence to support this point of view, and in Ref. 3 the laminar length of the jet for various upstream conditions is plotted against Reynolds Number. This diagram is reproduced in Fig. 3.5 and the decrease in length of the laminar part of the jet may be clearly seen.

Another interesting point is made by Chanaud and Powell concerning the reason for the stability of a jet of low Reynolds Numbers. By calculating the local values of Strouhal and Reynolds Numbers, based upon the actual width of the jet between the centre line and the inflection point and on the centre line velocity ($Re = Ub'/\nu$ and $S = fb'/U$), they were able to plot the locus of a disturbance as the distance downstream in the jet increases. This was superimposed on a diagram containing the theoretical neutral stability boundary, calculated by Curle as a solid line, together with boundaries in broken lines representing the general trend of spatial rate of amplification, the innermost one having the greatest degree of instability. This is given in Fig. 3.6, and it may be seen that for lower initial Reynolds Number the high degree of spreading of the jet causes the local conditions to move quickly into a stable region, so that the

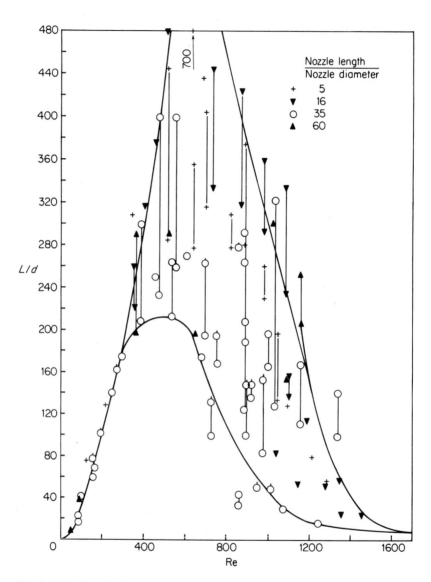

Fig. 3.5 The length of jet L in which the flow remains laminar plotted as a function of Reynolds Number for various circular nozzles; d is the nozzle diameter (Beatty and Markland, *Ref. 3*, Fig. 7)

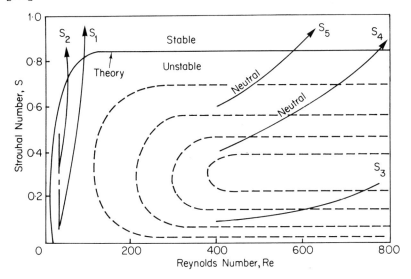

Fig. 3.6 Comparison of experimental results with theory based on the pseudo-laminar jet. Parameters are evaluated locally to show how the stability changes for a disturbance travelling downstream from the nozzle and how the change depends on starting conditions (curves S_1, S_2, S_3, S_4 and S_5) (Chanaud and Powell, *Ref. 12*, Fig. 8)

disturbance does not have the opportunity to be amplified sufficiently to be visible (curves S_1 and S_2). At higher Reynolds Numbers the slope of the locus S_3 is such that a long length of the jet lies in the region of large amplification, and therefore a disturbance results in transition of the jet. Some remarks are also made in the paper by Chanaud and Powell on the effects of the amplitude of the disturbance (S_5 large and S_4 small), and also on the generation of vortex pairs emanating from the point of inflection in the velocity profiles.

Sato[10] discusses the velocity fluctuations in a laminar jet at Reynolds Numbers close to the critical value and also the response characteristics of the jet when subjected to external excitation. From a series of careful measurements, the mean velocity on the centre line of the jet and the width of the jet are plotted against distance downstream, and the sharp rise in both values over a small range of distance is taken as an indication of the transition to turbulence. It was then confirmed that this transition was initiated by sinusoidal velocity fluctuations which were weak at first and later amplified as the distance downstream increased. A difference between these fluctuations and those in a boundary layer is that two modes of fluctuation are found in the jet, the first at a higher frequency and symmetrical in phase with respect to the centre line and the second at a lower frequency and anti-symmetrical in phase about the centre line. The latter is dominant when the $1/b$ ratio for the nozzle is large so that the profile of the jet issuing from the nozzle is parabolic, whilst the former is dominant for a more nearly square profile. In both cases the frequency of oscillation increases with slit velocity. For a particular geometry and slit velocity, the energy spectrum of the velocity fluctuations shows, on the centre line, one peak at the higher frequency and, off the centre line, two peaks, one of which is at the lower frequency. It appears, therefore, that natural fre-

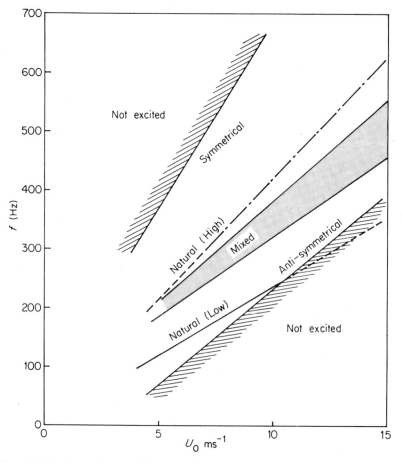

Fig. 3.7 The boundaries of the two types of velocity fluctuations in a jet as a function of jet velocity as the frequency of excitation is changed at various values of nozzle velocity. The labelling indicates the regions in which symmetrical and anti-symmetrical sinusoidal velocity fluctuations are induced. The slit width is 6 mm (Sato, *Ref. 10*, Fig. 13)

quencies occur in the laminar jet. When the intensity of velocity fluctuations along the centre line is plotted against distance from the nozzle, it is found to be initially proportional to that distance, but later levels off and then reduces. The jet is laminar over the initial linear portion, but further downstream it becomes unstable and turbulent.

When the jet is excited by a loudspeaker, the response of the jet contains frequencies in the two groups, and the diagram reproduced in Fig. 3.7 shows how these modes depend upon the nozzle velocity U_0 and frequency of excitation. It may be seen that frequency of excitation is important; more detail is given in this paper[10], and it is pointed out that whilst excitation at 330 Hz for a U_0 of 10 ms^{-1} (below the higher natural frequency) causes an increase in the mean velocity and the 'u' fluctuations in the outer part of the jet, it does not affect the dependence of 'u' fluctuations on distance downstream from the nozzle. On the other hand, excitation at 420 Hz (the natural frequency) has a very pronounced effect on the distribution of the fluctuations both across the jet and along its length.

Sato then goes on to modify the Orr–Sommerfeld calculations to include a flatter velocity profile than that due to Bickley, since he observed such profiles in practice. The chosen profile was

$$\frac{u}{U} = \left[1 + sky^2\left(\frac{2 - y^2 - y^4}{1 + ky^6}\right)\right] \operatorname{sech}^2 ay \qquad (3.10)$$

where k and s are parameters and a is a constant $= 0{\cdot}88136$. The ideal jet is given by $k = s = 0$, and Sato used three other combinations of k and s to give the appropriate profiles. The propagation velocity and wave number were calculated and were shown to be dependent on the parameters. It was found that the jet with $k = 0{\cdot}1977$ and $s = 0{\cdot}7$ gave a good correlation between theory and experiment for the variation in frequency of velocity fluctuations with Reynolds Number, assuming that the observed natural frequency corresponds to the maximum rate of amplification. The results are shown in Fig. 3.8, where it is seen that the ideal jet (Jet I) gives little correlation, whereas the modified profile (Jet II) is much better. Results for propagation velocity and spatial rate of amplification are also better.

In 1964 Sato, with a co-author Sakao, published an account of an investigation at lower Reynolds Numbers than in the previous instance[13]. First of all they observed that at these Reynolds Numbers the jet profile follows Bickley's solution fairly closely both for velocity variations across the stream and for variations along the stream. The momentum in the axial direction was found to be constant throughout the flow field, but the curious result is that the value of this momentum lay between 0·5 and 0·8

106 THE TRANSITION FROM LAMINAR TO TURBULENCE OF A JET [3.2]

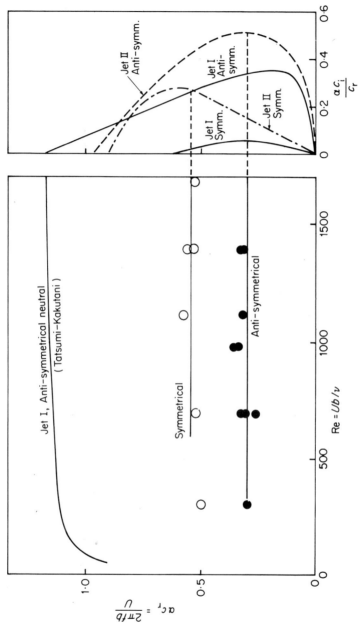

Fig. 3.8 Comparison of experimental natural frequencies, shown by the plotted points in the left-hand diagram, with those predicted by two calculations; the curves labelled Jet I are obtained using a commonly accepted profile due to Bickley and those labelled Jet II are obtained by Sato using a modification to this basic profile. The right-hand diagram shows the rate of amplification $\alpha c_i c_r$ for various frequencies αc_r at infinite Reynolds Number, and the maxima of these curves is assumed to predict natural frequencies; Jet II gives good correlation (Sato, *Ref. 10*, Fig. 29)

[3.2] THE TRANSITION FROM LAMINAR TO TURBULENCE OF A JET 107

of that at the nozzle exit; no explanation is offered except that a low pressure region exists near to the nozzle exit which may influence the flow.

Velocity fluctuations are then measured, and it is found that for very low Reynolds Number, i.e. between about 12 and 30, the jet is almost entirely laminar. Between about 30 and 60 the periodic fluctuations are found in a wide region of the jet, but these die out without developing into irregular fluctuations. Above Reynolds Numbers of between 40 and 60, irregular fluctuations are observed downstream of the regular ones. The non-dimensionalized frequency of the sinusoidal velocity fluctuations is plotted against Reynolds Number as shown in Fig. 3.9. It may be seen that the Strouhal Number is directly proportional to Reynolds Number for experiments at the lower Reynolds Numbers but that it is constant for

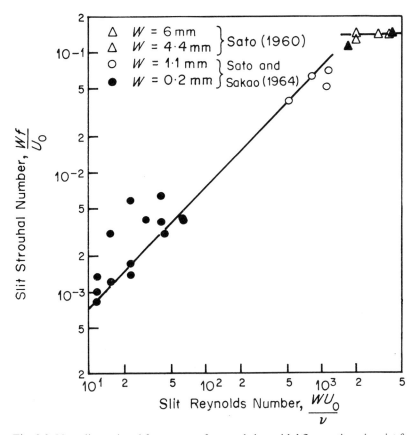

Fig. 3.9 Non-dimensional frequency of natural sinusoidal fluctuations in a jet for a range of Reynolds Numbers. Note the change of slope at higher values of Reynolds Number (Sato and Sakao, *Ref. 13*, Fig. 11)

the results of the previous experiments reported in Ref. 10. The explanation given is that at high Reynolds Numbers the breadth of the jet does not change much in the x-direction, so that the slit velocity and width may be taken as characteristic quantities, i.e.

$$f \propto \frac{U_0}{2W}$$

at low Reynolds Numbers

$$U \propto U_0^{\frac{4}{3}} \; W^{\frac{2}{3}} x^{-\frac{1}{3}} v^{-\frac{1}{3}}$$

$$b \propto U_0^{-\frac{1}{3}} W^{-\frac{1}{3}} x^{\frac{1}{3}} v^{\frac{1}{3}}$$

so that

$$f \propto \frac{U_0^2}{v} \frac{W}{x} \qquad (3.11)$$

However, the sudden change from one state to the other indicated in Fig. 3.9 is not explained.

The stability calculations based on the Orr–Sommerfeld equations are again examined in this paper[13], and it is pointed out that because of the relatively fast spread of the jet, the terms $\partial U/\partial x$ and $\partial^2 u/\partial x^2$, normally neglected at higher Reynolds Numbers, are not negligible at lower Reynolds Numbers. A solution is therefore produced which is based on the neutral curve of Tatsumi and Kakutani, but since for the calculation of this latter curve the time amplification factor c_i was assumed to be zero, the function of $\alpha c_i/c_r$ against αc_r was taken to be that calculated by Lessen and Fox. A spatial amplification A was then calculated where

$$A = \exp \left\{ \int_{Re_{b0}}^{Re_b} 0 \cdot 644 \left(\alpha \frac{c_i}{c_r} \right) dRe_b \right\} \qquad (3.12)$$

where Re_b is the local Reynolds Number in the jet where the width is b. Re_{b_0} is the value at the initial point of the disturbance where the width is b_0 (usually at the nozzle). Thus the effect of the change of Re_b and αc_r (similar to Strouhal Number) with distance downstream of the jet is calculated. Effectively this is the same idea as that suggested by Chanaud and Powell, where they plot the locus of Strouhal and Reynolds Numbers for a given disturbance as the jet is traversed. The difference is that Sato and Sakao define a modified or integrated neutral curve where the spatial amplification has become unity; disturbances are first amplified and then later damped after crossing a boundary where $c_i = 0$, but their amplitude will only have diminished back to the original value at some further distance downstream. This 'integrated' neutral curve will depend upon

[3.2] THE TRANSITION FROM LAMINAR TO TURBULENCE OF A JET 109

the initial starting conditions, and various curves can be plotted for a fixed initial Reynolds Number and varying initial αc_r. A set of such curves with experimental results superimposed on them is reproduced in Fig. 3.10 from Ref. 13, and it may be seen that the theoretical predictions are generally supported by experiment. Similar curves could be produced for

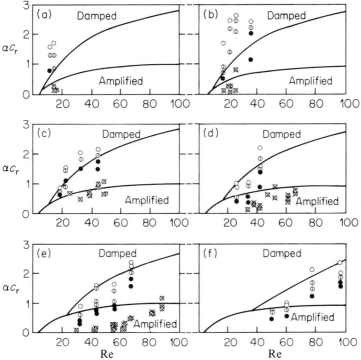

Fig. 3.10 Modified neutral stability curves due to Sato and Sakao. The upper 'integrated' theoretical curves are for a condition where a disturbance has been first amplified and then reduced back to its initial amplitude. The results depend on the initial starting conditions which are as follows:

	Slit velocity (cm s^{-1})	Slit Re
(a)	117	16
(b)	173	23
(c)	231	31
(d)	313	42
(e)	486	65
(f)	770	103

The lower curves are those due to Tatsumi and Kakutani. Experimental results are: ● amplified; ○ not amplified; ⊕ not clear; ⊗ natural fluctuation (Sato and Sakao, *Ref. 13*, Fig. 14)

higher Reynolds Numbers, but they would have less significance since the spread of the jet is much less and turbulence will have set in before the disturbance has the possibility of being damped; this is still a difficult point.

3.3 Noise Associated with a Laminar Jet. Edgetones

3.3.1 The Effect of a Wedge in the Jet Stream

It was shown in the previous section that laminar jets are sensitive to disturbances and that disturbances of certain frequencies, or to be more precise certain Strouhal Numbers, were amplified more than others. If the jet impinges on an obstacle, then an audible tone of a discrete frequency may often be heard. This is called an 'edgetone', although in the majority of experiments the obstacle is a wedge rather than an edge. The exact mechanism of the tone is the subject of considerable debate, and a great many papers have been written on the subject, many of which contain conjecture rather than convincing proof of the arguments put forward. Almost all the later references quote work by Brown[14, 15, 16] because of the meticulous experimental results presented by him, which are so detailed that the more mathematically minded workers are able to use them for comparison with the theoretical results. This was indeed his intention, and at the end of one paper he says: 'Every effort has been made to treat the matter as fully as possible in the hope that mathematicians may feel induced to address themselves to it. . . .' In the first of these papers (1935) he showed a series of excellent photographs of a free two-dimensional jet excited by an external sound source, taken by means of a stroboscope, in which the development of vortices is clearly shown. By means of this photographic technique, measurements were made of the velocity of the centres of the vortices, the angular velocity of the vortices and the wavelength of the vortex motion. Some of the photographs are shown in Fig. 3.11. At the higher jet speed, the stream breaks into strong vortex motion with the result that the jet spreads very quickly. As the jet speed is lowered, the spread caused by the vortex motion is lessened until at 100 cm s^{-1} very little spread is seen. Brown also shows that the effect of increasing the amplitude of the disturbance is similar to that of increasing stream velocity. Brown quotes the transition to turbulence as occurring at 150 cm s^{-1} so that very low Reynolds Numbers are not included in the results; if the jet velocity were lowered still further, the jet spread would be significantly greater because of the effect of shear stresses. The important conclusions from the photographs are that as the stream velocity increases for a given nozzle width and excitation frequency, the wavelength of the propagation of the disturbance increases, i.e. the

[3.3] NOISE ASSOCIATED WITH A LAMINAR JET. EDGETONES 111

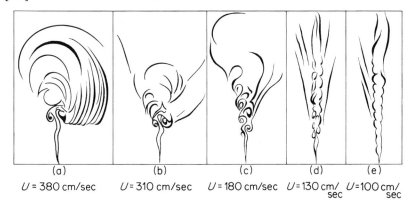

Fig. 3.11 Stroboscopic photographs by G. B. Brown of a jet excited by an external sound source at 126 Hz, for different stream velocities with a constant width of jet (Brown, *Ref. 14*, Plate 2)

velocity of propagation of the disturbance increases. However, the ratio of velocity propagation to stream velocity increases slightly at lower stream velocities, but decreases over most of the range considered. These are important observations in relation to the causes of edgetones. A graph of u_v/U_0 against stream velocity is reproduced in Fig. 3.12 for various

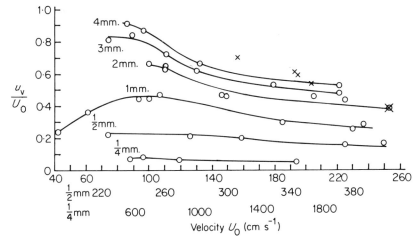

Fig. 3.12 Non-dimensional propagation velocity of vortices in a jet excited by an external sound source at 126 Hz against jet nozzle velocity for various nozzle widths. The initial flat or rising portion of the curve is where the vortices are changing from symmetrical to alternate. The portion where the curves fall with increasing nozzle velocity corresponds to the alternate vortex formation illustrated by (e) to (c) of Fig. 3.11, whilst the final flat portion of the curves occurs at the excessive-development phase of (a) and (b) of Fig. 3.11 (Brown, *Ref. 14*, Fig. 8)

nozzle widths, where u_v is the velocity of the vortices. A further important point is that Brown estimates the ratio d/λ, where d is the distance apart of the vortex rows and λ is the wavelength of the vortex motion, and shows that this ratio does *not* tend towards the constant value expected of a Karman vortex street.

The next important point is that the angular velocity of the vortices is directly related to the frequency, i.e. $\omega = \pi n$.

In the second reference by Brown, an attempt is made to explain the cause of the sensitivity of a jet to noise, but since his explanation is critically examined by later workers, this will be omitted here. His important contribution to the literature on edgetones is his thorough documentation of experimental results[14]. Very briefly the conclusions are, that if a wedge is placed in the jet, audible tones are produced whose frequency depends on the spacing of the wedge from the nozzle and on the speed of the jet. If the air stream is at a constant velocity, and the wedge is slowly moved away from the nozzle, the frequency reduces slowly and

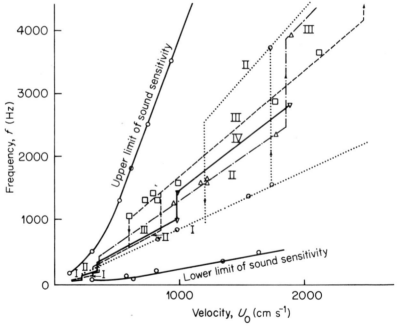

Fig. 3.13 The four stages in vortex formation in a jet-edge system showing the jumps in frequency of edgetones as the wedge distance and/or jet velocity changes. Note the hysteresis. —▽— $h = 1\cdot5$ cm; –☐– $h = 1\cdot0$ cm; —△— $h = 0\cdot75$ cm; ··⊙·· $h = 0\cdot50$ cm; where h is the distance of the wedge (or edge) from the nozzle (Brown, *Ref. 15*, Fig. 5)

[3.3] NOISE ASSOCIATED WITH A LAMINAR JET. EDGETONES

then abruptly jumps to a higher frequency, after which the frequency again decreases until a new 'jump' occurs. Photographs are presented[15] which indicate clearly that the first stage of tone has just over one wavelength of disturbance formed between the nozzle and the wedge, the second stage just over two wavelengths, the third just over three wavelengths, and the fourth just over four. The fifth stage is never reached because turbulence sets in. On reducing the distance, the jumps take place at a different frequency so that there is some hysteresis.

An alternative is to plot the change in frequency against jet velocity for several different wedge distances, and again jumps occur which exhibit hysteresis. Fig. 3.13 is a reproduction of Brown's results for this case. Each curve has several parts and these are labelled I, II, III or IV to indicate the stage, or number of wavelengths (or vortices) between nozzle and wedge. An important observation made by Brown is that when the boundaries for sound sensitivity formulated in his earlier paper (Ref. 14) are superimposed on the figure, the edgetones are generated only in the centre of the region thus demarcated. From the results obtained, Brown suggested a formula for the frequency of oscillation,

$$f = 0.466j(U_0 - 40)\left(\frac{1}{h} - 0.07\right) \quad (3.13)$$

where j has the value 1, 2·3, 3·8 and 5·4 for stages I, II, III and IV respectively.

Very much later than Brown's work, Curle[17] and Powell[18,19] critically reviewed the postulates for the mechanism of edgetones stemming from the results of previous research and produced more convincing reasons for the edgetone phenomenon, together with more plausible formulae for the frequency of the tones. Both these authors, working separately but concurrently, arrived at very similar conclusions, namely that the wavelength is not an integral number of slot widths but that a more realistic figure for it is

$$\frac{h}{\lambda} = \left(N + \frac{1}{4}\right) \quad (3.14)$$

where N is an integer, λ is the wavelength and h is the distance of the wedge from the slit.

Both authors conclude that a disturbance at the edge causes feedback to the slit, where the jet issuing from it is deflected. The vortex formed at the slit arrives at the wedge at a certain time later, dependent on the velocity of propagation of the disturbance. This then gives rise to another disturbance feeding back to the slit, and so on. Both Curle and Powell conclude that for maximum feedback, the jet must be in its mean position, i.e. passing the wedge with maximum velocity during its oscillation about the wedge. Curle then assumes that the vortex shed by the wedge immediately causes a hydrodynamic disturbance at the slit which generates another vortex at that point. Powell, on the other hand, assumes that the feedback occurs by means of an acoustic wave, and his results include an allowance for time delay.

By using experimental evidence of both Savic and Brown, a realistic formula for the propagation velocity of the turbulence in terms of the velocity at the nozzle and the nozzle width could be obtained, which, when coupled with the expression for the wavelength, gave a good prediction of edgetone frequency, i.e.

Curle:
$$f = 1 \cdot 024 \frac{U_0}{h^{\frac{3}{2}}} W^{\frac{1}{2}} \left(N + \frac{1}{4}\right)^{\frac{3}{2}} \qquad (3.15)$$

Powell:
$$f = 1 \cdot 024 \frac{U_0}{h^{\frac{3}{2}}} W^{\frac{1}{2}} \left(\frac{N + \frac{1}{4}}{1 + \frac{u_v}{c}}\right)^{\frac{3}{2}} \qquad (3.16)$$

where u_v is the velocity of propagation of the disturbance and c is the velocity of sound. The results compare well with experiment.

Curle uses theory by Lighthill[20] to show that the maximum amplitude of sound disturbance occurs at a Strouhal Number of about 0·55, which agrees with practice. In the later of the two papers by Powell[19] this theory is used to obtain the acoustic force on the wedge and to assume that the effect on the flow field must be equal and opposite. The velocity disturbance at the slit, due to this force, is calculated, and a mathematical proof is put forward to account for the mechanisms of the edgetones. This gives convincing quantitative support for the argument put forward to explain the jumps in frequency, the fact that a minimum distance h exists below which edgetones do not occur, and the reasons for hysteresis, and so on.

3.3.2 *The Effect of a Resonator Close to the Jet*

A further important aspect of the source of tones in jets was investigated by Nyborg *et al.*[21] who consider the effect of a resonator placed close to the jet. Here the cavity itself has a natural frequency which can be varied by changing the cavity length, from a little higher than the jet natural frequency (for a fixed edge position and jet velocity) to considerably lower. The results are shown in Fig. 3.14(a) and (b) and it may be seen that the whistle frequency is no longer the frequency of either of the contributing elements. Also as the frequency decreases the sound level rises up to a

[3.3] NOISE ASSOCIATED WITH A LAMINAR JET. EDGETONES

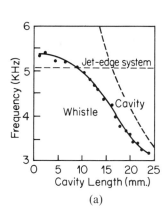
(a)

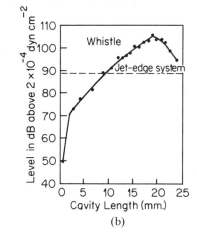
(b)

Fig. 3.14 Both sets of curves are plotted against cavity length for a jet width of 2·0 mm, distance between nozzle and edge of 1·59 mm and jet depth of 25·4 mm. The cavity height was 0·5 mm with the same depth of cavity as the jet. The mass flow was 0·27 g s^{-1}. (a) Natural frequency of a jet-edge system coupled to a cavity resonator. The broken lines show the natural frequency of the jet-edge system alone and also the lowest natural frequency of the cavity, which varies with cavity length. (b) The noise level of the jet-edge system and also that of the jet-edge system coupled to a cavity resonator, taken 15 cm away (Nyborg et al., Ref. 21, Fig. 2)

maximum and then falls again. Thus a tuned cavity by the side of a jet can produce a considerable amplification of the noise level by appropriate choice of the tuning frequency. It is also shown that tuning to any of the jet eigenfrequencies can be realized, and presumably jumps in frequency occur in the same way as with a jet by itself. Tuning to several of the cavity eigenfrequencies can also occur; some complicated maps are given of frequency against cavity length and also maps of different modes are plotted as contours on a graph of airflow through the slit against cavity length.

An earlier paper by Nyborg et al.[22] gives more information on the tones produced by the jet edge combinations, paying particular attention to directivity in polar patterns for the sound distribution. In a paper by Nyborg alone[23] another theory for the production of edgetones is presented, but Powell's later paper probably gives the fullest account of the problem.

3.3.3 Practical Application of the Edgetone Idea

Tamulis[24] applies the previous ideas to an investigation into the problem of noise in a 'knife-edge' proportional fluid amplifier. The configuration of the edge is different from that considered in the studies of edgetones because of the shape of the knife edge. Tamulis observed that the vortex shed by the knife edge is opposite in sense to that produced by the wedges of earlier studies, so that the frequencies are displaced by $\frac{1}{2}$ a wavelength. With this correction to the basic model, frequencies can be predicted to within 10 per cent. He then tested a proportional amplifier which incorporated some of the design features and cross-correlated the noise in the input and output channels. He found zero cross-correlation, suggesting that in a well designed amplifier noise due to turbulence is generally the dominating noise source, and that due to edgetones has insufficient strength to convert into the output channels.

The first to utilize the ideas developed on edgetones was Unfried[25, 26]. He describes an unsymmetrical wall-attachment device as shown in Fig. 3.15. As D_1 increases from a small value, the jet tends to be successively: stable on the lower wall; bistable with a preference for the lower

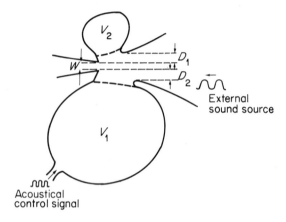

Fig. 3.15 Sketch of H. H. Unfried's acoustically sensitive element. The volumes on either side of the jet, the confines of which are indicated by broken lines, are effectively capacitances that combine with the inertia of the fluid in the openings close to the jet to create Helmholtz resonators. The static characteristics are fixed by the walls and the setback distances D_1 and D_2, whilst the dynamic characteristics are fixed by the resonators (Unfried, *Ref.* 25, Fig. 3)

wall; bistable; bistable with a preference for the upper wall; monostable on the upper wall. If further increased, the jet cannot properly attach to the upper wall and the affinity to the lower wall is repeated. By careful adjustment of the distance, the jet may be made just stable on the upper wall, but an acoustic disturbance will cause it to switch to the lower one.

By careful adjustment of volume V_1 in conjunction with the dimensions of the opening into V_1, a Helmholtz resonator may be created which is tuned to give a high impedance to an acoustical control signal at a specific frequency so that the device is selective. Alternatively, both the volumes V_1 and V_2 may be utilized to give a low impedance to an external sound source and allow switching, again at a sharply defined frequency.

In the second paper Unfried describes sound generators which again use jets and tuned cavities as oscillators to produce sound of a specific frequency. He also shows how a sound source can be frequency modulated and later demodulated to give sound transmission over a limited bandwidth. The experiments described are elegant in their utilization of ideas produced by basic research into the relatively academic problem of edgetones.

Acoustic sensors are now used in industry[27] where a signal at high frequency of about 50 kHz is beamed at a sound sensitive element. If an object interrupts the beam, the element changes state. The advantage of a high frequency is that the polar diagram of the sound is very narrow and good discrimination of the signal can be achieved at distances of several feet.

In a paper by Gradestsky and Dimitriev[28] turbulence amplifiers are described which are made sensitive to a particular frequency by locating a Helmholtz resonator close to the jet. Their behaviour is similar to that of the devices quoted in the previous references, but the frequency of operation is much lower.

Gottron[29] also considers the acoustic switching of bistable and monostable wall attachment elements and shows that they can switch with less acoustic power than pneumatic power. In another paper by the same author[30] noise reduction in amplifiers of up to 80 per cent is achieved by adjusting the resonator natural frequency to be the same as that of the jet-edge combination. He also points out that if the two natural frequencies are different, audible noise is present. This broadly confirms the result due to Nyborg et al.[22]

3.4 The Turbulent Jet

In the case of a turbulent jet, the equation for a two-dimensional boundary layer can again be used, but the difficulty is to give a value to the viscosity term. Because of this, the basic equation is given in terms of the shear stress τ, i.e.

$$\frac{\partial u}{\partial t} + u \frac{\partial u}{\partial x} + v \frac{\partial u}{\partial y} = \frac{1}{\rho} \frac{\partial \tau}{\partial y} \qquad (3.17)$$

with continuity again giving

$$\frac{\partial u}{\partial x} + \frac{\partial v}{\partial y} = 0 \qquad (3.18)$$

Use is now made of Prandtl's mixing length theory which argues that because turbulence moves particles from one level in the stream to another, the turbulent velocity fluctuations are related to distance displaced and the rate at which the velocity changes along the velocity profile, i.e. $|\bar{u}'| = l|d\bar{u}/dy|$ and by a further argument this equals $|\bar{v}'|$ where $\bar{u}'$ and $\bar{v}'$ are the average values of the turbulent velocity fluctuations. The length l is called the 'mixing length' and is a parameter which varies from situation to situation. The apparent shear stress can be shown to be proportional $\rho\overline{|u'v'|}$ and this is assumed to be proportional to $\rho|\bar{u}'|.|\bar{v}'|$. In order to keep the sign correct for the shear stress, one modulus sign is dropped and

$$\tau = \rho l^2 \left|\frac{\partial u}{\partial y}\right|\frac{\partial u}{\partial y} \qquad (3.19)$$

Thus the apparent viscosity is $\rho l^2 |\partial u/\partial y|$, which gives the apparent kinematic viscosity as

$$\varepsilon = \rho l^2 \left|\frac{\partial u}{\partial y}\right| \qquad (3.20)$$

and of the various possible values of l, the one which has been proved to be useful in the free turbulent flow case is that which assumes

$$\varepsilon = \kappa_1 b(\bar{u}_{\max} - \bar{u}_{\min}) \qquad (3.21)$$

For one analysis by Goertler, this is simplified in turn by ignoring $\bar{u}_{\min}$ as being negligible and taking $\bar{u}_{\max}$ as U, the centre line velocity. As κ_1 is a constant,

$$\frac{l}{b} \text{ is assumed constant}$$

and

$$\frac{\partial u}{\partial y} \propto \frac{U}{b}$$

Schlichting[1] gives an account of the procedure for establishing the jet equations by first stating a rule which has stood the test of time, i.e.

$$\frac{Db}{Dt} \propto v'$$

[3.4] THE TURBULENT JET

and since

$$v' \propto l \frac{\partial u}{\partial y}$$

then

$$\frac{Db}{Dt} \propto l \frac{\partial u}{\partial y} = \text{const.} \frac{l}{b} U$$

For a jet boundary, the rate of change of b with time equals the rate of change with distance times the maximum velocity and

$$\frac{Db}{Dt} \propto U \frac{db}{dx}$$

Hence

$$\frac{db}{dx} = \text{const.} \frac{l}{b} = \text{const.}$$

so that

$$b = \text{const.} \, x \tag{3.22}$$

For the free jet, the momentum is assumed constant throughout the flow field so that the momentum per unit length of a two-dimensional jet is

$$J' = \rho \int u^2 \, dy = \text{const.} \, \rho U^2 b = \text{const.}$$

Hence

$$U \propto \frac{1}{x^{\frac{1}{2}}} \left(\frac{J'}{\rho} \right)^{\frac{1}{2}}$$

U and b are now written in terms of the values at an arbitrary distance s from the slit,

$$U = U_s \left(\frac{x}{s} \right)^{-\frac{1}{2}}; \quad b = b_s \frac{x}{s} \tag{3.23}$$

and

$$\varepsilon = \varepsilon_s \left(\frac{x}{s} \right)^{\frac{1}{2}} \quad \text{with} \quad \varepsilon_s = \kappa_1 b_s U_s \tag{3.24}$$

In order to satisfy the previously derived variation in U with x, a stream function is now defined to be

$$\psi = \sigma^{-1} U x F(\eta) \tag{3.25}$$

$$\eta = \sigma \frac{y}{x} \tag{3.26}$$

where σ is a constant to be determined empirically. Therefore

$$U = U_s \left(\frac{x}{s}\right)^{-\frac{1}{2}} F' \quad \text{and} \quad v' = \sigma^{-1} U_s s^{\frac{1}{2}} x^{-\frac{1}{2}} (\eta F' - \tfrac{1}{2} F)$$

and on substituting these values in the equation (3.17), the following differential equation for F results,

$$\frac{1}{2} F' + \frac{1}{2} FF'' + \frac{\varepsilon_s}{U_s s} \sigma^2 F''' = 0 \tag{3.27}$$

with the boundary conditions that $F = 0$ and $F' = 0$ at $\eta = 0$ and $F' = 0$ at $\eta = \infty$.

Because ε_s contains the free constant κ_1, it is possible to put

$$\sigma = \frac{1}{2} \sqrt{\frac{U_s s}{\varepsilon_s}}$$

which simplifies the differential equation so that on integrating twice it gives

$$F^2 + F' = 1 \tag{3.28}$$

which is exactly the same as the equations for the two-dimensional laminar jet. The characteristic velocity U_s can now be expressed in terms of the momentum per unit length because the velocity profile is known, and

$$J = \frac{4}{3} \rho \frac{U_s^2 s}{\sigma} \quad \text{with} \quad \frac{J}{\rho} = K$$

so that the final equations for the jet are

$$u = \frac{\sqrt{3}}{2} \cdot \sqrt{\frac{K\sigma}{x}} \cdot (1 - \tanh^2 \eta) \tag{3.29}$$

$$v = \frac{\sqrt{3}}{4} \sqrt{\frac{K}{x\sigma}} [2\eta(1 - \tanh^2 \eta) - \tanh \eta] \tag{3.30}$$

$$\eta = \sigma \frac{y}{x} \tag{3.31}$$

[3.4] THE TURBULENT JET

The volume flow rate is

$$Q = \sqrt{3}\sqrt{\frac{Kx}{\sigma}} \tag{3.32}$$

From these results it is possible to obtain ε in the following form

$$\varepsilon = \frac{U}{4\sigma}\cdot\frac{y}{\eta}$$

If y is taken to be $b_{\frac{1}{2}}$, the point at which the velocity is $\frac{1}{2}U$, then y/η may be evaluated and

$$\varepsilon = \frac{1\cdot 125}{4\sigma} b_{\frac{1}{2}} U \tag{3.33}$$

Reichardt, by careful experiments, found that in practice for a fully developed free jet

$$\sigma = 7\cdot 67$$

which gives that

$$\varepsilon = 0\cdot 037 b_{\frac{1}{2}} u \tag{3.34}$$

Alternative solutions may be obtained by different assumptions for the turbulent shear stress. In addition to the previous result, Newman[31] lists two others as follows:

(a) Based on Prandtl's first hypothesis and constant eddy viscosity across the jet (the result given above is based on Prandtl's second hypothesis and constant eddy viscosity across the jet)

$$\varepsilon = \kappa_1 b(u_{max} - u_{min}) \quad \text{gives} \quad F = \exp(-\eta)^2 \tag{3.35}$$

(b) Prandtl's first hypothesis assuming the mixing length l constant across the jet and proportional to its width

$$l \propto b \quad \text{gives} \quad F = (1 - \eta^{\frac{3}{2}})^2 \tag{3.36}$$

The main disadvantage of these profiles is that the jet does not become fully turbulent immediately on leaving the nozzle, since turbulence begins in the shear layer between the jet and the surroundings and requires time to spread towards the centre of the jet. The real nature of the velocity distribution in the jet is given in Fig. 3.16, where a wedge of laminar flow, having a uniform velocity equal to that at the slit, extends for several slit widths downstream of the nozzle. From the point where the laminar

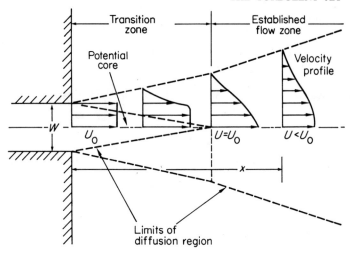

Fig. 3.16 Velocity distribution in a turbulent jet showing the laminar core region

region ends, the jet spreads more rapidly. An alternative set of expressions for the velocity profile is given by Albertson et al.[32], in which the laminar core and the fully turbulent region are treated separately, i.e.

(a) For the zone of establishment, i.e. where there is a laminar core

$$\frac{u}{U} = \exp\left[-\frac{\left(y + \sqrt{\pi}\, C_1 \frac{x}{2} - \frac{W}{2}\right)^2}{2(C_1 x)^2}\right] \quad (3.37)$$

(b) For the zone of established flow

$$\frac{u}{U} = \sqrt{\frac{1}{\sqrt{\pi}\, C_1}\frac{W}{x}} \exp\left[-\frac{1}{2C_1^2}\frac{y^2}{x^2}\right] \quad (3.38)$$

where C_1 is defined by

$$\frac{x_0}{W} = \frac{1}{\sqrt{\pi}\, C_1} \quad (3.39)$$

in which x_0 represents the length of the constant velocity core and U is the magnitude of the velocity in this core. The chief departures from experiment are: (a) at the edge of the jet and (b) in the region of transition. The first point does not necessarily introduce much error into an analysis based on the equation, and in any case it still gives a slight improvement on the Goertler profile. The second point is important, but no simple

[3.4] THE TURBULENT JET   123

equations exist to describe this region precisely, and generally speaking it occupies such a short length of the jet that again the errors resulting in calculations based on the profile are not too great. The core region is rather more difficult to specify because there is evidence to indicate that this depends strongly on aspect ratio. Jones et al.[33] give the result shown in Fig. 3.17 where it may be seen that for the aspect ratios used in fluid

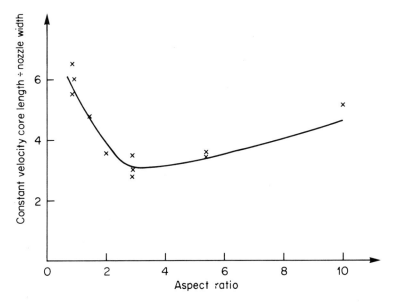

Fig. 3.17 Variation of constant velocity core length with aspect ratio
(Jones, Foster and Mitchell, Ref. 33, Fig. 12)

elements the core is only between 3 and 4 nozzle widths long compared with the normally quoted value of 5·2—the latter being approached only at higher Reynolds Numbers. This difference is small, but when it is remembered that the re-attachment bubble of a wall attachment element may be only 5 or 6 nozzle widths long, the importance may be realized.

Kirschner[34] gives an interesting justification of the equation for the established zone which was obtained empirically by Albertson et al.[32] By using Reichhardt's inductive theory for turbulence, he first of all finds a velocity distribution in Gaussian form issuing from a slit of infinitesimal width, i.e.

$$u = \left(\frac{J}{\rho\sigma_1 x}\right)^{\frac{1}{2}} \left(\frac{1}{\pi}\right)^{\frac{1}{4}} \exp\left(-\frac{y^2}{2\sigma_1^2 x^2}\right) \quad (3.40)$$

By using a convolution integral over the slit width, i.e.

$$u^2 = \frac{J}{\rho\sigma_1 x}\left(\frac{1}{\pi}\right)^{\frac{1}{2}} \int_{-\infty}^{\infty} f^2(\alpha) \exp\left\{-\frac{(y-\alpha)^2}{\sigma_1^2 x^2}\right\} d\alpha \qquad (3.41)$$

the equation for the profile of a jet may be obtained for any velocity distribution at the exit from the nozzle merely by modifying $f(\alpha)$. For a uniform velocity at the nozzle, $f(\alpha)$ must be $1/W$ to make the exit momentum equal to J and the integration must be for from $\alpha = -W/2$ to $\alpha = +W/2$. The integration gives the result that

$$u = \left(\frac{J}{\rho\sigma_1 x}\right)^{\frac{1}{2}}\left\{\text{erf}\left(\frac{\frac{W}{2}+y}{\sigma_1 x}\right) + \text{erf}\left(\frac{\frac{W}{2}-y}{\sigma_1 x}\right)\right\}^{\frac{1}{2}} \qquad (3.42)$$

For $x \gg W$, the equation reduces to

$$u = \left(\frac{J}{\rho\sigma_1 x}\right)^{\frac{1}{2}}\left(\frac{1}{\pi}\right)^{\frac{1}{4}} \exp\left(\frac{y^2}{2\sigma_1^2 x^2}\right) \qquad (3.43)$$

which is of the same form as the equation for the established jet given by Albertson[32] if $\sigma_1 = C_1$.

Kirschner then points out that the centre line velocity, which is given by

$$U = \left(\frac{J}{W\rho}\right)^{\frac{1}{2}}\left\{\text{erf}\left(\frac{W}{2C_1 x}\right)\right\}^{\frac{1}{2}} \qquad (3.44)$$

decays immediately on leaving the nozzle, but that there is a knee in the centre line velocity decay curve. This knee is interpreted as being the end of the core. Analysis establishes that this is at the point

$$x_0 = \frac{W}{\sqrt{2}\,C_1} \qquad (3.45)$$

which gives a 25 per cent greater value for x_0 than the empirical value given by Albertson. Altogether the paper by Kirschner gives an excellent review of the theoretical arguments used in calculating jet flows.

There are, however, two main disadvantages in using the Albertson model. The first is that there is no exact integral for the function (3.40) and hence the mass flow and momentum could only be found by numerical integration, although on a computer this is not a problem. Secondly, the function is again asymptotic, as was the Goertler profile, and so an assumption has to be made for the width of the jet.

An alternative profile due to Simson[35] overcomes these difficulties, for which he assumes a form similar to the third of the theoretical ones

[3.4] THE TURBULENT JET

described by Newman. The velocity in the potential core is U_0 and one equation describes the velocity in the jet, i.e.

$$\frac{u}{U} = \left\{1 - \left(\frac{y}{y_{max}}\right)^{\frac{7}{4}}\right\}^2 \tag{3.46}$$

where $y_{max} = Cx$, and

(a) in the zone of establishment y is measured from the edge of the potential core

$$U = U_0$$

(b) in the zone of established flow y is measured from the centre line

$$U = U_0 \left(\frac{x_0}{x}\right)^{\frac{1}{2}}$$

where x_0 is the length of the laminar core from the nozzle exit.

Miller and Comings[36] show that if the data towards the edge of the jet is corrected for the transverse velocity component, then the exponential expression for the velocity profiles is a better fit. They also measured turbulent velocity fluctuations, static pressure and shear stress in the jet. Some of their results are given in Fig. 3.18. Within the core region it can be seen that the static pressure has a sharp sub-ambient minimum on either side of the centre line and that the velocity fluctuations have a maximum value at about the same point. Beyond the core region, the static pressure is a minimum at the centre line and the shear stress has a maximum value near to the point of maximum velocity fluctuations, probably around the point of inflection on the velocity profile.

In order to make a comparison between the turbulent two-dimensional jet and the laminar jet, it is convenient to use the Goertler profile for the turbulent jet because the equations for u are then of the same form in both cases. First of all, it should be pointed out that since K, the kinematic momentum, is equal to $U_0^2 W$ for a uniform nozzle exit velocity,

$$\frac{u}{U_0} = \frac{\sqrt{3}}{2} \sqrt{\frac{W\sigma}{x}} (1 - \tanh^2 \eta) \tag{3.47}$$

i.e. is independent of Reynolds Number. Similarly for v, so that the jet spread is independent of Reynolds Number. This is not so with a laminar jet in which u is proportional to $Re^{\frac{1}{3}}$.

If a ratio of the centre line velocities for the laminar and turbulent cases is taken, one obtains:

$$\frac{U_{lam}}{U_{turb}} = 0.189 \left(\frac{x}{W}\right)^{\frac{1}{6}} Re^{\frac{1}{3}} \tag{3.48}$$

which clearly depends on x/W and Re. For a typical distance downstream

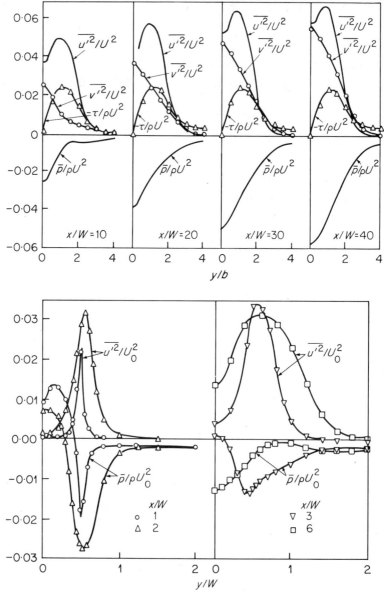

Fig. 3.18 Velocity fluctuations, shear stress and static pressure in a turbulent jet (a) in the fully developed region and (b) in the core region (Miller and Comings, *Ref. 36*, Figs. 9 and 10)

[3.4] THE TURBULENT JET

from the nozzle, say $x = 10W$,

$$\frac{U_{\text{lam}}}{U_{\text{turb}}} = 0.277\, \text{Re}^{\frac{1}{3}} \tag{3.49}$$

Hence the jet widths at that point are the same for $\text{Re} = (1/0.277)^3 = 47$. However, for Reynolds Numbers around the critical region, say $\text{Re} = 2000$, $U_{\text{lam}}/U_{\text{turb}} = 4.4$, a relatively large ratio, giving a significant ratio of energies in the two jets at that downstream position.

This analysis is not strictly accurate because laminar jets are normally operated with the flow from the slit having a parabolic profile for greater stability. Also, the origin is at a distance x_0 behind the nozzle which is different in the two cases, so that 10 nozzle widths from the origin does not give the same point downstream from the nozzle. Nevertheless, the analysis gives an indication of the difference in spreading rates for the jets and also indicates that the laminar jet can spread as much as the turbulent only, but only at very low Reynolds Numbers.

3.5 Power Consumption of a Jet

The power consumption of a jet is given by PU_0A where P is the supply pressure, U_0 is the velocity of the fluid in the nozzle and A is the cross-sectional area of the nozzle. Since $A \propto W^2$ for a laminar jet, $U_0 \propto P$, and therefore the power consumption is

$$\propto U_0^2 A = \frac{\text{Re}_{\text{crit}}^2\, v^2 A}{W^2} \tag{3.50}$$

W being the width of the nozzle. Hence, for the laminar jet, the power consumption is independent of the width of the jet.

For the turbulent case, we have approximately that $U_0 \propto \sqrt{P}$ and so the power consumption is

$$\propto U_0^3 A = \frac{\text{Re}_{\text{crit}}^3\, A^3 v}{W^3} \tag{3.51}$$

i.e.

$$\frac{\text{Re}_{\text{crit}}^3}{W}$$

Therefore, in this case, for a given Critical Reynolds Number, the smaller the element the higher is the power consumption. At the same time, the smaller the nozzle the higher is the velocity for a given Reynolds Number and therefore the faster is the switching speed.

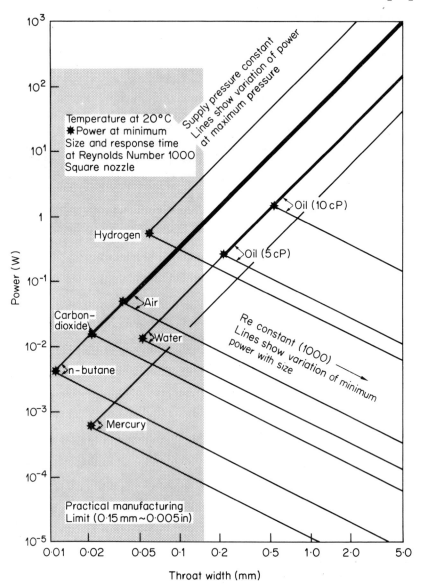

Fig. 3.19 Estimated variation of power with linear size for a wall attachment element working on various fluids (Comparin et al., Ref. 37, Fig. 9)

[3.5] POWER CONSUMPTION OF A JET   129

Curves of power consumption of turbulent jets for various sizes of element and different fluids are given in Fig. 3.19. The minimum power consumption is determined by the Critical Reynolds Number which, as has already been stated, is not a very clearly defined number for a two-dimensional jet.

For a three-dimensional jet there is a further variation, because if the actual minimum Reynolds Number is accepted to be experimentally based on the hydraulic mean depth, then the minimum Reynolds Number expressed in terms of nozzle width varies with aspect ratio of the nozzle. It may be shown that the Critical Reynolds Number at an aspect ratio σ divided by the Critical Reynolds Number at an infinite aspect ratio is

$$\frac{\text{Re}}{\text{Re}_\infty} = 1 + \frac{1}{\sigma} \tag{3.52}$$

Thus the minimum power consumption depends on the aspect ratios of the device. Similarly, the lines showing the power consumption at maximum pressure are arbitrary because this pressure will depend on the element type and is usually fixed by a Mach Number criterion. A Mach Number of unity is a reasonable upper limit for common fluid amplifiers.

Clearly, also, a bottom limit will be the ability to manufacture the very miniature nozzles dictated by the Critical Reynolds Numbers for some fluids, and the shaded area indicates the region in which this is most likely to present a serious problem.

3.6 The Deflexion of a Jet by Another Transverse Jet

The most comprehensive information to date on the deflexion of laminar jets is given by Beatty and Markland[3]. Two problems were examined. The first was the effect of the transverse jet on the length of laminar flow in the combined jet, and the second, the amount of deflexion compared to that predicted by theory from momentum considerations. It is pointed out that maximum deflexion will occur with the jets inclined at 90° to each other but that this is most likely to disturb the laminar flow—if the jets are inclined at a more acute angle they may merge without instability. A number of photographs are given of the deflexion of one jet by another at various angles of inclination, and the results plotted in such a way that the Reynolds Number may be found for a given laminar length and jet deflexion at the receiver. It does appear that it is possible to work at higher Reynolds Numbers in the main jet if the acute angle between the main jet and the control jet is small, but it is difficult to choose the best compromise.

It was further pointed out that the laminar length was longer for a long supply channel than for a short supply channel, because of the parabolic profile in the jet. However, the deflexion was lower than that predicted by momentum considerations for a parabolic profile by as much as 20 per cent, whereas for the square profile the correlation was good. It was thought that since the momentum was concentrated towards the centre of the parabolic jet, a proper transfer of momentum was in some way prevented by the slower velocity at the edges. It was also found that the pressure recovery in a deflected jet was very much less than in the same undeflected case—below 50 per cent—and this was thought due to the distortion of the cross-section of the jet so that it no longer conformed to the shape of the receiver. With parabolic profiles, the two jets do not appear to coalesce properly and the jet spreads very rapidly for large values of the control flow. Probably with turbulent jets this problem is not so severe, because of the relatively square profile, but clearly work still needs to be done on the mechanics of interacting jets.

However, for square profiles, it is shown that momentum considerations apply, i.e. if the jet deflection is ϕ, then by resolving perpendicular to the deflected jet,

$$J_c \sin(\theta - \phi) = J_s \sin \phi \tag{3.53}$$

where J_c is the control momentum directed at an angle θ to the main jet momentum J_s.

Note that pressure forces are neglected because of the open configurations of the nodel, i.e.

$$\frac{\sin \phi}{\sin(\theta - \phi)} = \frac{J_c}{J_s} = \frac{\beta_c \rho_c \bar{u}_c^2 d_c^2}{\beta_s \rho_s \bar{u}_s^2 d_s^2} \tag{3.54}$$

where β_c, β_s are factors depending on the velocity profiles and $\bar{u}_c$, $\bar{u}_s$ are average velocities in the channels. If the jet profiles are similar,

$$\frac{\sin \phi}{\sin(\theta - \phi)} = \frac{\mathrm{Re}_c^2}{\mathrm{Re}_s^2} \tag{3.55}$$

Olson[38] points out that when control jets impinge on a main jet, there is a virtual origin of the resultant jet deflexion which is located somewhere downstream of the power jet nozzle exit. This distance exists because of the finite momentum exchange rate between the two jets, and this may well depend on the velocity profiles. Following this region, there is an adjustment distance in which the velocity profile changes from asymmetrical to symmetrical, and beyond this the jet spreads normally. These various corrections to an ideal theory must be determined experimentally.

Olson gives more details of his fairly elaborate theoretical–experimental method predicting the deflexion angle, and this is worth reading. In Section 4.2 of this book the work of Reilly and Moynihan is summarized since this is a frequently quoted reference and the method appears to give adequate results.

Two papers appearing more recently provide a more analytical approach to the jet deflexion problem. Begg[39] uses the conception of integrated pressure defects to estimate the change in jet angle immediately on leaving an orifice due to an asymmetry in the pressure field downstream of the orifice. Theoretical results are shown in which the jet angle is predicted to change by as much as 25 per cent, but the accuracy of the method has yet to be determined, and the amount of pressure asymmetry to be expected is not yet well established. Olsen in his analysis considers the pressure field in the vicinity of the main nozzle and control ports and also concludes that this has a considerable influence. Zalmanzon *et al.* in the other paper[40], use potential flow methods to predict the jet angle resulting from the interaction of two jets at an angle, but points out that the real situation in fluid elements varies considerably depending on the geometry, again confirming the general impression that much work needs to be done before all the possible variations between practice and theory can be fully explained.

We will finish this brief section merely by saying that all amplifiers, including wall re-attachment devices, contain an element of jet deflexion in their operation and it is necessary to estimate this before applying the methods for calculation of the wall attachment parameters if a reasonable description of the phenomenon is to be achieved.

3.7 The Turbulent Re-attachment Process

3.7.1 The Basic Mathematics of the Re-attachment Effect

When a turbulent jet passes close to an adjacent wall, the entrainment of fluid into the jet causes a depression between the jet and the wall. The resulting pressure difference across the jet causes it to bend towards the wall and the depression is increased. This process is cumulative and finally the jet re-attaches to the wall as in Fig. 3.20. The pressure difference across the bubble balances the force due to the radial acceleration v^2/r, which in turn is due to the fluid flowing in a curved path. The pressure difference acting on faces AB and DE of the control column ABDE causes fluid to be reversed into the jet in order to supply the entrained fluid. The situation is stable because a smaller radius of curvature would require a lower bubble pressure to balance the increased acceleration, but this would draw in more fluid than required for entrainment so the bubble would grow

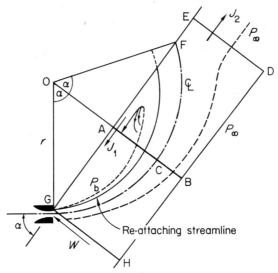

Fig. 3.20 The model for the re-attached jet used in the theory of Bourque and Newman (Bourque and Newman, *Ref. 41*, Fig. 4)

back to its original curvature. The converse is true if the radius of curvature increases from the equilibrium position.

An analytical model for this process was first put forward by Bourque and Newman[41] and relies on the following assumptions:

(a) That a re-attachment streamline may be defined extending from the edge of the nozzle to the re-attachment point. There is no mean flow across this line so that all fluid on the outside of it proceeds downstream while fluid on the inside is reversed into the re-attachment bubble.

(b) That the velocity distribution in the re-attachment jet is essentially the same as that of a free jet for the same distance from the nozzle along the centre line.

(c) That the radius of curvature of the centre line of the jet is constant.

(d) That the pressure in the bubble is constant up to the centre line, after which point it changes abruptly to the pressure on the outside of the jet.

(e) That the width of the jet is small compared to the radius of the curvature.

(f) That the skin friction of the forward and reverse flow along the wall is negligible compared to the momentum.

(g) That the flow is incompressible and two-dimensional, i.e. there are no boundary layer effects due to the top and bottom cover plates, and consequently no secondary flow into the bubble.

Bourque and Newman use the Goertler profile and it has the advantage of being relatively easy to integrate. The mass flow at any section is

$$2\int_0^\infty \rho u \, dy = 2\rho \left[\frac{3J(x+x_0)}{4\rho\sigma}\right]^{\frac{1}{2}} \tag{3.56}$$

But the mass flow at the nozzle must be given by the velocity at the nozzle and hence

$$2\rho \left[\frac{3Jx_0}{4\rho\sigma}\right]^{\frac{1}{2}} = (\rho WJ)^{\frac{1}{2}} \tag{3.57}$$

$$\therefore x_0 = \tfrac{1}{3}\sigma W$$

Now for the re-attaching streamline

$$\frac{1}{2}(\rho WJ)^{\frac{1}{2}} = \int_0^y u\rho \, dy = \rho \left[\frac{3J(x+x_0)}{4\rho\sigma}\right]^{\frac{1}{2}} \tanh \frac{\sigma y}{x+x_0} \tag{3.58}$$

If we let

$$t = \tanh \frac{\sigma y}{x+x_0}$$

the result becomes

$$\frac{1}{t^2} - 1 = \frac{3x}{\sigma W} \tag{3.59}$$

a parametric equation for the re-attaching streamline.

Two points are now taken in the jet; Point (1) is where the centre line is parallel to the wall and Point (2) where it cuts the wall. Because of the assumption of constant radius of curvature, at point (1)

$$\frac{1}{t_1^2} - 1 = \frac{3x}{\sigma W} = \frac{3r\alpha}{\sigma W} \tag{3.60}$$

at point (2)

$$\frac{1}{t_2^2} - 1 = \frac{6r\alpha}{\sigma W} \tag{3.61}$$

The latter equation is obtained on the assumption that the jet width is small compared with the radius of curvature so that $\widehat{BOF} = \alpha$, which gives $x = 2\alpha r$. In a real situation where α is small, this may lead to serious inaccuracies.

Combining the two previous equations gives

$$\frac{1}{t_2^2} = \frac{2}{t_1^2} - 1 \tag{3.62}$$

Now considering the momentum change across the control volume ABCD parallel to the wall,

$$(P_\infty - P_b)AC = J + J_1 - J_2 \tag{3.63}$$

where

$$J_1 = \int_{y_1}^{\infty} \rho u^2 \, dy = \frac{3}{4} J\left(\frac{2}{3} - t_1 + \frac{t_1^3}{3}\right) \tag{3.64}$$

and

$$J_2 = \int_{-\infty}^{y_2} \rho u^2 \, dy = \frac{3}{4} J\left(\frac{2}{3} + t_2 - \frac{t_2^3}{3}\right) \tag{3.65}$$

But, by balancing the pressure forces across the jet with the centrifugal force of the circular path of radius r,

$$P_\infty - P_b = \frac{\int_{-\infty}^{+\infty} \rho u^2 \, dy}{r} = \frac{J}{r} \tag{3.66}$$

Hence

$$\cos \alpha = \tfrac{3}{4}(t_1 - t_2) - \tfrac{1}{4}(t_1^3 + t_2^3) \tag{3.67}$$

We therefore have two equations for t_1 and t_2, so that these parameters may be found.

The equation for the point at which the re-attaching streamline cuts the wall is given by geometrical considerations as

$$l = 2r \sin \alpha - \frac{y_2}{\sin \alpha} \tag{3.68}$$

where l is the distance of the re-attachment point from the nozzle. This gives

$$\frac{l}{W} = \frac{2}{3} \sigma \frac{\left(\frac{1}{t_1^2} - 1\right)}{\alpha} \sin \alpha - \frac{\tanh^{-1} t_2}{3 t_2^2 \sin \alpha} \tag{3.69}$$

and for the bubble pressure the following equation is found by dividing

[3.7] THE TURBULENT RE-ATTACHMENT PROCESS

$P_\infty - P_b$ by the dynamic head and using (3.60)

$$\frac{P_\infty - P_b}{P - P_\infty} = \frac{2W}{r} = \frac{6}{\sigma\left(\dfrac{1}{t_1^2} - 1\right)} \qquad (3.70)$$

Consequently, P_b and r may be evaluated.

The results are shown in Fig. 3.21, where theoretical predictions of Bourque and Newman are given, for various values of σ, by the continuous lines. The symbols denote various experimental points. It may be seen that moderate agreement is obtained but that this is achieved by using values of σ considerably different from the 7·67 of the free jet. The broken line gives the theory due to Sawyer which is discussed a little later.

This method has also been used by Comparin, Moore and Jenkins[42] to predict the re-attachment of a laminar jet because the form of the velocity profile is exactly the same as that for a turbulent jet. The resulting parametric equations are evaluated in the same way, but because the profile is a function of $(x + x_0)^{-\frac{2}{3}}$ the resulting equations are slightly different, and because the centre line velocity is a function of $Re^{\frac{1}{3}}$ the final solutions for the bubble pressure and length of the bubble are Reynolds Number dependent. The relevant parametric equations are:

$$\cos\alpha = \tfrac{1}{4}[3(t + t_1) - (t^3 + t_1^3)] \qquad (3.71)$$

and

$$t^3 = \frac{t_1^3}{2 - \dfrac{36x_0 t_1^3}{Re\, W}} \qquad (3.72)$$

where again, for the second equation, the assumption that the jet is narrow compared with the radius of curvature must be made. From these two equations, t_1 may be eliminated and t substituted in the equation for the re-attachment distance x_r which was again arrived at geometrically, i.e.

$$\frac{1}{W} = \left(\frac{Re}{36t^3} - \frac{x_0}{W}\right)\frac{\sin\alpha}{\alpha} - \frac{1}{3t^2}\frac{\tanh^{-1}t}{\sin\alpha} \qquad (3.73)$$

where

$$\frac{x_0}{W} = \frac{5}{64}Re$$

In this case the value of x_0 takes into account the fact that the fully developed profile does not start till the end of the core region. An expression for bubble pressure may also be found as before.

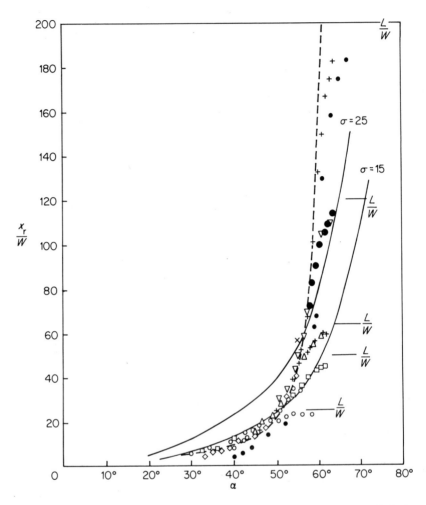

Fig. 3.21(a) Comparison of Bourque's measurements with Sawyer's analysis for a jet attaching to an adjacent inclined plate (Sawyer, *Ref. 44*, Figs. 7 and 8):

---- Analysis with $E = 0{\cdot}130$ and $2E_1/E = 0{\cdot}638$;
——— Bourque and Newman's analysis using various values of σ.

Dimensionless distance to re-attachment point x_r/W against wall angle. L is the length of the plate onto which the jet attaches and the figure indicates how different values of L/W limit the observed x_r/W. The various symbols denoting measured points refer to different Reynolds Numbers

[3.7] THE TURBULENT RE-ATTACHMENT PROCESS 137

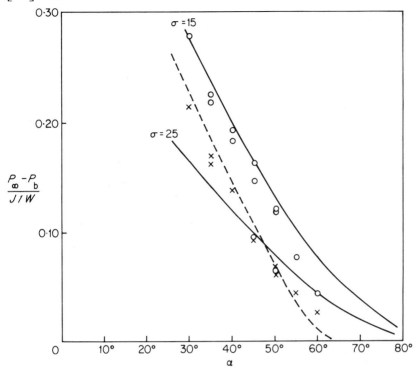

Fig. 3.21(b) Comparison of Bourque's measurements with Sawyer's analysis for a jet attaching to an adjacent inclined plate (Sawyer, *Ref. 44*, Figs. 7 and 8):
---- Analysis with $E = 0.130$ and $2E_1/E = 0.638$;
—— Bourque and Newman's analysis using various values of σ.

The variation of bubble pressure with wall angle
○ corresponds to pressure minimum on the plate.
× corresponds to pressure near the jet exit.

This theory proved moderately successful for wall angles greater than 25° but gave physically impossible solutions below that figure. This fact is similar to the theory for the turbulent jet and may be in part due to the neglect of the thickness of the jet in deriving the parametric equation. The results for wall angles of 30°, 40° and 50° and for aspect ratios of 10, 15 and 30 are reproduced from Ref. 42 in Fig. 3.22.

3.7.2 An Alternative Control Volume

Sawyer[43] considered the attachment of a jet to a wall parallel to the jet and distance h from it. He used, however, after some discussion, a control volume which was introduced by Bourque in work earlier than that reported here and which gives a simpler result than that already quoted.

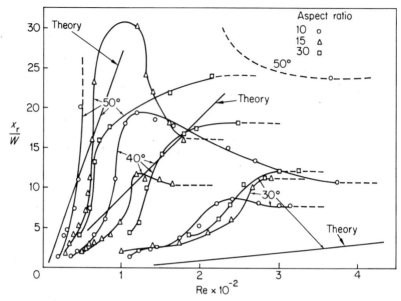

Fig. 3.22 Bubble length for the re-attachment of a laminar jet at wall angles of 50°, 40° and 30°, showing the dependence on Reynolds Number (Comparin *et al.*, Ref. 42, Fig. 4)

Briefly, the momentum of the jet at the re-attachment point is resolved parallel to the wall and put equal to the difference forward along the plate and that reversing back into the bubble, i.e.

$$J \cos \theta = J_2 - J_1$$

The expression $J_2 - J_1$ may be evaluated using (3.64) and (3.65), but in this case both are calculated at the same point so that $t_2 = t_1$. Hence

$$\cos \theta = \tfrac{3}{2}t_1 - \tfrac{1}{2}t_1^3 \qquad (3.74)$$

Sawyer further takes the re-attachment point to be defined as the position at which $u/U = 0.1$ streamline cuts the wall. If b_1 is the thickness of the jet at re-attachment, defined as the distance between the two 0·1 velocity points, and x_1 is the coordinate along the jet centre line of that point

$$\frac{b_1}{2(x_1 + x_0)} = \frac{1}{\sigma} \tanh^{-1} \sqrt{0 \cdot 9} = \frac{1 \cdot 825}{\sigma} \qquad (3.75)$$

Using (3.57) and (3.59) together with $x_1 = r\theta$,

$$(x_1 + x_0) = \frac{r\theta}{1 - t_1^2} \qquad (3.76)$$

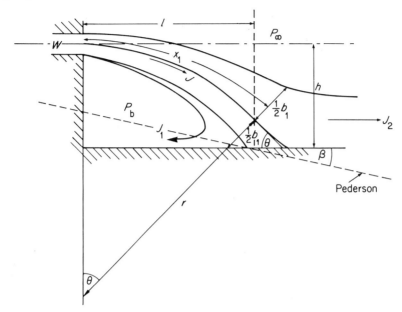

Fig. 3.23 Sawyer's first jet attachment model. Pederson considers the case where the wall may be put at an angle β to the nozzle centre line, but still locates the re-attachment point by measurements l and h (Sawyer, Ref. 43, Fig. 1)

From the geometry of the attachment model indicated in Fig. 3.23, a relation is obtained between the height of the re-attachment point measured from the nozzle centre line (h), the radius of curvature of the centre line (r), the width of the jet at re-attachment and θ. A second relation is also obtained for the distance of the re-attachment point from the nozzle measured along the nozzle centre line, i.e.

$$h = r(1 - \cos\theta) + \tfrac{1}{2}b_1 \cos\theta \qquad (3.77)$$

and $l = r \sin\theta$. From (3.75), (3.76) and (3.77) the following expression may be obtained

$$\frac{l}{h} = \frac{\sin\theta}{1 - \cos\theta + \dfrac{1 \cdot 825 \theta \cos\theta}{\sigma(1 - t_1^2)}} \qquad (3.78)$$

By choosing values of t_1, $\cos\theta$ may be found in two ways from (3.74) and (3.78), and by trial and error a complete solution is obtained.

In a later paper[44], Sawyer modified the definition of the re-attachment point to be where the 0·1 velocity line meets the edge of the reversed flow. The reversed flow profile was assumed to be identical with that between the re-attachment streamline and the 0·1 velocity line.

3.7.3 The Effect of Curvature of the Flow on the Turbulent Mixing

In an additional important observation Sawyer points out that the effect of curvature should be to modify the mixing coefficient. If Prandtl's arguments are correct, then they may be modified to account for the curvature of the jet, i.e. if the eddy shear stress is due to the displacement of a block of fluid from one level to another across the stream, then the geometrical effect of curvature of the jet will be such that at the new position of the block of fluid the radius of curvature of the streamline will be different. For movement to the outside, the radius of curvature will be larger and therefore the centripetal acceleration less, so that the particle moves out further than it might otherwise have done. Conversely, if a disturbance is towards the inside of the curved jet, it will move less than expected. Consequently, one would expect more mixing on the outside and less on the inside, and the difference is proportional to $1/r$. Sawyer uses an entrainment parameter E defined by

$$E = \frac{1}{U}\frac{d}{dx}\int_{-\infty}^{+\infty} u \, dy = \frac{1}{\sigma} \tag{3.79}$$

and takes parameters E_1 and E_2 for the inner and outer edges of the jet so that $E = E_1 + E_2 = 0.130$ for the plane jet. The difference between the two sides is taken to be $1 - 2(E_1/E)$. Sawyer points out that this should be $\propto (b/r)$, but for the results given in Fig. 3.21 E_1 and E_2 are taken to be constants of value 0·0413 and 0·0884, which gives values of σ of 12·1 and 5·65 for the inner and outer edges of the jet respectively.

This difference in entrainment rates means that the profile on each side of the jet spreads differently. The interesting point, however, is that it is shown that the resulting complete profile is very little different from that of the free jet, but that the centre line is merely displaced from the original centre line. Sawyer shows that this is because the difference in entrainment rates is made up for by a cross flow in the jet at an almost identical rate.

3.7.4 The Effect of Control Flow

Pederson[45] modified Sawyer's method to allow for a control flow into the bubble region and also to allow for a wall which was both set back from the slot (or nozzle) and also inclined to the nozzle axis. This results in modifications of two of Sawyer's equations, i.e. the parametric equation for the constant mass flow line becomes

$$\frac{(1+B)^2}{t_1^2} = \frac{3x_1}{\sigma W} \tag{3.80}$$

where B is the ratio of the control flow to half the flow leaving the slot.

Also the momentum balance at the re-attachment point yields:

$$\cos(\theta - \beta) = \tfrac{3}{2}t_1 - \tfrac{1}{2}t_1^3 \tag{3.81}$$

The results show moderate agreement with experiment for various wall angles, set-back and control flow. Further, by the assumption that the dynamic behaviour is governed by the ventilation process (i.e. by the control flow) a relation for switching time was obtained. This approach neglects inertia effects on the assumption that the switching time is generally long enough for them to be neglected. By differentiating half the volume flow rate at a distance x downstream Pederson obtains

$$\frac{d}{dx}\left(\frac{Q}{2}\right) = \frac{1}{4}\sqrt{\frac{3J}{\sigma x \rho}} \tag{3.82}$$

The jet is then assumed to be moving laterally with velocity v over its length x, so that the rate of volume to be filled is vx. The two quantities are equated to give a value for v, and for full switching the jet is assumed to move approximately the distance h, which gives a switching time of 22 times the transport time of a particle in the jet over the distance h. This is a very crude analysis but gives an order of magnitude of switching time, which indicates indeed that the 'ventilation' of the bubble region must control the switching.

3.7.5 Modifications to the Basic Re-attachment Model

In a later publication[46] Bourque points out that these methods for the turbulent jet generally give poor agreement for an inclined wall adjacent to the nozzle, but gives reasonably good agreement for the case of the wall which is offset from the nozzle and parallel to the nozzle centre line. He replaces the assumption of a constant radius of curvature for the centre line with the assumption that the dividing (or re-attachment) streamline has a sinusoidal shape. His argument for this is that the pressure difference across the jet in practice increases for 70 per cent of the bubble length so that the radius of curvature gradually increases and then decreases for the last 30 per cent of the bubble. This point was considered earlier by Jones, Foster and Mitchell[33] and will be discussed later. He also points out that for the inclined wall case, the limit of attachment is for a wall angle α of $67°$ and under this condition the jet should theoretically hit the wall at right angles at infinity. A profile in polar coordinates appropriate to this is

$$\frac{r}{W} = \frac{a}{W}\sin\frac{\pi}{2}\frac{\theta}{\theta_m} \tag{3.83}$$

where r is the radius vector at an angle θ with the direction of the jet at the nozzle exit and θ_m is the maximum possible value for θ which is 67°. The scale factor a is determined from the continuity and momentum equations. Bourque uses the original parametric equation for the re-attachment streamline by assuming constant mass flow between it and the centre line. However, he replaces his original control volume concept of Ref. 41 with that due to Sawyer which equates momenta at the re-attachment point. Further geometrical relations are required in addition to (3.82) so that the solution is now rather more complex, and the values for the re-attachment point can only be solved by trial and error. Using a value for σ of 10·5 to allow for reduced entrainment close to the wall, his results give good agreement with experiment. A comparison with the experimental results of Levin and Manion[47] is reproduced from the paper and shown in Fig. 3.24, because the experiments are often quoted in the literature. (For their theory, Levin and Manion used the method of Bourque and Newman.)

Perry[48] proposes a simple modification to the 'control-volume' model which is shown to give better results using only a single value of the jet-spread parameter σ. He takes the control volume EDGH in Fig. 3.20 for the momentum equation, so that if J is the momentum issuing out of the nozzle,

$$J \cos \alpha - J_2 = (P - P'_b)\left(D + \frac{W}{2}\right) \cos \alpha \qquad (3.84)$$

where D is the setback of the wall from the edge of the nozzle (not shown in Fig. 3.20). P'_b is assumed different from P_b, the bubble pressure, because experiment shows this to have a slightly higher value, and it is this modification that is Perry's contribution. He further suggests that for simplicity

$$P_\infty - P'_b = \frac{1}{C'_p}(P_\infty - P_b) \qquad (3.85)$$

He further does not assume that $\widehat{BOF} = \alpha$, but gives it a symbol θ, which is then evaluated from an equation obtained from the momentum equation, i.e.

$$\cos \theta = \cos \alpha (1 - C'_p) + C'_p \left(\frac{1}{2} + \frac{3}{4} t' - \frac{t'^3}{4}\right) \qquad (3.86)$$

Again because of the reduction of entrainment close to a wall, Perry takes a value of σ of 10, which is larger than the 7·67 quoted for a free jet and is close to the value used by Bourque. C'_p is then adjusted to accom-

[3.7] THE TURBULENT RE-ATTACHMENT PROCESS

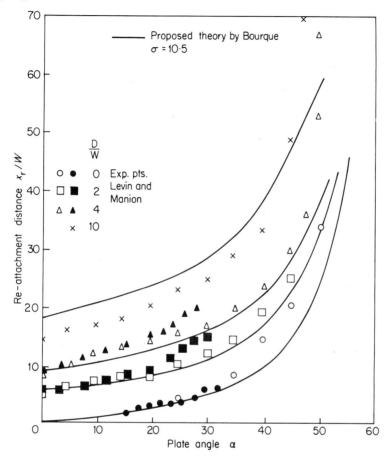

Fig. 3.24 The theoretical bubble length for attachment of a turbulent jet on to an inclined plate due to Bourque, compared with experiments by Levin and Manion. D is setback of wall (Bourque, Ref. 46, Fig. 9)

modate other errors and a value of 2 is found to give good agreement with experiment. It is also found that if this value is applied to the case of a wall parallel to the centre line of the nozzle, the equation reduces to that given in Sawyer's analysis, thus ensuring equal experimental agreement for that particular case and better agreement for the inclined wall case. His results for bubble length are given in Fig. 3.25 for various wall angles and setback.

A point worth noting is that he did not find any noticeable change in the experimental results over an aspect ratio range of 1·5 to 100 (other workers' experimental results are quoted in order to cover this range).

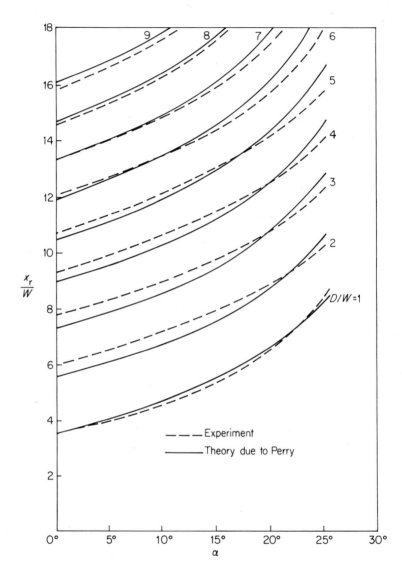

Fig. 3.25 Comparison of theory with experiment for attachment of a turbulent jet on to an inclined wall with offset. Modification to basic theory is due to Perry (*Ref. 48*, Fig. 8)

[3.7] THE TURBULENT RE-ATTACHMENT PROCESS

This point is of some interest because the argument is still open. First of all the core length changes with aspect ratio, and secondly more flow enters the bubble region through the boundary layer as the aspect ratio reduces, thus causing the bubble to lengthen. The implication is that the various effects cancel each other out, but this is by no means proven.

Olson reports work on the wall attachment phenomenon[38,49]. He expresses the jet parameters in terms of the Mach number of the stream emerging from the slit. He also considers the initial deflexion of the jet due to a transverse control flow, a point missed by other theoretical approaches. Another important point is that the jet centre line is assumed to be offset from the nozzle centre line because of the differing shear stress constant on the inner and outer edges of the jet. If the distance from the top of the nozzle and the bottom of the nozzle to the centre line are h_{oc} and h_{ic} respectively,

$$\frac{h_{ic}}{W} = \frac{\kappa_{ci}}{\kappa_{ci} + \kappa_{co}} \quad \text{and} \quad \frac{h_{oc}}{W} = \frac{\kappa_{co}}{\kappa_{ci} + \kappa_{co}} \quad (3.87)$$

where κ is defined by Prandtl's second hypothesis with constant mixing length assumed across the jet, i.e.

$$\varepsilon = \kappa \xi \frac{U_{\text{shear}}}{2} \quad (3.88)$$

where

$$\kappa = 2C_2 f(1) \frac{d\xi^*}{dx}$$

ξ^* is the distance from the edge of the shear layer to the point of half the shear layer velocity (U_{shear}). C_2 and $f_4(1)$ which are functions of Mach Number (M_0) are defined in the references. For fully developed flow, U_{shear} is U, the centre line velocity.

A Gaussian profile is then assumed and a numerical integration gives the position of the re-attaching streamline for both the core region and fully developed region. These are corrected for the fact that in the experiments the centre line velocity decay was found to be a power of x slightly different from $-\frac{1}{2}$.

Olson then comments on the momentum balance used by other authors to give the further equation necessary for a solution of the problem, and states: 'The difficulty found with using this procedure in the present technique was that the predicted re-attachment location was quite sensitive to small changes in the factors contributing to the momentum balance, particularly the pressure force contribution which most investigators

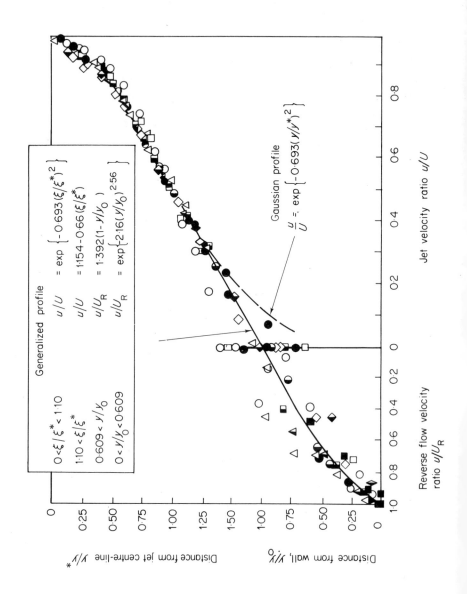

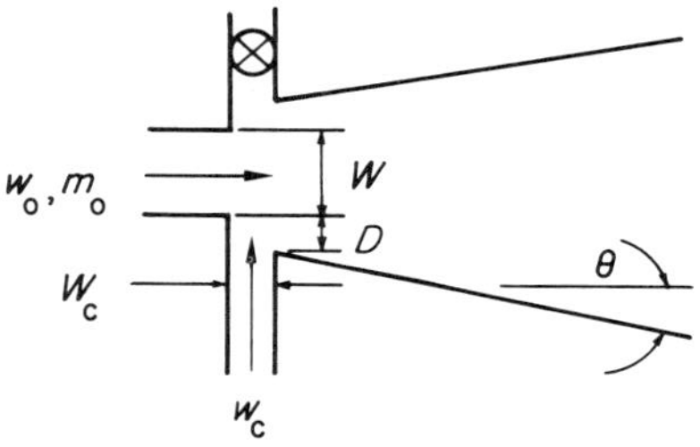

[3.7] THE TURBULENT RE-ATTACHMENT PROCESS 147

Symbol	△	◇	○	□
D/W	0.5	1.0	1.0	1.0
θ (deg)	10	6	10	13.5

w_c/w_o	Symbol
0	Open
0.05	Half-closed
0.15	Closed

$W = 0.5$ in
$M_o = 0.66$
$W_c/W = 0.5$

Fig. 3.26 Olson's generalized velocity profile for the bubble region at the beginning of the recompression. Note that ξ is the distance of a point in the y direction measured from the edge of the shear layer which is difficult to define in the region of developing flow. The diagram describes the flow near the re-attachment point where the edge of the shear layer is the jet centre line, so that ξ becomes the same as y. There is a slight confusion because for the lower part of the curve, y/y_o expresses the distance from the wall perpendicular to the jet centre line (Olsen, Ref. 38, Fig. 3)

have ignored.' Although in Ref. 49 he uses Sawyer's control volume, this is discarded in Ref. 38, where he postulates that the velocity profiles in the jet boundary, enclosing the separation bubble and the reverse flow in the separation bubble, can be generalized in terms of the percentage distance to re-attachment. The evidence for this is given in Ref. 49. Effectively this must mean that by equating forward and reverse flow inside the re-attachment line, computed from empirical expressions for velocity profiles, it is possible to fix a point relative to the re-attachment point which has a suitable profile for forward and reverse flow, with a smooth gradation between the two. This turns out to be at a distance 0·65 of that of the re-attachment point itself. Details are not given of the actual procedure in the references, but the generalized velocity profile for this point (apparently the point of the initiation of recompression) is reproduced in Fig. 3.26.

Ref. 49 also gives considerable information on the effect of a finite wall length on the re-attachment point and also on the effect of a second wall. Probably the most elegant part of the paper, however, is the measurement of spreading rates for attached jets at various Mach Numbers, expressed as an effective shear stress constant κ. Olson points out that it is difficult to define the edge of the shear layer and therefore not easy to determine ζ^* directly. He therefore plots the experimental profile logarithmically so that the curve becomes almost a straight line which can be matched to the theoretical Gaussian profile for forward flow (indicated in Fig. 3.26) by suitable choice of ζ^*. By repeating this for various values of x, $d\zeta^*/dx$ may be obtained and used in (3.88). Some of the results are given in Fig. 3.27 where it may be seen that the constant for the inner part of the jet does not appear to vary with jet curvature but is more a function of the distance to the re-attachment point. On the other hand, on the outer part of the jet, the results shown in Fig. 3.28 indicate that κ in this region is a function of jet curvature. The results may be compared with Sawyer's predictions, since for a Goertler profile the eddy viscosity may be expressed as in (3.33), which may then be compared with Olson's expression which includes the constant κ, i.e.

$$\varepsilon = \zeta^* \frac{U}{2} \qquad (3.89)$$

where

$$\zeta^* = 2b_{\frac{1}{2}}$$

Hence

$$\kappa = \frac{1 \cdot 125}{4\sigma} = \frac{0 \cdot 28}{\sigma} \qquad (3.90)$$

[3.7] THE TURBULENT RE-ATTACHMENT PROCESS

The values of E_1 and E_2 chosen by Sawyer give values of κ of 0·023 and 0·05 on the inner and outer boundaries of the jet respectively, thus agreeing in principle with the experimental results of Olson.

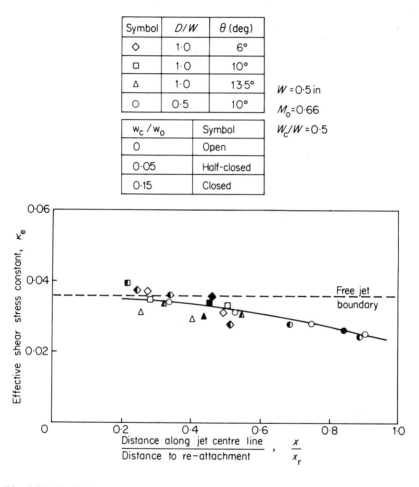

Fig. 3.27 Variation of shear constant with distance from re-attachment for inner boundary of re-attaching jet for single-boundary-wall configurations (Olson and Chin, *Ref. 49*, Fig. 28)

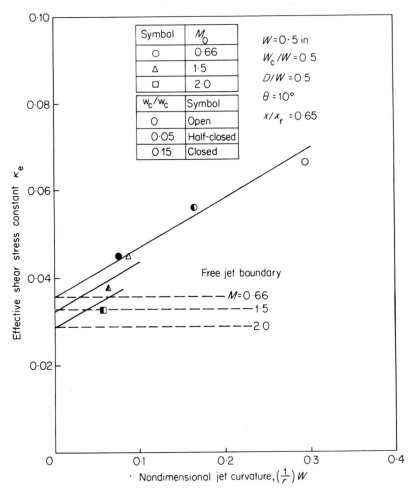

Fig. 3.28 Variation of shear stress constant with jet curvature for outer boundary of re-attaching jet (Olson and Chin, *Ref. 49*, Fig. 31)

3.7.6 An Estimate of Wall Pressure Distribution

One point about all the results quoted here is that a constant bubble pressure is assumed, although there is considerable experimental evidence to show that this is not realistic for a wall very close to the jet. For the correct design and positioning of control channels in an element, some knowledge of the pressure distribution along the re-attachment wall is desirable. Jones, Foster and Mitchell[33] reported on a step-by-step calculation in which the bubble was split up into small control volumes

151

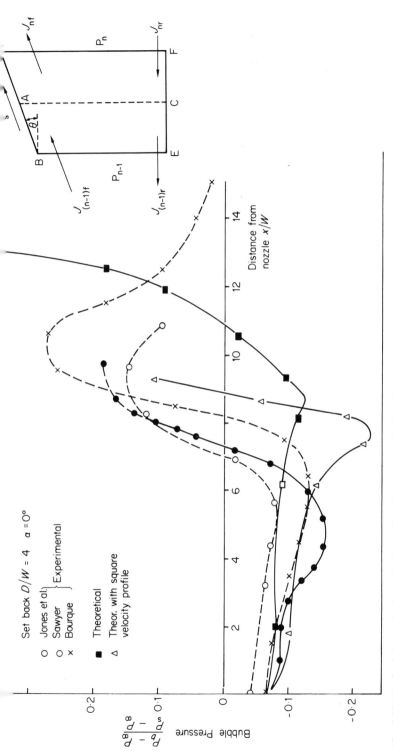

Fig. 3.29 Typical theoretical predictions of wall pressure distribution compared with experiment for calculations in which the bubble is split up into a large number of small control volumes, one of which is shown below the curves. BD is a length of the constant mass flow line (or re-attachment streamline) and J_s, the momentum entrained into the jet over this length, is assumed equal to the shear stress along BD. J_{nf}, $J_{(n-1)f}$ are estimates of the momentum of the forward flowing fluid outside the re-attachment streamline at stations n and (n − 1) and J_{nr}, $J_{(n-1)r}$ are the estimated reversed flow momentum. P_n and P_{n-1} are average pressures across the boundaries DF and BE respectively, so that a momentum equation for each volume may be written:

$$AC(P_{n-1} - P_n) = (J_{nf} - J_{(n-1)f}) \cos \theta - (J_{nf} - J_{(n-1)r}) - J_s \cos \theta$$

(Jones, Foster and Mitchell, *Ref. 33*, Fig. 11)

as shown in Fig. 3.29. Here the important assumption is that the shear stress on BD is the same as in the free jet case, and therefore can be calculated as a momentum difference in that case. A momentum balance is now carried out for each small control volume, based on Albertson's profile and some straight line approximations as the position comes close to the re-attachment point. The average pressure for the volume is obtained, the radius of curvature computed, and an angle of inclination for the next control volume boundary BD is then calculated. For the calculation an initial pressure must be chosen, either on the basis of experiment or by other means yet unproven.

From the analysis it is clear that after the point of maximum distance from the wall, the radius of curvature must decrease sharply. This is accompanied by the rapid rise of pressure up to the re-attachment point and clearly indicates that the assumption of constant pressure within the bubble is not valid and that the centre line of the jet does not follow a circular path. This is particularly so for a wall with no offset and small angle of inclination, so that the other methods are least likely to be successful for such a case. This comment gives some justification for Bourque's later assumption of a sinusoidal shape for the centre line of the jet. A typical result is shown in Fig. 3.29 where two theoretical curves are indicated. One is for normally accepted velocity profiles and the other for a rather more extreme velocity distribution in the jet, which would more clearly describe the high velocities occurring near the centre of the vortex in the bubble region.

3.7.7 The Stability of Attachment

Two other papers are of interest in making observations about the re-attachment problem. Brown[50] points out that if a jet attaches to a short wall, the normal momentum balance at the re-attachment point becomes invalid and the jet tends to 'lock on' to the corner at the end of the wall itself. This is because a slight deflexion causes a large change in the flow reversed into the bubble, resulting in a very stable situation. He then goes on to describe a pressure controlled amplifier in which this effect is utilized. A sharp edge is placed near the jet which causes some flow to reverse into the bubble, as shown in Fig. 3.30. The situation may be stable or unstable, depending on the geometry and on the impedance of the control channel and load. An important difference between this design and a conventional wall attachment device is that here the bubble has a constant length; this simplifies receiver and vent design.

On the other hand, Sawyer[51] points out that it is possible to design a cavity such that the jet is neutrally stable and the re-attachment point can take up any position along the cavity wall, depending on the initial

[3.7] THE TURBULENT RE-ATTACHMENT PROCESS

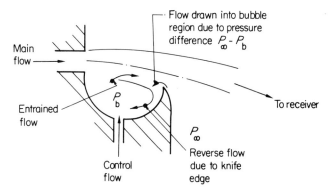

Fig. 3.30 Brown's single control 'knife-edge' amplifier without receiver ports shown. It can also be symmetrical and bistable. The deflexion of the jet is determined primarily by the position of the knife edge, the jet taking up a position relative to the edge so that the reversed flow replaces that which is entrained. The pressure difference is fixed by the acceleration of the fluid due to curvature of the jet (Brown, *Ref. 50*, Fig. 6)

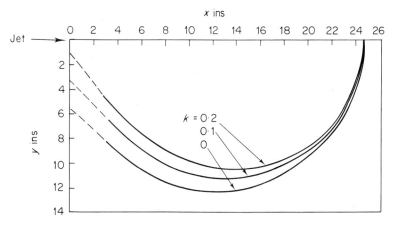

Fig. 3.31 Wall shapes calculated by Sawyer to give neutral stability of the jet so that the re-attachment point could take up any position along the wall. k is an empirical constant that depends on the ratio of jet thickness at re-attachment to radius of curvature of the jet (Sawyer, *Ref. 51*, Fig. 9)

conditions. Using a similar momentum equation to that given in Ref. 44, he produces a relation between the angle θ subtended by the arc of the bubble and the angle ϕ between the re-attaching streamline and the wall. This is plotted for various values of t^* the value of the parameter t for the re-attaching streamline. An equation is then derived for the rate of bubble growth based on the assumption of a quasistatic situation, i.e. that the rate of increase of volume of the bubble depends upon the surplus of flow into the bubble over that entrained out of it by the turbulent mixing in the jet. The shapes for zero rate of growth, calculated for three values of an empirical constant k, are given in Fig. 3.31. Clearly this is of importance for the design of wall attachment elements because a shape of this kind should give greatest sensitivity to control flow. However, this is an idealized situation and will be strongly modified by the rest of the geometry of the device and the loading conditions on the element, as pointed out in the next paragraph.

3.7.8 The Effect of Downstream Loading on the Attachment Bubble

All of these attempts at explaining wall attachment are concerned with attachment to a single adjacent wall. The effect of other boundaries such as an opposite wall or a splitter may be considerable. Several experimenters have tested the effect of a second wall on the attachment process and have found little effect. Such observations are erroneous however, because they neglect the effect of the splitter, and it is this coupled with the second wall which may be shown to have a dominating influence. Foster et al.[52] show the effect of the pressure field in the other side of the jet from the re-attachment wall by introducing flow into the mixing region from the splitter in two ways as shown in Fig. 3.32. The first was from the side of the splitter in the receiver into which the jet was flowing and the second was from the centre of the splitter; in both cases this flow was parallel to the nozzle centre line, but opposing the main flow. Referring back to Fig. 3.20 if the flow is from the *side* of the splitter, the pressure is raised in the region opposite the re-attachment point F. This causes more flow to be reversed into the bubble and the bubble lengthens. Theoretical results are given following the method of Bourque and Newman but with the assumption that the pressure on the control volume boundary ED is greater than ambient. In addition the geometrical equation is modified to take into account the finite width of the jet (this is important for small setback and wall angle, and the original equation gave impossible results). Fig. 3.32 is reproduced from the paper and shows the very strong influence of downstream pressure on the bubble length.

In the second way of changing the pressure distribution in the mixing zone[52], the flow was introduced from the centre of the splitter, thus raising

[3.7] THE TURBULENT RE-ATTACHMENT PROCESS 155

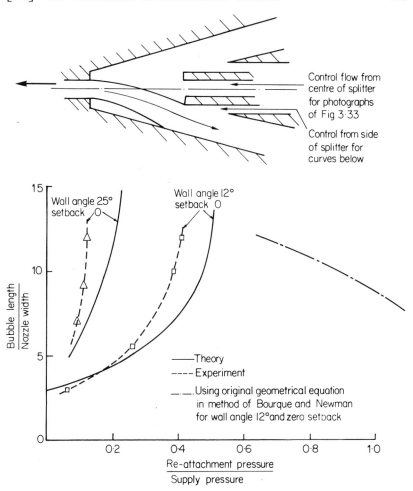

Fig. 3.32 Comparison between theory and experiment for the effect of downstream loading on the stability of an attached jet. Control flow is introduced from the side of the splitter past which the jet is flowing, as indicated in the diagram above the curves (Foster, Misra and Mitchell, *Ref. 52*, Fig. 3)

the pressure around the area of line CB in Fig. 3.20. This raises the pressure on boundary AB of the control volume and causes more flow out of the bubble than is reversed back, so that the bubble length shortens. Finally, the element switches due to interaction of the two opposing jets. Fig. 3.33 shows photographs of this taking place. Thus it is clear that to concentrate theoretical methods to the case of attachment to a single wall is not likely to result in an adequate description of a real amplifier and

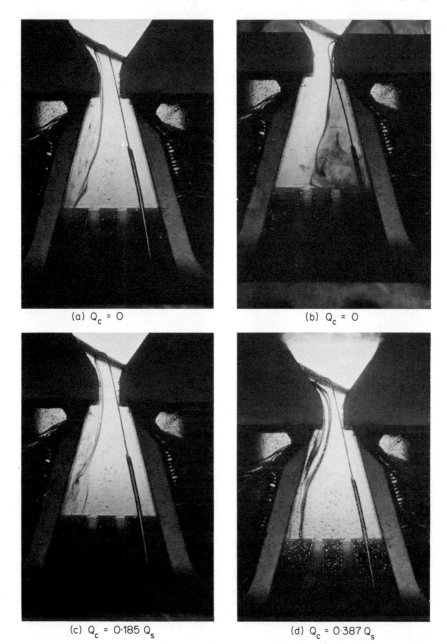

(a) $Q_c = 0$

(b) $Q_c = 0$

(c) $Q_c = 0.185\, Q_s$

(d) $Q_c = 0.387\, Q_s$

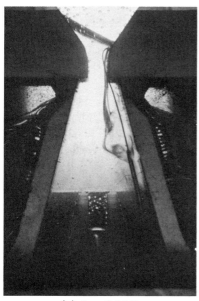 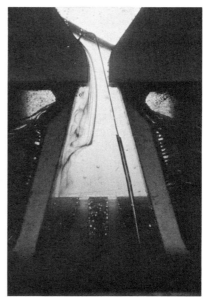

(e) $Q_c = 0.46 Q_s$ (f) $Q_c = 0.46 Q_s$

Fig. 3.33 Photographs showing the re-attachment point moving upstream as the control flow from the centre of the splinter is increased. Fig. (b) is for the control flow as (a) but showing the outer edge of the jet. Similarly for Figs. (f) and (e). $Q_s = 0.2$ g.p.m. (Foster, Misra and Mitchell, *Ref. 52*, Fig. 9)

more sophisticated techniques are necessary to take into account the complete pressure or vortex field in both sides of the jet.

3.7.9 Boundary Layer Separation

An alternative form of attachment is that in which the flow contacts a curved surface and then continues to flow around it without any bubble formation. This form of attachment is a boundary layer phenomenon and theoretical methods are available to determine the separation point. Kadosch[53] calculates the pressure distribution along a curved surface for both turbulent and laminar boundary layers, although he concludes that the wall effect is not likely to occur in the latter condition. This work is used to design proportional amplifiers which are reported on by Pavlin and Facon[54].

Thomson[55] describes the separation of a boundary layer in a convergent-divergent duct at pressure ratios above critical. This is followed by re-attachment on one side of the duct so that the main stream may be effectively diverted. A compression wave is formed at the separation point and this alters the direction of the main stream, causing it to move back towards the wall. A further secondary compression occurs at the re-attachment point. No theory is given but a number of experimental results for the wall pressure distribution pressure ratio and separation distance are given.

3.7.10 Supersonic Flow

Page[56] gives a useful summary of the re-attachment of supersonic flow separating from an edge. The difference between this phenomenon and the subsonic equivalent is that there is a Prandtl–Meyer expansion at the edge from which the flow separates, and the stream then heads towards the wall with less curvature of flow than in the subsonic case. Mixing conditions still exist, but in this case heat transfer is important in addition to turbulent shear. Page discusses in simple terms the effect of reflexions of shock waves from a second wall, or from the outer edge of a re-attaching jet. He also gives a list of references, and since it is difficult to summarize an already clear concise paper, the reader is referred to the paper itself.

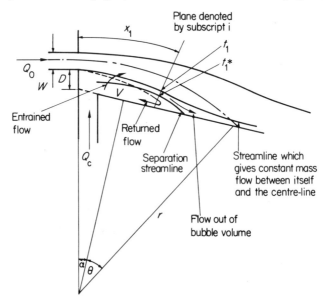

Fig. 3.34 Separation bubble model used by Lush to calculate rate of change of bubble volume. The constant mass flow line is assumed different from the separation streamline and the flow between the two is the flow into or out of the control volume. As shown, the bubble will decay, but with the separation streamline above the constant mass flow line it would increase (Lush, *Ref. 57*, Fig. 2b)

3.7.11 The Rate of Growth of the Re-attachment Bubble

At this stage, it is worth examining the ideas on the mechanism of switching due to Sawyer and given initially in Ref. 51 in connection with the design of wall shapes for neutral stability of the jet. Lush[57] amplifies on the method and compares the estimated switching times with those obtained experimentally for a large scale model of a wall attachment fluid amplifier. A theoretical model is used which is similar to the model due to Sawyer for the static case, but the assumption that the re-attachment streamline is the same as the line of constant mass flow from the nozzle is dropped, and two separate lines for re-attachment and constant mass flow are assumed, represented by t^* and t respectively. The flow between these two lines represents the net flow into and out of the bubble respectively at the re-attachment point, as indicated in Fig. 3.34. If there is some flow in the control channel, a differential for the rate of change of bubble volume V may be written down assuming that inertia forces are negligible, i.e.

$$\frac{dV}{dt} = Q_c - \frac{1}{2} Q_0 \sqrt{\frac{x_1}{x_0}} (t_1^* - t_1) \tag{3.91}$$

where subscript 1 denotes the conditions at the re-attachment point.

An equation for x_1/x_0 is found from entrainment considerations, i.e.

$$\sqrt{\frac{x_1}{x_0}} = \frac{\frac{2E_1}{E}}{\frac{2E_1}{E} - 1 + t_1} \tag{3.92}$$

and a momentum equation is obtained as before, i.e.

$$\cos\theta = \tfrac{3}{2}t_1^* - \tfrac{1}{2}t_1^{*3} \tag{3.93}$$

Three geometrical equations are required to be able to calculate the rate of growth of the jet, i.e.

$$\frac{x_1}{x_0} = 1 + \frac{6E(\alpha + \theta)}{C_{\Delta p}} \tag{3.94}$$

$$\cos\theta = \cos\alpha \left\{ 1 - \frac{1}{2} \frac{\left(D + \frac{W}{2}\right)}{W} C_{\Delta p} \right\} \tag{3.95}$$

$$\frac{x}{W} = \frac{2}{C_{\Delta p}} \frac{\sin(\alpha + \theta)}{\cos\alpha} \tag{3.96}$$

and the non-dimensional bubble volume is

$$V = \frac{2W^2}{C_{\Delta p}^2}\left\{(\alpha + \theta) - \frac{\cos\theta}{\cos\alpha}\sin(\alpha + \theta)\right\} \qquad (3.97)$$

where $C_{\Delta p}$ is the pressure difference across the bubble divided by $\frac{1}{2}\rho U^2$. Angles α and θ are defined as indicated in the diagram.

The equations were solved graphically and bubble switching times estimated as a function of control flow by calculating the time for the bubble to grow a certain length under the action of control flow and then for it to shrink back to size on the other wall. The time between leaving one wall and arriving at the other could not be estimated with the present theory. The results are shown in Fig. 3.35, and it may be seen that the

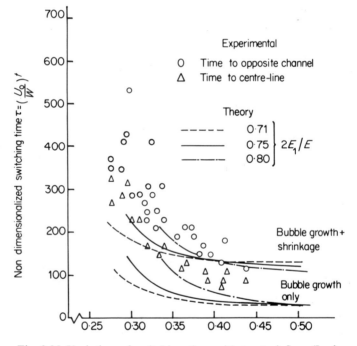

Fig. 3.35 Variation of switching time with control flow (Lush, *Ref. 57*, Fig. 4)

theory underestimates the switching time, as expected. However, the trend of decreasing switching time with increasing control flow is clearly predicted and supports the experimental results due to Muller which are quoted in Chapter 7. The switching time is expressed non-dimensionally

[3.7] THE TURBULENT RE-ATTACHMENT PROCESS

as $\tau = (U_0/W)t$, and can only be related to transport time of the element if the length of the wall is known.

In the model used, the splitter distance was $12W$, so that τ should perhaps be divided by 12 in order to express switching time as the ratio of real time to transport time. Clearly the experimental switching times were between 10 and 50 times the transport time and were very dependent on control flow.

3.8 The Trapped Vortex

When a jet attaches to a wall, the assumptions concerning the nature of the flow inside the bubble region are very much simplified, and it is assumed that fluid which is returned recirculates in a smooth manner along the boundary wall and back into the jet. This is not always so and the flow pattern may be observed to have broken up into at least two vortices with opposite rotations, the one closest to the nozzle actually opposing the jet flow. It is also well known that a vortex on the side of the jet opposite the re-attachment point has a marked influence on the stability of the jet. These two points have led to more interest in the problem of the flow due to vortex formation, and in particular into the conditions for a single vortex to be stable between a uniform stream and a cavity.

Rosenhead[58] is a much quoted reference on the stability of a vortex formed at a surface of discontinuity. He carried out much work on vortex streets which was extended and improved by Hama and Burke[59].

However, the more relevant theoretical treatment is that due to Ringleb[60, 61, 62]. Ringleb first points to the flow over a snow cornice where the air separates from the downstream edge of a high mountain and the snow is caused to form in a shape as shown in Fig. 3.36. A naturally

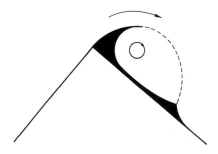

Fig. 3.36 Flow over a snow cornice (Ringleb, *Ref. 60*, Fig. 1)

rounded cavity is formed which is probably the best shape for the minimum energy loss in the air flow. Vortices may easily be represented singly or in pairs in potential flow fields, but generally these may be shown to move with the stream, as in the Karman vortex street. The problem in a 'snow cornice case' is to put in a condition that the centre of the vortex should have zero velocity. Ringleb shows how a transformation of the form used for the Joukowski airfoil may be used to obtain the conditions for stability of vortices in a cusped diffuser. A transformation

$$Z = \tau + \frac{\gamma}{\tau - c} + \frac{\bar{\gamma}}{\tau - \bar{c}} \tag{3.98}$$

with $\zeta = ie^\tau$ is used to transform a simple duct given by

$$Z = \ln \zeta - i\frac{\pi}{2}$$

in the upper half of the ζ plane. (Two parallel lines a distance π apart.)

For a uniform flow field represented by a source C_Q with two vortices of strength C_Γ symmetrically placed with centres at

$$\zeta_0 = \xi_0 + i\eta_0$$

and conjugate point, the condition for stability reduces to

$$\frac{C_Q}{C_\Gamma} = \frac{1}{4} \frac{\xi_0^2 - \eta_0^2}{\xi_0 - \eta_0} \tag{3.99}$$

The condition for finite velocity at the cusps also gives

$$\frac{C_Q}{C_\Gamma} = \frac{4\xi_0 \eta_0}{(1 + \xi_0^2 + \eta_0^2) - 4\xi_0^2} \tag{3.100}$$

Hence for various values of (C_Q/C_Γ), ξ_0 and η_0 can be found. In other words, a locus for the centres of the vortices can be obtained.

The shape of one set of cusps is shown in Fig. 3.37 in which the chain dotted lines are the loci of the centres of the vortices. The actual position of the centres of the vortices along the lines depends upon the ratio of strength of the source representing main flow to the strength of the vortex. In practice, this will clearly depend upon the shearing forces between the main flow and the trapped volume, and so far little work has been done in bringing together the idealized situation of potential flow theory and the theories of attachment based on turbulent mixing ideas.

The idea of a cusped diffuser has been used in the elements manufactured by the Aviation Electric Company and is described further in Chapter 7. An additional use of the vortex in this case is that a vent is

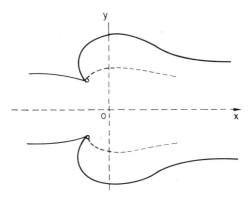

Fig. 3.37 Locus of equilibrium points for centre of vortex within a cusped diffuser (Ringleb, *Ref. 60*, Fig. 10)

placed at the centre of the vortex; when the output is loaded, the flow passes out via the vortex to atmosphere and so builds up a back pressure. Rechten[63] considers the idea from a different point of view and uses the Joukowski transformation to design both the shape of the re-attachment wall and the cusp of the element. The theory is not rigorous because there is no guarantee that the conditions for a standing vortex will be met. However, in other work described in Chapter 7, he reports on good characteristics obtained from elements embodying this design method.

More experimental results for the flow in a cusp and on flow in diffusers with sudden enlargements are given by Frey[64, 65].

An alternative theoretical approach is given by Duggins[66] who uses two Schwarz–Christoffel transformations to obtain the streamlines for the separation from the internal corner of a 90° elbow in a pipe.

3.9 Theoretical Approaches to the Problem of Vortex Flow

3.9.1 Neglecting the Boundary Layer

Some reports of work on vortex valves is reported in Chapter 4, but in the main this is semi-empirical. Very few publications are concerned with the theory of the fluid flow in the vortex, probably because it is such a complex phenomenon. The vortex cannot be assumed to be two-dimensional because of the effect of the side plates, and the boundary layers along these plates have the effect of dominating the flow situation. The reason for the strong influence of the boundary layer is that is slows down the tangential velocity of the fluid close to the walls of the chamber and so reduces its centripetal acceleration. Consequently the pressure difference, which still exists on account of the rotation of the mass of the fluid in the centre of the vortex, causes the slow fluid to move in to the centre of the vortex. The result is that most of the flow through a vortex device takes place in the boundary layer, so that an accurate theoretical description is difficult.

Royle[67] gives some indication of the theoretical approach to the axisymmetric flow problem, but the boundary layer is not considered. The Navier–Stokes equations for axially symmetrical, incompressible, two-dimensional flow are:

$$\frac{\partial v_\phi}{\partial t} + v_r \frac{\partial v_\phi}{\partial r} + \frac{v_r v_\phi}{r} = \nu \left[\frac{\partial^2 v_\phi}{\partial r^2} + \frac{\partial}{\partial r} \frac{v_\phi}{r} \right] \quad (3.101)$$

$$\rho \frac{\partial v_r}{\partial t} + v_r \frac{\partial v_r}{\partial r} - \frac{v_\phi^2}{r} = -\frac{\partial P}{\partial r} \quad (3.102)$$

$$\frac{\partial v_r}{\partial r} + \frac{v_r}{r} = 0 \quad (3.103)$$

A moment of momentum per unit mass is defined as $\phi = v_\phi r$ and a radial Reynolds Number

$$N = -\frac{v_{r_o} R_o}{\nu} \quad (3.104)$$

where v_{ϕ_o} is the tangential velocity and v_{r_o} the radial velocity at the outer radius of the vortex. (3.101) and (3.103) give

$$\frac{\partial \phi}{\partial t} = \frac{\partial^2 \phi}{\partial r^2} + \frac{N-1}{r} \frac{\partial \phi}{\partial r} \quad (3.105)$$

For a sink of radius R_e the solution of equation (3.105) becomes

$$\frac{v_\phi r}{v_{\phi_o} R_o} = \frac{\left(\frac{r}{R_e}\right)^{2-N} - \frac{N}{2}}{\left(\frac{R_o}{R_e}\right)^{2-N} - \frac{N}{2}} \quad \text{for } N = 2 \quad (3.106)$$

where $v_{\phi_o} R_o$ is the inlet moment of momentum.

It is then pointed out that for $N > 2$ and $(R_o/R_e) \gg 1$ or for $N \gg 2$ and $(R_o/R_e) > 1$ the flow approximates to that of an inviscid free vortex, i.e. $v_\phi r = $ const., whereas for small N, it tends to the solid core vortex $v_\phi/r = $ const.

The analysis is then confined to an inviscid vortex flow, but compressibility is assumed and all flow velocities are put in terms of Mach Number with the conditions for continuity giving $\rho v_r = K_1$ and conservation of angular momentum giving $v_\phi r = \text{const.} = K_2$. The ratio of those constants gives the angle α which the velocity at any point makes with the radius vector, i.e. $\tan \alpha = (K_2/K_1)\rho = \Gamma/Q$, where Γ is the circulation and Q the main flow rate.

The pressure P at radius r divided by the supply pressure P_0 is plotted against the radius r divided by the radius r^* at which the Mach Number is unity for various angles α, and the figure reproduced in Fig. 3.38. The

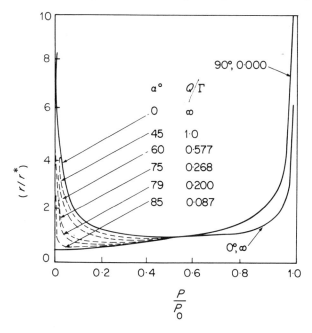

Fig. 3.38 Non-dimensional radius versus non-dimensional pressure for various ratios of strengths of sink and vortex in a vortex valve (Royle and Hassan, *Ref. 67*, Fig. 1b)

effect of changes in the rates of vortex to sink strength is clearly seen. Royle shows that there is good correlation with experiment providing that the initial angle α is correctly evaluated as the ratio strength of the vortex to that of the sink. This is not the same as the inlet angles of the jets used to create the vortex.

Kwok[68] split the vortex into two parts. The first is for the flow in the annular outer region where the axial velocity is zero and the second is in the inner part of the vortex where this component is significant. The two parts are matched at the appropriate radius by assuming that the tangential velocity v_ϕ, the tangential shear and the static pressure are continuous. Kwok non-dimensionalizes the equations using the following parameters:

$$X = \frac{r}{R_o}, \quad V_r = \frac{v_r}{v_{\phi o}}, \quad V_Z = \frac{v_z}{v_{\phi o}}$$

$$Z = \frac{z}{R_o}, \quad V_\phi = \frac{v_\phi}{v_{\phi o}}, \quad P = \frac{p}{v_{\phi o}^2}$$

(3.107)

where $v_{\phi o}$ is the peripheral tangential velocity. A further parameter $\alpha = (Q/2\pi v_a h)$ is defined where v_a is the apparent viscosity and h is the height of the chamber. An apparent Reynolds Number is also used, i.e. $\mathrm{Re} = v_{\phi o} R_o / v_a = \alpha/\beta$ where $\beta = -NW/2\pi R_o$ with N and W being the number and width of the tangential inlet nozzles.

In the outer annular region, the analysis and result is essentially similar to that of Royle but restricted to compressible flow. As the flow enters the inner core region, the axial velocity component is suddenly increased as a result of pressure differences across the central orifice. The axial velocity V_Z is now assumed to be of the form

$$V_Z = Zf(X) \qquad (3.108)$$

The two-dimensional equations are now considered and an expression for V_r is obtained which contains terms in N. It is shown that for a sensible solution, N must be either 0 or 1. For low swirl conditions within a vortex rate sensor, $N = 0$ has apparently been used, but in the analysis reported on here $N = 1$ is used. The resulting equations are

$$V_r \frac{dV_r}{dX} - \frac{V_\phi^2}{X} + \frac{\beta}{\alpha} \left\{ \frac{d}{dX}\left(\frac{V_r}{X} + \frac{dV_r}{dX}\right) \right\} = -\frac{\partial P}{\partial X} \qquad (3.109)$$

$$V_r \frac{dV_\phi}{dX} + \frac{V_r V_\phi}{X} + \frac{\beta}{\alpha} \left\{ \frac{d}{dX}\left(\frac{V_\phi}{X} + \frac{dV_\phi}{dX}\right) \right\} = 0 \qquad (3.110)$$

$$V_r \frac{df}{dX} + f \frac{df}{dX} + \frac{\beta}{\alpha} \left\{ \frac{d^2 f}{dX^2} + \frac{1}{X}\frac{df}{dX} \right\} = 4C^2 \qquad (3.111)$$

$$\frac{dV_r}{dX} + \frac{V_r}{X} + f = 0 \qquad (3.112)$$

The continuity equation is used to eliminate f from the Z momentum equation and a third order differential equation results which has an exact

[3.9] THEORETICAL APPROACHES TO THE PROBLEM OF VORTEX FLOW 167

solution, which when put into the continuity equation yields an expression for f

$$f = -2C(1 - 3 e^{-mX^2}) \qquad (3.113)$$

where $m = C\alpha/2\beta$.

By satisfying the conditions for matching at the boundary of the two flows, an expression for C is obtained, i.e.

$$C = \frac{\beta + \dfrac{6\beta}{\alpha}(1 - e^{-mX_e^2})}{X_e^2} \qquad (3.114)$$

The non-dimensional tangential velocity distributions and wall static pressure distribution resulting from this analysis applied to a thin vortex chamber with given inlet conditions are reproduced from Ref. 65 and given in Figs. 3.39 and 3.40. In the reference, curves are also given of the axial velocity which predict reversed flow towards the centre of the sink tube—a prediction confirmed by experiment.

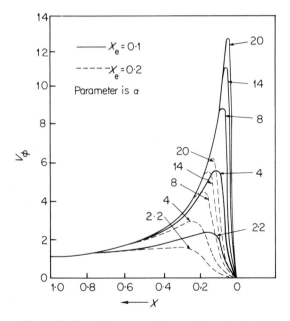

Fig. 3.39 Non-dimensional tangential velocity distributions for various values of α and two values of X_e (the value of X at the edge of the outlet) (Kwok, *Ref. 68*, Fig. 2)

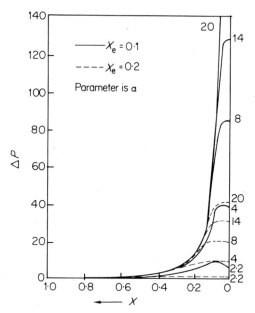

Fig. 3.40 Non-dimensional wall static pressure distributions for various values of α and two values of X_e (the value of X at the edge of the outlet) (Kwok, *Ref. 68*, Fig. 3)

The important aspect of the figures presented here is the very strong influence of the size of the outlet tube on the performance of the vortex valve. The effect of α, which is inversely proportional to effective viscosity, is also strong. It is also clear from this work and that of Royle that for high pressure drop across the vortex, the strength of the vortex compared with that of the sink must be high, i.e. the radial flow should be kept as low as possible.

3.9.2 Taking the Boundary Layer into Account

Sarpkaya *et al.*[69, 70] do not assume one-dimensional flow in the annular region, but say in their introduction 'the radial flow between the two coaxial discs is modified by the viscous shear and by a vortex created by the rotation of the unit about an axis parallel to its symmetry axis'. Consequently, the determination of the viscous efficiency of the sensor, or in other words the fraction of the angular momentum which the fluid retains at the pick-off location, requires a detailed analysis of the resulting boundary layers. The type of flow present in a rate sensor is a combination of a strong sink flow and a weak confined vortex flow.

[3.9] THEORETICAL APPROACHES TO THE PROBLEM OF VORTEX FLOW 169

Because of the difficulty of obtaining a closed solution of the Navier–Stokes equations for the confined vortex, the problem is divided into analysing the radial flow by use of the boundary layer and integral-momentum equations and then extending the solutions to flows with a weak swirl by assuming similarity between the radial and tangential boundary layers.

The results of the boundary layer analysis for purely radial flow indicate that the boundary layer thickness should rise quickly initially from the outer coupling to a maximum and then fall steadily to zero at the centre of the vortex. If this actually is so, it may suggest that the methods of Royle and Kwok yield reasonable results because the boundary layer thickness is small in the region of largest velocity and pressure changes. Sarpkaya then defines a 'viscous efficiency' E which is the circulation at radius $r(\Gamma_r)$ divided by the initial circulation at the outer coupling (Γ_o), i.e.

$$E = \frac{\Gamma_r}{\Gamma_o} \quad (3.115)$$

The relative tangential velocity of the fluid at radius r can be shown by the assumption of similarity between tangential and radial boundary layer flows to be

$$v_\phi = \frac{\omega R_o}{v_{r_o}} \left(\frac{\Gamma_r}{\Gamma_o} - \frac{r^2}{R_o^2} \right) v_r \quad (3.116)$$

when v_r is the inlet velocity at radius R_o.

The circumferential shear stress on an annular ring of fluid is

$$\tau_\phi = \mu \left| \frac{dv_\phi}{dz} \right|_{z=0} \quad (3.117)$$

$$\therefore \tau_\phi = \mu \frac{\omega R_o}{v_{r_o}} \left(\frac{\Gamma_r}{\Gamma_o} - \frac{r^2}{R_o^2} \right) \left| \frac{dv_r}{dz} \right|_{z=0} \quad (3.118)$$

a momentum thickness is now defined, i.e.

$$\frac{\tau_r \theta}{\mu v_r} = f_2(K) \quad (3.119)$$

where θ is the momentum thickness and $\tau_r = |dv_r/dz|_{z=0}$. $f_2(K)$ is a tabulated function which depends upon the form of the polynomial assumed for the velocity distribution in the boundary layer. It is stated that the viscous efficiency can also be written

$$E = \frac{1}{\rho h \omega R_o^2} \int_r^{R_o} \frac{\tau_\phi r}{v_r} dr \quad (3.120)$$

where h is the disc spacing. Hence it may be shown that

$$\frac{dE}{d(r/R_o)} = 4\pi \sqrt{\frac{2vR_o^2}{Qh}} \left(E - \frac{r^2}{R_o^2} \right) \frac{f_2(K)}{\sqrt{v_r\theta^2/vr}} \qquad (3.121)$$

where

$$\frac{v_r\theta^2}{v} = 0\cdot 4r \left(1 - \frac{r}{R_o} \right) \qquad (3.122)$$

These equations may be integrated numerically to give the viscous efficiency plotted as a function of Qh/vR_o^2 and the results are shown in Fig. 3.41. It is clear from the diagram that the efficiency is poor for very

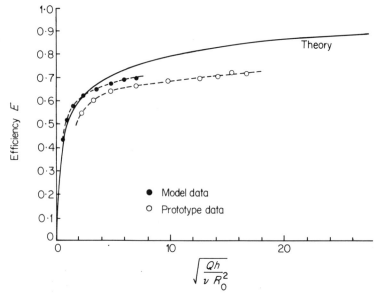

Fig. 3.41 Theoretical and experimental 'viscous efficiency' of a vortex rate sensor (Sarpkaya et al., Ref. 70, Fig. 3)

low values of the parameter, i.e. if v or R_o are large and if Q or h are small. Sarpkaya also discusses in the references a particular kind of pick-off which sense the tangential velocity in the centre of the vortex.

3.9.3 Combination of Ideal Fluid Theory and Boundary Layer Concepts

In Ref. 71, Wormley and Richardson make the very useful point that for low swirl rates the vortex valve behaves in a manner predictable by inviscid flow theory. At sufficiently high swirl rates, the chamber flow pattern consists of a central spinning toroid or 'doughnut' of otherwise stationary fluid bounded by end wall boundary layers; under these conditions all the flow through the valve is contained in the boundary layer.

Using inviscid-flow analysis, a relation is obtained between the control weight flow and the output flow, both non-dimensionalized by dividing by the maximum supply weight flow, which is simplified by approximation to

$$\overline{W}_o = 0{\cdot}707 \sqrt{1 \pm \sqrt{1 - \frac{\overline{W}_c^4}{G^4}}}$$

where

$$G = \sqrt{\frac{\dfrac{1}{2C_e}\left(\dfrac{r_e}{r_o}\right)\left(\dfrac{A_c}{A_e}\right)}{\sqrt{1 - \dfrac{r_e^2}{r_o^2}}}} \qquad (3.123)$$

where $\overline{W}_o$ and $\overline{W}_c$ are the normalized output flow and control flow respectively; r_e, r_o are the exit hole and chamber radii; and A_c, A_e are the control port and exit hole areas. C_e is the discharge coefficient of the exit port.

Such a function shows $\overline{W}_o$ increasing first as $\overline{W}_c$ increases and then as $\overline{W}_c$ reduces, thus indicating a maximum value for $\overline{W}_c$ which is found to occur always for $\overline{W}_o = 0{\cdot}707$. This value $\overline{W}_c^*$ is given by

$$\overline{W}_c^* = \overline{W}_c|_{\overline{W}_o = 0{\cdot}707} = G$$

The authors then show experimentally that the real vortex valve behaves as predicted by the above equation as far as the maximum value for $\overline{W}_c$. Thereafter, the characteristic may start to follow the reducing trend for $\overline{W}_c$ and later start increasing, in which case three values of $\overline{W}_o$ exist for a single value of $\overline{W}_c$, or $\overline{W}_c$ may continue to increase. The former situation is unstable, whilst the latter is proportional. Fig. 3.42 shows the two alternative situations.

Using this as a guide, the authors then carry out a considerable program of experimental work to determine the value of $\overline{W}_c$ for cut-off of the valve ($\overline{W}_{cc}$). By cut-off it is understood that the main flow is reduced to zero but the control flow is still passing through the valve, so that then $\overline{W}_o = \overline{W}_{cc}$. Values of $\overline{W}_{cc}$ are plotted for various values of R_e/R_o and (A_c/A_e) in Fig. 3.43, and furthermore, a locus is plotted of the cases when $\overline{W}_c^* = \overline{W}_{cc}$; this is important because it gives an indication of the

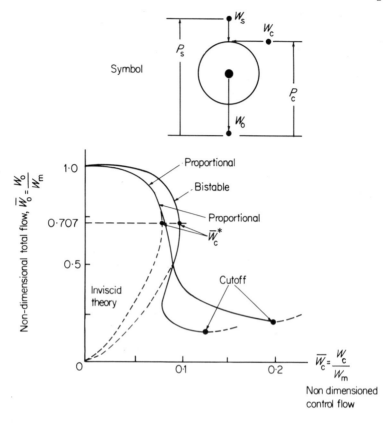

Fig. 3.42 Flow characteristics of typical vortex amplifiers. W_m is the maximum supply weight flow (Wormley and Richardson, *Ref. 71*, Fig. 2)

boundary between bistable and proportional action. It is then pointed out that three points can be found on the characteristic, i.e. $\overline{W}_o = 1$, $\overline{W}_c = 0$; $\overline{W}_o = 0.707$, $\overline{W}_c^* = G$ and $\overline{W}_{cc}$ from Fig. 3.43. By correct choice of $\overline{W}_{cc}$, a proportional amplifier with any desired gain may be roughly designed— the area to the left of the locus of $\overline{W}_c^* = \overline{W}_{cc}$ gives proportional action. For convenience, the required control pressure for cut-off, $\overline{P}_{cc}$ is given, where the control pressure is normalized by dividing by the supply pressure. The supply pressure is related to the maximum flow which occurs with zero control flow. This is because conventional vortex valves are designed so that all the pressure drop under zero swirl conditions occurs across the exit port; the normal hydraulic orifice equation then holds with a discharge coefficient of between 0·7 and 0·8.

[3.9] THEORETICAL APPROACHES TO THE PROBLEM OF VORTEX FLOW

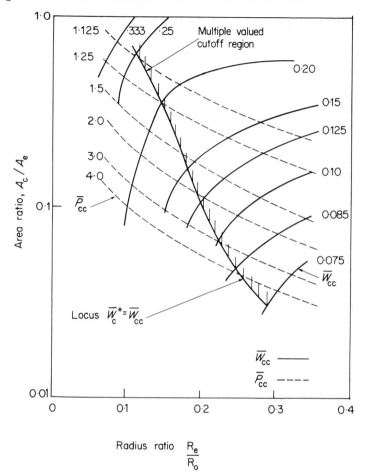

Fig. 3.43 Values of non-dimensional flow and pressure at cut-off ($\overline{W}_{cc}$ and $\overline{P}_{cc}$) for various geometries of vortex device. The locus of $\overline{W}_c^* = \overline{W}_{cc}$ gives an approximate indication of the dividing line between proportional and bistable action, the left of the boundary giving the former characteristic (Wormley and Richardson, *Ref. 71*, Fig. 4)

Wormley[72] examines theoretically the effect of the boundary layer on the end plates of the vortex chamber. He obtains a solution for the circulation (Γ) at any radius in the chamber as a function of two parameters; the ratio of radial to tangential velocity at the outer radius of the chamber and a boundary layer coefficient B.L.C. He then finds that these two may be combined together to provide only one parameter on which the distribution of circulation depends, i.e.

$$\text{B.L.C.}^* = \frac{2fR_o v_{\phi_o}}{hv_{r_o}} \left(\frac{v_{r_o} h}{2v_a} \right)^{-\frac{1}{4}}$$

In Fig. 3.44 a plot is given of Γ against X for various values of B.L.C.*, where $X = 1 - r/R_o$ and $\Gamma = v_{\phi_s} r / v_{\phi_o} R_o$. Here v_{ϕ_s} is the tangential velocity at the edge of the boundary layer of thickness S and f is an end wall friction coefficient.

There is an interesting discussion at the end of Ref. 72 concerning the validity of the various vortex analyses, to which the reader is referred.

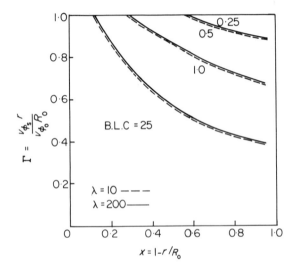

Fig. 3.44 Circulation distribution in a vortex chamber. Wormley shows that it is a function of only one parameter, B.L.C.* (Wormley, *Ref. 72*, Fig. 11)

References

1. Schlichting, H., *Boundary Layer Theory*, McGraw-Hill, New York, 1960.
2. Bickley, W., The plane jet. *Phil. Mag.*, Ser. 7, **23**, 727 (1939).
3. Beatty, E. K. and Markland, E., Feasibility study of laminar jet deflection fluidic elements, C.F.C. (3), Paper H1.
4. Savic, P. On acoustically effective vortex motion in gaseous jets. *Phil. Mag.*, Ser. 7, **32**, 245 (1941).

5. Goertler, G. and Tollmien, W. (Eds.), *50 Jahre Grenzschichtforschung*, Wieweg and Sohn, Braunschweig, 1955.
6. Curle, N., On the hydrodynamic stability in unlimited fields of viscous flow. *Proc. Roy. Soc. (London)*, **238**, 489 (1957).
7. Tatsumi, T. and Katutani, T., The stability of a two-dimensional jet. *J. Fluid Mech.*, **4**, 261 (1958).
8. Howard, L., Hydrodynamic stability of a jet. *J. Math. Phys.*, **37**, 283 (1958).
9. Clenshaw, C. W. and Elliott, D., Numerical treatment of Orr–Sommerfeld equations in case of laminar jet. *J. Mech. Appl. Math.*, **13**, 300 (1960).
10. Sato, H., The stability and transition of a two-dimensional jet. *J. Fluid Mech.*, **7**, Pt. 1, 53 (1960).
11. Andrade, E. N. da C., The sensitive flame. *Proc. Phys. Soc. (London)*, **53**, 329 (1941).
12. Chanaud, R. C. and Powell, A. Experiments concerning the sound-sensitive jet. *J. Acoust. Soc. Am.*, **34**, No. 7, 908 (1962).
13. Sato, H. and Sakao, An experimental investigation of the instability of a two-dimensional jet at low Reynolds Numbers. *J. Fluid Mech.*, **20**, Pt. 2, 337 (1964).
14. Brown, G. B., On vortex motion in gaseous jets and the origin of their sensitivity to sound. *Proc. Phys. Soc.*, **47**, Pt. 4, 703 (1935).
15. Brown, G. B., The vortex motion causing edgetones. *Proc. Phys. Soc.*, **49**, 493 (1937).
16. Brown, G. B., The mechanism of edgetone production. *Proc. Phys. Soc.*, **49**, 508 (1937).
17. Curle, N., The mechanics of edgetones. *Proc. Roy. Soc.*, A216, 412 (1953).
18. Powell, A., On edgetones and associated phenomena. *Acoustica*, **3**, 233 (1953).
19. Powell, A., On the edgetone. *J. Acoust. Soc. Am.*, **33**, 4 (1961).
20. Lighthill, M. J., On sound generated aerodynamically. *Proc. Roy. Soc.*, A211, 564 (1952).
21. Nyborg, W. L., Woodbridge, C. L. and Schilling, H. K., Characteristics of jet-edge-resonator whistles. *J. Acoust. Soc. Am.*, **25**, 1 (1953).
22. Nyborg, W. L., Burkhard, M. D. and Schilling, H. K., Acoustical characteristics of jet-edge and jet-edge-resonator systems. *J. Acoust. Soc. Am.*, **24**, 1 (1952).
23. Nyborg, W. L., Self-maintained oscillations of the jet in a jet-edge system. *J. Acoust. Soc. Am.*, **26**, 2 (1954).
24. Tamulis, J. C., Jet-knife edge interaction as a source of signal noise in proportional fluid amplifiers. *Proc. I.F.A.C. Symp. Fluidics, London*, Paper A8 (1968).
25. Unfried, H. H., Experiment and theory of acoustically controlled fluid switches. *H.D.L.* (3), **II**.
26. Unfried, H. H., An approach to broad band fluid amplification at acoustic frequencies. *H.D.L.* (3), **II**.
27. Becken, B. B., An acoustic fluidic sensor. Advances in Instrumentation, **23**, Pt. II, *Proc. 3rd Annual Conf. of I.S.A.* (1968).
28. Gradestsky, V. and Dimitriev, V., Design problems of fluidic digital elements and devices. *Fluidics International* (July 1968).
29. Gottron, R. N., Acoustic control of pneumatic digital amplifiers. *H.D.L.* (2), **I**.
30. Gottron, R. N., Noise reduction by jet-edge and resonator coupling. *H.D.L.* (2), **I**.

31. Newman, B. G., Turbulent jets and wakes. *Fluid Mechanics of Internal Flow*, Ed. Sovran, Elsevier Pub. Co. (1961).
32. Albertson, M. L., Dai, Y. B., Jensen, R. A. and Hunter Rouse, Diffusion of submerged jets. *A.S.C.E. Papers*, **74** (1948).
33. Jones, N. S., Foster, K. and Mitchell, D. G., A method of calculating the wall re-attachment pressure distribution in a turbulent re-attachment bubble. *C.F.C.* (2), Paper A3.
34. Kirschner, J. M., Jet flows. *Fluidics Quarterly*, **1**, 3, 33 (1968).
35. Simson, A. K., Gain characteristics of subsonic pressure controlled proportional fluid jet amplifiers. *Trans. A.S.M.E.* (June 1966).
36. Miller, D. E. and Comings, E. W., Static pressure distribution in the free turbulent jet. *J. Fluid Mech.*, **3**, Pt. 1 (Oct. 1957).
37. Comparin, R. A., Mitchell, A. E., Mueller, H. R. and Glaettli, H. M., On the limitations and special effects in fluid jet amplifiers. *A.S.M.E. Symp. on Fluid Jet Control Devices* (Winter 1962).
38. Olson, R. E., Analytical techniques for predicting the characteristics of jet flows in fluidic devices. *Fluidics Quart.*, **1**, 1 (1968).
39. Begg, R. D., The effect of downstream pressure asymmetry on the flow from two-dimensional orifices, *Proc. I.F.A.C. Symp. Fluidics*, Paper A5 (1968).
40. Zalmanzon, L. A., Ivanov, N. N. and Limonova, M. E., Theoretical and experimental investigation of fluidic elements. *Proc. I.F.A.C. Symp. on Fluidics*, Paper A4 (1968).
41. Bourque, C. and Newman, B. G., Re-attachment of a two-dimensional jet issuing parallel to a flat plate. *J. Fluid Mech.*, **9** (1960).
42. Comparin, R. A., Moore, R. B. and Jenkins, W. C., Jet re-attachment at low Reynolds Numbers and moderate aspect ratios. *A.S.M.E.*, Paper 67-FE-25.
43. Sawyer, R. A., The flow due to a two-dimensional jet issuing parallel to a flat plate. *J. Fluid Mech.*, **9**, Pt. 4, 543 (1960).
44. Sawyer, R. A., Two-dimensional re-attachment jet flows including the effects of curvature on entrainment. *J. Fluid Mech.*, **17**, Pt. 4, 481 (1963).
45. Pederson, J. R. C., The flow of turbulent incompressible two-dimensional jets over ventilated cavities. *C.F.C.*(1).
46. Bourque, C., Re-attachment of a two-dimensional jet to an adjacent flat plate. *Proc. A.S.M.E. Fluidics Symp.*, **1967**, 192.
47. Levin, I. and Manion, F. M., Fluid amplification 5. Jet attachment distance as a function of adjacent wall offset and angle. *H.D.L.* (1), TR-1087.
48. Perry, C. C., Two-dimensional jet attachment. *Proc. A.S.M.E. Fluidics Symp.*, **1967**, 205.
49. Olson, R. E. and Chin, Y. T., Fluid amplification 17. Studies of re-attachment jet-flows in fluid-state wall attachment devices. *H.D.L.* (3), AD-623 911.
50. Brown, F. T., A combined analytical and experimental approach to the development of fluid-jet amplifiers. *A.S.M.E. Basic Eng.* (June 1964).
51. Sawyer, R. A., Analysis of time dependent jet attachment processes and comparison with experiment. *C.F.C.* (2), Paper A2.
52. Foster, K., Misra, A. K. and Mitchell, D. G., A note on the effect of a downstream pressure rise on a re-attachment jet. *C.F.C.* (3), Paper F9.
53. Kadosch, M., The curved wall effect. *C.F.C.* (2), Paper A4.
54. Pavlin, C. and Facon, P., The curved wall amplifier operational characteristics and applications. *Proc. I.F.A.C. Symp. Fluidics*, Paper B5 (1968).

REFERENCES

55. Thomson, R. V., The switching of supersonic gas jets by atmospheric venting. *C.F.C.* (2), Paper A5.
56. Page, R. H., Fluid mechanics of supersonic separation and re-attachment. *Proc. I.F.A.C. Symp. Fluidics*, Paper A7 (1968).
57. Lush, P. A., Investigation of the switching mechanism in a large scale model of a turbulent re-attachment amplifier. *G.F.C.* (2), Paper A1.
58. Rosenhead, L., The formation of vortices from a surface of discontinuity. *Proc. Roy. Soc. (London)*, A134, 170 (1931).
59. Hama, F. R. and Burke, E. R., On the rolling up of a vortex sheet. *A.F.O.S.R.*, TN 60-1069 (Sept. 1960).
60. Ringleb, F. O., Two-dimensional flow with standing vortices in ducts and diffusers. *A.S.M.E. J. of Basic Eng.* (Dec. 1960).
61. Ringleb, F. O., Discussion of problems associated with standing vortices and their applications. *A.S.M.E. Symp. Fully Separated Flows* (1964).
62. Ringleb, F. O., *Boundary Layer and Flow Control*, V. Lachmann (Ed.), Vol. 1, p. 265, Pergamon Press, 1961.
63. Rechten, A., Stabilität und belastbareit bistabiler fluid elemente. *I.F.A.C./I.F.I.P. Symp.* (Oct. 1965).
64. Frey, K. P. H. and Vasuki, N. C., Flow stability for two-dimensional cusp devices. *H.D.L.* (3), I, 111.
65. Frey, K. P. H., New comprehensive studies on sudden enlargements, *H.D.L.* (3), I, 119.
66. Duggins, R. K., A potential flow model for a closed separation region. *Inst. Mech. Eng., Thermodynamics and Fluid Mechanics Convention*, Paper 11 (March 1968).
67. Royle, J. K. and Hassan, M. A., Characteristics of vortex devices. *C.F.C.* (2).
68. Kwok, C. C. K., A theoretical study of vortex flow in a thin cylindrical chamber. *Bendix Technical Journal*, 1, 4 (1969).
69. Sarpkaya, T., A theoretical and experimental investigation of the vortex-sink angular rate sensor. *H.D.L.* (3), II.
70. Sarpkaya, T., Goto, J. M. and Kirschner, J. M., A theoretical and experimental study of vortex rate gyro. *Proc. A.S.M.E. Fluidics Symp.* **1967**, 218.
71. Wormley, D. N. and Richardson, H. H., Experimental investigation and design for vortex amplifiers operating in the incompressible flow region. *H.D.L. Tech. rep* DSR 70167-1 (Dec. 1968).
72. Wormley, D. N., An analytical model for the incompressible flow in short vortex chambers. *A.S.M.E. J. of Basic Eng.* (June 1969).

4
Pure Fluid Analog Amplifiers

4.1 General

This class of amplifier is characterized by the continual modulation of the output by changes of the control signal. As this involves proportionality between the two quantities, a fluid analog amplifier must avoid fluid flow phenomena which are inherently non-linear in action. For instance, the sudden changes in flow pattern produced by the attachment of a jet to an adjacent wall in digital devices must be eliminated. Momentum interaction of two or more fluid jets is the most common effect exploited in these devices, although in some cases this is modified by controlled viscous effects.

Two important forces associated with a fluid jet are due to pressure energy and changes in momentum. For incompressible jet flow into a relatively large chamber with a uniform pressure distribution, the jet and chamber pressure rapidly equalize and the jet behaviour is determined by its momentum. By impinging two or more jets transversely or in opposition a variety of analog fluid amplifiers may be constructed with a proportional action largely as a result of momentum changes. Fig. 4.1 illustrates the simplest case of two mutually perpendicular jets impinging on each other. Assuming that the two supply nozzles are not too close together thereby impeding the flow, the direction of the resultant jet centre line, θ, is determined by the relative momenta of the two jets. Thus, if we consider a fluid jet of density ρ, cross-section area A and average velocity $\bar{u}$, the mass rate of fluid flow from the jet is $\rho A \bar{u}$ and the corresponding momentum is $\rho A \bar{u}^2$. Hence the momentum of nozzles 'a' and 'b' are respectively

$$J_a = \rho_a A_a \bar{u}_a^2$$
$$J_b = \rho_b A_b \bar{u}_b^2$$

[4.1] GENERAL

As the two flows are initially at right angles and momentum is conserved, the resultant jet has a momentum J_r given by the triangle of forces

$$J_r = \sqrt{(J_a^2 + J_b^2)}$$

and

$$\tan \theta = \frac{\rho_a A_a \bar{u}_a^2}{\rho_b A_b \bar{u}_b^2}$$

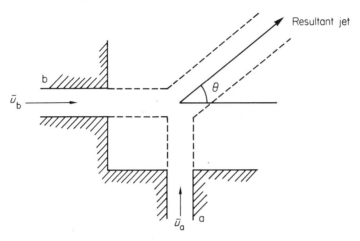

Fig. 4.1 Momentum interaction of two jets

For example, if a receiver is placed axi-symmetrically downstream, then the flow or pressure signal it records should be a maximum when the flow momentum of the signals from nozzles 'a' and 'b' are equal. The output signal falls when the flow through one of the input nozzles is increased as the resultant jet is deflected away from the receiver centre line. Care must be taken in the use of the simple equations given above, as the local pressure may significantly alter the jet momentum. If, say, the momentum of the flow through nozzle 'a' increases and the upstream pressure in 'b' remains unaltered, then the flow momentum through nozzle 'b' may in fact reduce, due to the higher local pressure at the point of impact of the two jets. This gives a larger jet deflexion θ than otherwise expected.

The most widely used analog elements are the Beam-deflexion and Vortex amplifiers. The Impact-modulator amplifiers and the more complex Double-Leg amplifier have found less general acceptance. As all these devices are active, the ratio of output to control signal is important giving a measure of the amplification or gain. The gain may be based on the flow, pressure or power quantities although many different ways of defining each type exists. Pressure gain, G_P is usually defined as the slope of the curve of differential output static pressure ΔP_o against differential input static pressure ΔP_c, for a two output device, i.e.

$$G_P = \frac{\Delta P_o}{\Delta P_c} \qquad (4.1)$$

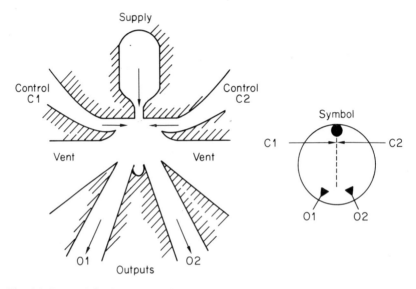

Fig. 4.2 Beam-deflexion proportional amplifier (Parker and Jones, C.F.C.(2), Paper C.4, 60)

However, it is necessary to specify the load conditions in the output channels as this may substantially modify the flow patterns within the device thereby altering the gain obtainable. Pressure gain is usually quoted for zero output flow and is referred to as the blocked load condition. Similarly, flow gain G_f for proportional amplifiers is usually expressed as the slope of the curve of differential output flow ΔQ_o, against differential control flow, ΔQ_c, i.e.

$$G_f = \frac{\Delta Q_o}{\Delta Q_c} \qquad (4.2)$$

The no load or maximum flow condition is usually specified for the flow gain, although this does not usually correspond to zero output pressure due to the finite flow resistance of the output channels.

The stage gain of a single fluid amplifier is usually reduced when con-

[4.1] GENERAL 181

nected in an array of other elements. This is because optimizing the design for good signal matching between the output of one amplifier and the input of the next is a compromise between their channel resistance. Another difficulty encountered by the amplifier design is the noise generated in the element itself. Most amplifiers operate with turbulent jets giving rise to random flow fluctuations within the jet stream which may be amplified in the receiver to give noise levels of appreciable magnitude. The receiver entrance itself is also a source of noise (see Section 5.1.1).

4.2 Beam-deflexion Amplifier

4.2.1 Internal fluid mechanics

The Beam-deflexion amplifier uses an array of three impinging two-dimensional jets, one of which is the supply jet and the other two are opposing control jets arranged as shown in Fig. 4.2. The main jet, in the absence of control signal, flows undeflected across the interaction chamber and divides equally between two receiver channels set symmetrically about the jet centre line. In order to provide uniform pressure in the interaction region the amplifier has large vents on either side of the supply jet which also take any flow overspill from the output channels. The output channels may be separated by a splitter or alternatively a further vent may be used at this point to improve the output characteristics.

The control signals are directed into the interaction region from opposing nozzles perpendicular to the supply jet. The momentum flux of the supply jet and the forces acting on it from the control flows determines the resultant direction of the main jet as it leaves the interaction region. That is to say, as flow is introduced into the control channels, they interact with the supply jet and deflect it across the output channels. As two output and two input channels are available, the amplifier may be used as a single input-output device or alternatively as a two input-output device employing positive and negative valued differential signals.

For control nozzle widths comparable to the supply nozzle width, deflexion of the main jet is usually, but not necessarily, achieved by momentum interaction of the control and main jets. On the other hand amplifiers have been constructed by Simson[1] using large control nozzle widths to achieve pressure instead of momentum control of the jet deflexion. The interaction of the three jets of a beam deflexion amplifier have been investigated in some detail by Reilly and Moynihan[2] who define a control volume around the interaction region as shown in Fig. 4.3. The face BC of the control volume is taken sufficiently far downstream so that the static pressure is equal to the ambient pressure and all the exit flow is assumed to flow uniformly through it. If, in addition, it is assumed

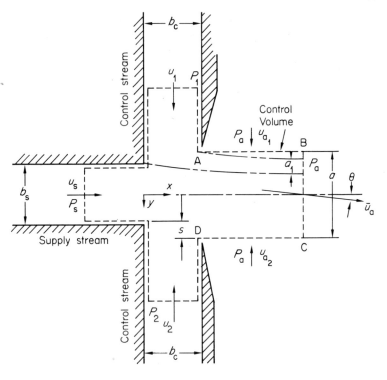

Fig. 4.3 Interaction region flow model (Reilly and Moynihan, *Ref. 2*, 133)

that the entrainment flows through AB and DC are equal in magnitude and opposite in direction, while maintaining equal static pressure distribution along the same lines, a force balance can be obtained in the x and y directions. Equating the difference in momentum leaving and entering the control volume to the sum of the external forces acting on the volume in the y direction

$$M_a \bar{u}_a \sin \theta - M_1 u_1 + M_2 u_2 = b_c p_1 - b_c p_2 + \sum F_{WY} \quad (4.3)$$

and in the x direction:

$$M_a \bar{u}_a \cos \theta - M_s u_s = b_s p_s - a p_a + \sum F_{WX} \quad (4.4)$$

where M = mass flow rate
 u = velocity (bar is average value)
 p = static pressure
 a = effective channel width across AD
 b = channel width

[4.2] BEAM-DEFLEXION AMPLIFIER

F_W = pressure force on the walls in the respective directions due to the internal centrifugal forces of the deflected jet streams.

Subscripts:
- a = interact region
- c = control port
- s = supply port
- 1, 2 = control streams.

Combining (4.3) and (4.4) and putting $p_a = 0$.

$$\tan \theta = \frac{M_1 u_1 - M_2 u_2 + b_c(p_1 - p_2) + \sum F_{WY}}{M_s u_s + b_s p_s + \sum F_{WX}} \quad (4.5)$$

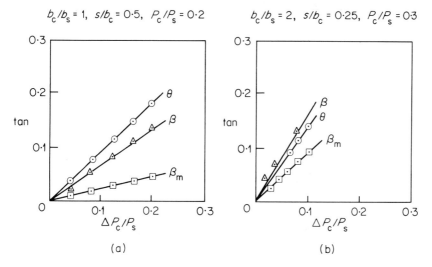

Fig. 4.4 Effect of momentum, static pressure and centrifugal forces on supply jet deflexion (Reilly and Moynihan, *Ref. 2*, 142)

where θ = combined jet deflexion angle. Neglecting elevation changes, the total pressure P may be expressed as:

$$P = \frac{\rho u^2}{2} + p \quad (4.6)$$

It is also convenient to express the dynamic pressure in dimensionless form as

$$E = \frac{\rho u^2}{2P} \quad (4.7)$$

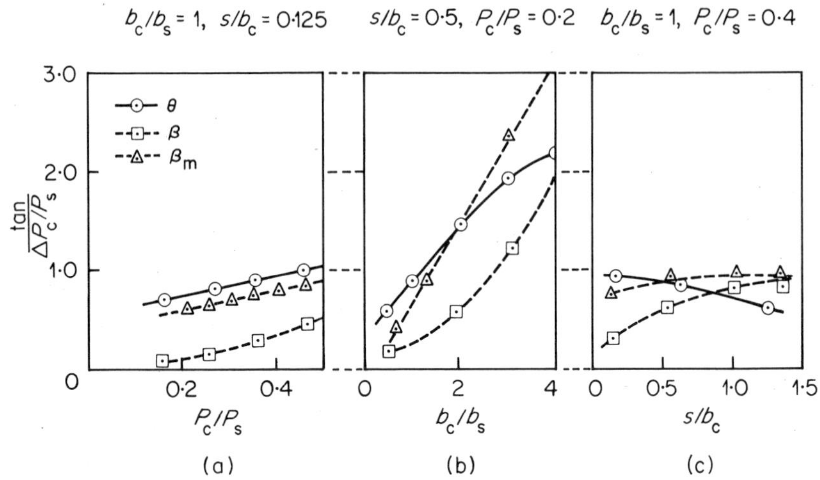

$b_c/b_s = 1$, $s/b_c = 0.125$ $s/b_c = 0.5$, $P_c/P_s = 0.2$ $b_c/b_s = 1$, $P_c/P_s = 0.4$

Fig. 4.5 Variation of supply jet deflexion slope with pressure (Reilly and Moynihan, Ref. 2, 144)

Using (4.6) and (4.7), the jet deflexion angle may be rewritten as

$$\tan \theta = \frac{b_c[(1 + E_1)P_1 - (1 + E_2)P_2] + \Sigma F_{WY}}{b_s P_s (1 + E_s) + \Sigma F_{WX}} \quad (4.8)$$

The deflection β_m of the main jet due to the momentum of the entering fluid alone is simply:

$$\tan \beta_m = \frac{b_c}{b_s} \left[\frac{E_1 P_1 - E_2 P_2}{E_s P_s} \right] \quad (4.9)$$

and the deflexion, β, caused by the combination of entering fluid momenta and static pressure forces is:

$$\tan \beta = \frac{b_c}{b_s} \left[\left(\frac{1 + E_1}{1 + E_s} \right) \frac{P_1}{P_s} - \left(\frac{1 + E_2}{1 + E_s} \right) \frac{P_2}{P_s} \right] \quad (4.10)$$

The net effect of the centrifugal forces is the difference between θ and β. The experimental results presented by Reilly and Moynihan show that the jet deflexion θ is essentially a linear function of the non-dimensional control pressure change $\Delta P_c/P_s$ for a wide range of jet nozzle configurations. Fig. 4.4 shows two extreme cases of the relative effects of momentum, static pressure and centrifugal forces on the total jet deflexion: (a) shows a case where centrifugal forces are aiding deflexion ($\theta > \beta$) while (b) illustrates the opposite effect in a geometry which includes wider

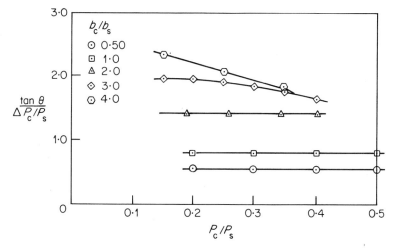

(a) For various control port areas, all with $s/b_c = 0.50$

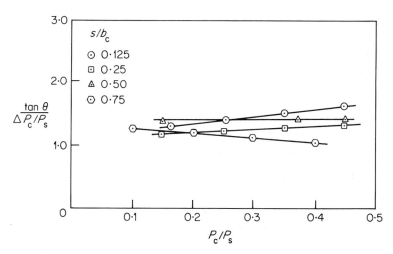

(b) For various control port off-sets, all with $b_c/b_s = 2.0$

Fig. 4.6 Variation of supply jet deflexion with geometry changes (Reilly and Moynihan, *Ref.* 2, 144)

control ports. If the difference in jet deflexion between β and β_m is due to static pressure forces, then in configuration (a) the control static pressure forces are greater than those due to momentum despite the use of a narrow control port channel. The geometry of (b) might be expected to produce greater pressure control, due to the increased control port width, but instead the momentum forces are dominant. This experimental work shows the danger in considering an amplifier as either momentum or pressure controlled, based purely on the control port width. This is because the geometric setback of the control nozzle (distance s in Fig. 4.3) and the average control pressure are inter-related and have a complex effect on the jet deflexion. Figs. 4.5 and 4.6 show the variation in the jet deflexion slope with changes in pressure and geometry which gives a measure of the gain of the amplifier (the greater the slope the higher the expected gain). Reilly and Moynihan conclude that, in general, large control port areas and small setback distances increase the jet deflexion slope, but the choice of average control pressure is dependent on the particular geometry.

Although most Beam-deflexion amplifiers are designed to operate at subsonic velocities, it is possible to obtain proportional control at much higher velocities provided care is taken with the position of the resulting shock waves. Sheeran and Dosanjh[3, 4] have investigated this aspect of jet impingement in some detail while Render[5] has given some experimental data on the jet pressure distribution for the interaction of a supersonic jet with a single control jet.

Apart from a sensitive interaction region geometry, a high-gain proportional amplifier design requires an efficient receiver channel configuration to minimize signal degradation and noise generation. Kallevig[6] has concluded that the receiver geometry, except for raising the pressure level of the amplifier, does not affect the velocity and total pressure profiles in the interaction region for a variety of vented and unvented amplifier designs. Reid[7] considers in more general terms the receiver diffusion effect resulting from the non-uniform impingement of a jet on the mouth of the receiver. Fig. 4.7 shows experimental velocity profiles obtained by Reid for various load conditions when a non-aligned jet discharges into a receiver. The constant-area mixing section provides rapid diffusion of the non-uniform flow. The channel length required before the profile approaches the fully developed state in the constant area section depends on the distance downstream of the receiver, the receiver alignment, and the receiver back pressure P_b due to the amplifier loading. The figure indicates that decreasing the receiver back pressure smooths the velocity profile more slowly so that, for the worst case in the lower illustration, the profile in the diffuser section entrance is still not symmetrical. This leads to early separation and a significant reduction in efficiency.

The shape and function of a receiver is also important in reducing unwanted noise associated with the output signal. Three major causes of

[4.2] BEAM-DEFLEXION AMPLIFIER

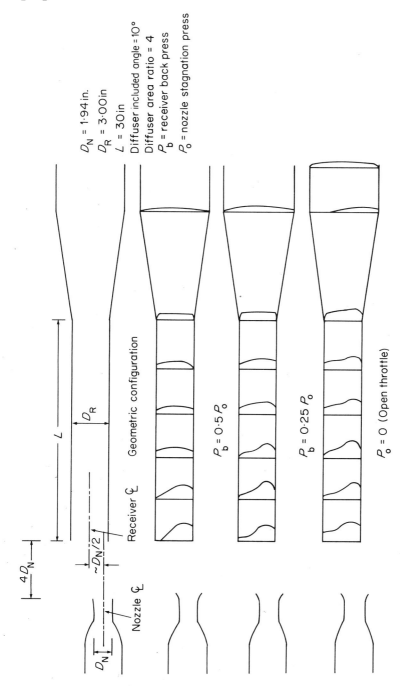

Fig. 4.7 Receiver diffuser characteristics (Reid, *Ref.* 7, 96)

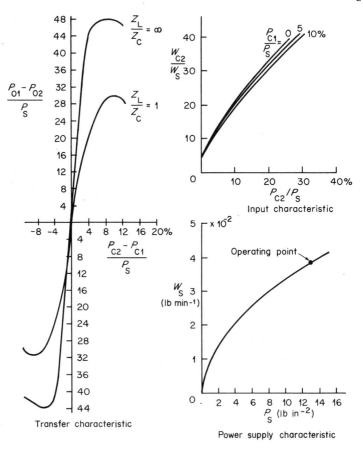

Fig. 4.8(a) Static characteristics of a Beam-deflexion amplifier (McCabe and Hughes, *Ref. 16*, 88)

noise within an amplifier appear to be turbulent fluctuations within the main jet, high shear stresses at the junction of main and control flows, and edgetone noise at the receiver entrance. Prosser and Fisher[8] consider a single jet-receiver combination and conclude that the noise-to-signal ratio is smallest when the receiver is close to the edge of the jet potential core, or alternatively, on the jet centre line. For any fixed position of a receiver, the ratio may be also expected to decrease as the channel width is made larger. Moynihan[9] shows that the noise level increases in the downstream direction and may be in the order of 4 to 6 per cent of the signal for a single jet discharging into a receiver placed 8 to 10 nozzle widths downstream. Addition of a control stream does not appreciably

[4.2] BEAM-DEFLEXION AMPLIFIER

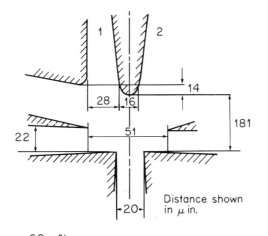

Fluid is dry air. Ambient temperature = 21·8° c
Barometric pressure = 29·64 in. Hg. $W_s = 3·85 \times 10^2$ lb min^{-1}
$P_s = 12·9$ lb in^{-2}

Distance shown in μ in.

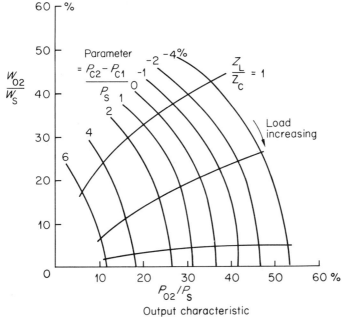

Output characteristic

Fig. 4.8(b) Static characteristics of a Beam-deflexion amplifier (McCabe and Hughes, *Ref. 16*, 89)

increase the noise level. Weinger[10] has recently shown that the aspect ratio (nozzle depth/nozzle width) has an effect on the noise level in a practical Beam-deflexion amplifier. Experimental results show a general increase in noise as the aspect ratio is increased in the range 0·5 to 7 and a marked reduction as the main jet velocity falls. Recent significant noise experiments on operational amplifiers is discussed in 5.1.1.

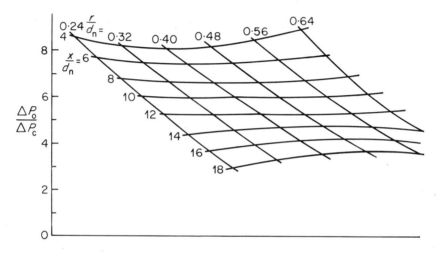

Fig. 4.8(c) Static characteristics of a Beam-deflexion amplifier (McCabe and Hughes, *Ref. 16*, 90)

4.2.2 Static Characteristics

Various workers have obtained static characteristics of Beam-deflexion amplifiers for a variety of geometries[11-18]. The results of McCabe and Hughes[16] shown in Fig. 4.8 are typical of the characteristics expected for an amplifier with its main jet operating at subsonic velocities. The linearity of the transfer characteristic, between the differential control signal and the differential output signal, may be as good as ± 0.5 per cent up to 80 per cent of the range under design load conditions although this is not usually maintained under increased load conditions. Many amplifier designs, while exhibiting increasing pressure gain with load on the output, become unstable under blocked load conditions. Two methods of overcoming this difficulty are to use a central vent between the receiver channels or alternatively to use a vortex vent in the output channel, as described by Hayes and Kwok[17]. Typical blocked load pressure gains are in the range 6 to 10 which drop to 2 to 5 when the amplifier is loaded with another identical element.

[4.2] BEAM-DEFLEXION AMPLIFIER

No load flow gains similar to the values for pressure are possible although it is to be noted that the amplifier design requirements for the two gains are different. Paperone et al.[13] have shown that a wider receiver is required for maximum flow gain than for maximum pressure gain, and that all gains are at a maximum when the power stream is evenly divided by the two output channels. Van Tilburg and Cockran[14] give data showing that, for certain configurations, the amplifier pressure gain is a maximum when using main nozzle aspect ratios in the range 2·0 to 2·5. In addition, the pressure gain is roughly independent of supply pressure and also increases with decreasing scale size. The control bias pressure level may also significantly alter the pressure gain of the amplifier.

Commercially available Beam-deflexion amplifiers operating with gaseous fluids use supply nozzle widths in the range 0·010 in. to 0·020 in. and aspect ratios in the range 2·0 to 2·5. Control nozzles tend to be wider than the supply nozzle, 1·5 to 2·0 times being the most used range. This would suggest, from the work of Reilly and Moynihan discussed earlier, that a combination of momentum and pressure control forces are used in the beam deflexion for optimized practical designs.

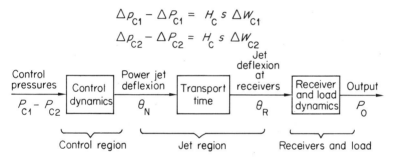

Fig. 4.9 Block diagram of a Beam-deflexion amplifier (Boothe, *Ref. 21*, 85)

4.2.3 Dynamic Characteristics

Analysis of the dynamic characteristics of Beam-deflexion amplifiers for sinusoidal input signals has been made by Belsterling and Tsui[19] and Manion[20] based on the use of lumped parameter representations of the constituent parts of the amplifier. However, the most concise analysis has been given by Boothe[21], who obtained good correlation with experimental frequency response results, for a range of operating conditions. Boothe considers three separate regions for the amplifier, as shown in Fig. 4.9, and derives transfer functions for each based on linearized small amplitude theory. The dynamics of the control ports are represented on a lumped parameter basis by the inertia of the fluid in the control channel, giving a transfer function between the control pressures and the supply jet deflexion (θ_N). The dynamic behaviour of the power jet is treated as a pure time delay so that θ_R, the deflexion of the jet at the receiver, represents the value of θ_N that existed one transport time-period earlier. The receiver dynamics are also represented by a fluid inertia term but the transfer function between θ_R and the output pressure P_o also includes the effect of the nature of the load on the amplifier output.

Denoting P_{c1} and P_{c2} as the steady state control port pressures and p_{c1}, p_{c2} the respective dynamic values, then the steady state control characteristics, which are dependent on the weight flow W, can be linearized in the following way:

$$P_{c1} = f(W_{c1} P_{c2}) \qquad P_{c2} = f(W_{c2} P_{c1}) \qquad (4.11)$$

Thus

$$\Delta P_{c1} = K_1 \Delta W_{c1} + K_2 \Delta P_{c2}$$
$$\Delta P_{c2} = K_1 \Delta W_{c2} + K_2 \Delta P_{c1} \qquad (4.12)$$

where K_1, K_2 are constants determined from steady state control characteristics.

Assuming that the jet deflexion is dependent on the difference in control flows:

$$\Delta \theta_N = K_3 (\Delta W_{c1} - \Delta W_{c2}) \qquad (4.13)$$

Representing the control ports in the dynamic case purely by the inertia or inductance of the fluid in the channel:

$$\Delta p_{c1} - \Delta P_{c1} = H_c \frac{d}{dt}(\Delta W_{c1}) \qquad (4.14)$$

$$\Delta p_{c2} - \Delta P_{c2} = H_c \frac{d}{dt}(\Delta W_{c2}) \qquad (4.15)$$

where H_c is the control port fluid inductance.

Combining (4.11) and (4.15) and using (4.13) the control region transfer function becomes

$$\frac{\Delta \theta_N}{\Delta p_{c1} - \Delta p_{c2}} = \frac{G_c}{1 + \tau_c s}$$

where gain

$$G_c = \frac{K_3}{K_1}(1 + K_2)$$

[4.2] BEAM-DEFLEXION AMPLIFIER

and the time constant

$$\tau_c = (1 + K_2)\frac{H_c}{K_1} \qquad (4.16)$$

The time delay, or transport time, for a signal to be transmitted by the jet, from the nozzle to the receiver, is assumed to depend purely on the jet velocity. Since this produces a phase lag but no amplitude change in the sinusoidal signal, the jet transfer function may be expressed as

$$\frac{\Delta\theta_R}{\Delta\theta_N} = e^{-\tau_t s} \qquad (4.17)$$

where τ_t = transport time from nozzle to receiver.

Considering one output channel, the steady state tests by Boothe show that, for small changes, the receiver pressure is related to jet deflexion and receiver flow by

$$\Delta P_{o1} = K_4 \Delta\theta_R - K_5 \Delta W_{o1} \qquad (4.18)$$

Where K_4, K_5 are constants and suffix o refers to the output. Representing the output channels by inductive terms in a similar manner to the control channels, but noting that the pressure signals are measured downstream of the channel (rather than upstream in the control channel case) thereby altering the signs, we have

$$\Delta p_{o1} - \Delta P_{o1} = -H_R \frac{d}{dt}(\Delta W_{o1}) \qquad (4.19)$$

A further relationship is required between the output flow W_{o1} and the output pressure P_{o1} to specify the system. This is dependent on the output load, which may in general be defined as the load impedance $Z_{o1}(j\omega)$ where:

$$Z_{o1}(j\omega) = \frac{\Delta P_{o1}}{\Delta W_{o1}} \qquad (4.20)$$

Combining (4.18), (4.19) and (4.20) the transfer function of the receiver and load is

$$\frac{\Delta p_{o1}}{\Delta\theta_R} = \frac{K_4}{1 + (K_5/Z_{o1}) + (H_R/Z_{o1})s} \qquad (4.20)$$

The overall transfer function of the amplifier is obtained from the combination of all three regions given by (4.16), (4.17) and (4.21)

$$\frac{\Delta p_{o1}}{(\Delta p_{c1} - \Delta p_{c2})} = \frac{G_c K_4 \, e^{-\tau_t s}}{(1 + \tau_c s)\{1 + (K_5/Z_{o1}) + (H_R/Z_{o1})s\}} \qquad (4.21)$$

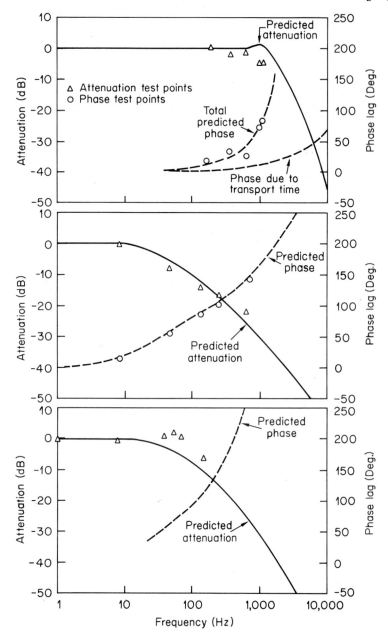

Fig. 4.10 Comparison of theoretical and practical frequency response curves for a Beam-deflexion amplifier (Table 4.1) (Boothe, *Ref. 21*, 90)

[4.2] BEAM-DEFLEXION AMPLIFIER 195

Table 4.1—Beam-deflexion amplifier test results[21]

		Case I	Case II	Case III
	FLUID	AIR	AIR	WATER
P_s	lb in^{-2}	15	10	5
W_s	lb s^{-1}	9.56×10^{-4}	7.84×10^{-4}	1.89×10^{-2}
Receiver vol.	in^3	4.92×10^{-3}	4.92×10^{-3}	4.92×10^{-3}
Load vol.	in^3	0.00	3.05×10^{-1}	0.00
C_R	in^2	1.06×10^{-8}	6.66×10^{-7}	7.05×10^{-10}
Load orifice diam.	in.	1.4×10^{-2}	—	1.4×10^{-2}
H_c	s^2 in^{-2}	4.26×10^{-1}	4.62×10^{-1}	4.62×10^{-1}
H_R	s^2 in^{-2}	9.91×10^{-1}	9.91×10^{-1}	9.91×10^{-1}
K_1	s in^{-2}	14.0	11.4	236
K_2		7.5×10^{-1}	7.5×10^{-1}	7.5×10^{-1}
K_5	s in^{-2}	9.86×10^{3}	7.04×10^{3}	1.66×10^{2}
K_6	s^2 in^{-2}	5.60×10^{-6}	0.00	1.82×10^{-4}
f_c	(Hz)	2.76×10^{3}	2.24×10^{3}	46.5
ξ_R		5.25×10^{-1}	2.88	3.35
f_R	(Hz)	1.61×10^{3}	1.96×10^{2}	6.12×10^{3}

Boothe considers three experimental cases (Table 4.1) and obtains satisfactory correlation of frequency response tests with his theoretical model as shown in Fig. 4.10.

Commercial Beam-deflexion amplifiers using supply jet widths in the range 0.010 in.–0.015 in. typically have a break frequency response in the range 100–1000 Hz when operating with gases, depending on the loading conditions. As with other analog fluidic amplifiers, a capacitive load causes degradation in the frequency response, the increased phase lag being particularly significant. Because the geometry of this amplifier may be reproduced in miniaturized form, it provides the best frequency response characteristics of all the proportional fluidic amplifiers being developed at the present time. It is for this reason that it has been used as the basic element in the operational-amplifier and active shaping network designs.

4.3 Impact Modulator

4.3.1 Static Characteristics

This device employs two axially opposed circular nozzles through which supply fluid flows causing direct impingement of the two jets. If the momenta of the two supply jets are equal, the impact or balance point occurs midway between the nozzles and is very sensitive to small changes in axial momentum. A wide range of operating conditions is possible using both incompressible and compressible jet flow. Two practical devices, known as the *Transverse Impact Modulator* (TIM) and *Direct Impact Modulator* (DIM) have been described by Bjornsen[23]. Fig. 4.11 gives dimensions of the Transverse type which has an annular chamber

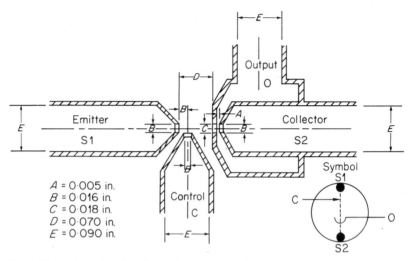

Fig. 4.11 Design details of an impact modulator (N.A.S.A. Rep., *Ref.* 22, Fig. 4.3.2-1)

surrounding one of the supply tubes, known as the collector, which senses the output flow and pressure. The other supply tube is called the emitter and reacts with control flow from a nozzle situated perpendicular to the supply flow axis. Normally with no control flow present, the momentum of the emitter jet is arranged to be slightly greater than that of the collector jet so that the balance point occurs just outside the collector orifice. This produces the maximum output pressure or flow signal. When control flow occurs, the impingement of the control and emitter jets is similar to the action occurring in a Beam-deflexion amplifier with momentum, pressure and centrifugal forces modifying the jet deflexion. The flow field then becomes asymmetric, allowing an increased flow and momentum condition to be established in the collector tube. The slight reduction in the axial momentum of the emitter nozzle and the corresponding rise in the collector nozzle results in a shift of balance point away from the collector, thereby reducing the output signal level. The phenomenon is very sensitive to changes in control flow and has a proportional action, resulting in an amplifier of high negative gain, i.e., device giving a reduced output signal as the input is increased. Unfortunately, Bjornsen gives no

[4.3] IMPACT MODULATOR

physical dimensions of the Impact Modulators described, but static characteristics[22] for the configuration shows in Fig. 4.11 are very similar to those of Bjornsen. The gap D between the emitter jet and the collector is very short (less than five nozzle widths) to achieve high sensitivity. It is reported that the device is sensitive to slight mis-alignment of the supply

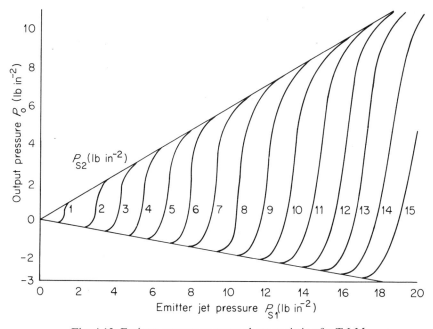

Fig. 4.12 Emitter-output pressure characteristic of a T.I.M.

jets, and changes in the collector chamber dimensions, such as A and C.

Many combinations of signals are possible with this device; for example, it is possible to make the emitter jet into a control signal. However, the commonest mode of operation is to hold the emitter pressure, P_{s1} and the collector pressure, P_{s2} constant at the optimum point on the characteristic curves to give the necessary gain and operating range for the input and output signals. Bjornsen describes experiments to establish the effect of different supply jet combinations on the output pressure for blocked load conditions. Fig. 4.12 shows the results of changes in the emitter pressure on the output pressure for fixed collector pressures due to the jet balance point being forced towards the output chamber. Values quoted are for operation using air as the working medium and show that the device functions over a wide range of supply pressures. The reverse characteristics with the jet balance point being forced away from the out-

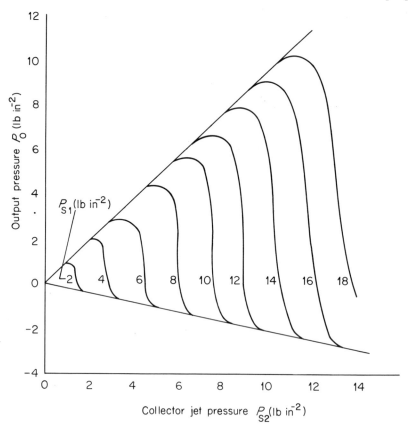

Fig. 4.13 Collector-output pressure characteristics of a T.I.M.

put chamber is similar in form as shown in Fig. 4.13, although it gives rise to a negative instead of a positive gain. The curves illustrated were obtained by maintaining the emitter jet pressure constant, and varying the collector pressure.

In both modes of operation it would appear, from published results, that higher pressure gains are possible as the supply pressure level is decreased at the expense of a decreasing operating range. Lechner and Sorenson[24] quote blocked load pressure gains ranging from 20 to 100 and no load flow gains from 5 to 30. Linearity is reported to be very good, as illustrated in Fig. 4.14 by a typical input-output pressure characteristic for two Transverse Impact Modulators in cascade, and is often in excess of 80 per cent of the output range. In this case, the two stage amplifier has a pressure gain of nearly 300, or about 17 for each stage. When four

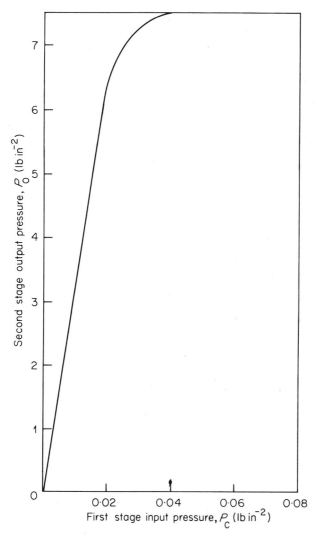

Fig. 4.14 Pressure characteristic of a two stage T.I.M.

or five such amplifiers are cascaded together, the gain is sufficiently high, and the output linear enough, to form a fluid operational-amplifier.

The output flow-pressure characteristic in Fig. 4.15 given by Lechner and Sorenson[24], shows that for high supply pressure the constant control pressure curves are remarkably linear. By suitable shaping, it has been shown to be possible to induce the output chamber to act as though it were a linear resistor, and is a characteristic apparently unobtainable from the receiver of other proportional fluidic amplifiers.

The second arrangement of this type of device, known as the Direct Impact Modulator was shown in Fig. 1.33. The geometry differs from the Transverse type in the method of controlling the changes in axial momentum. Instead of a control nozzle perpendicular to the jet axis, an annular chamber concentric with the emitter nozzle is used. As an

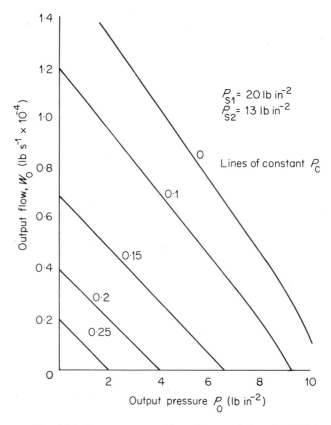

Fig. 4.15 Output pressure-flow characteristics of a T.I.M.

increase in control flow also increases the emitter jet momentum, the jet balance-point is initially arranged to be outside the output chamber. Increasing control flow then moves the balance point towards the chamber, resulting in a positive gain amplifier with similar or rather greater sensitivity than the transverse type. One difference of interest pointed out by Bjornsen, is that the Direct type control chamber has a pressure-flow characteristic indicating a region of operation involving very small constant suction control flow, as shown in Fig. 1.34. This would suggest that the emitter jet is almost entirely pressure controlled under these conditions although the fluid flow mechanism is not as yet fully understood.

As with other types of proportional fluidic-amplifiers, noise has been reported to be a problem, particularly under dynamic operating conditions. Static matching of one element to another has not appeared to be so difficult as for the Beam-deflexion amplifier due to the greater degree of isolation between input and output channels, and the reduced sensitivity to loading effects. However, it is probable that manufacturing difficulties are severe in miniaturized designs leading to significant element characteristic variations.

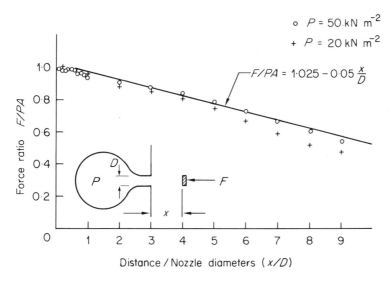

Fig. 4.16 Force exerted by a jet impinging on a disc (Katz, Ref. 25, 121)

4.3.2 The Mechanism of Opposed Jet Impact

The fluid mechanics of impact modulators is complex, and little attempt has been made to provide a complete model description of the device. In particular, the geometry of the collector chamber and direct-impact control chamber, have a significant effect on the jet flow which has not been clarified. However, some understanding of the operating mechanism of the amplifier can be gained from a study of the impacting of two axially opposed circular jets. Katz[25] has attempted to explain the impact phenomenon by assuming that the amplifier operates in a similar fashion to the well-known flapper valve. He measured the force exerted on a small disc by one of the jets, as shown in Fig. 4.16, and obtained an empirical formula for the reduction in this force with separational distance. The equilibrium-force balance position for the two jets can then be determined from this data as a function of the jet pressures. The correlation of experimental and theoretical pressure gains by this method was found to be poor, possibly due to using a fixed area disc.

An investigation of the effect of impact on the structure of an incompressible jet and the determination of the jet balance-point has been made by Misevich[26]. Normally, a circular free jet maintains its centre line velocity for a certain distance downstream before it starts to degrade. This

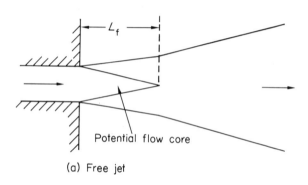

(a) Free jet

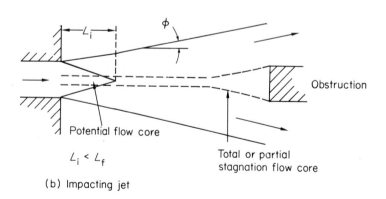

(b) Impacting jet

Fig. 4.17 Structure of an impacting jet

[4.3] IMPACT MODULATOR

is known as the core length L_f, as illustrated in Fig. 4.17. This varies slightly with jet velocity for incompressible flow but significantly for compressible conditions (see Chapter 3). Misevich gives the core length as 6 nozzle diameters at low velocities reducing to 2 nozzle diameters near sonic conditions for air. However, under impacting conditions with a circular obstruction of the same diameter as the nozzle the jet is subjected to deceleration forces and the new core length L_i is appreciably different from L_f. Fig. 4.18 shows the variation of L_i, with reducing nozzle pressure

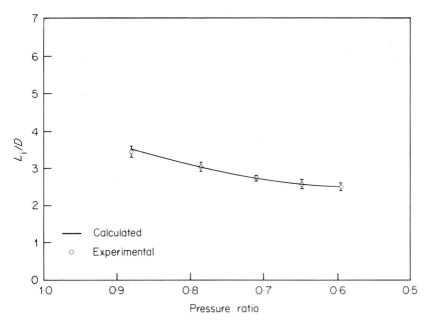

Fig. 4.18 Variation of potential core length for different pressure ratios (Misevich, *Ref. 26*, 112)

ratio or increasing jet velocity. L_i is approximately 3 nozzle diameters over a wide range of velocities and Misevich gives an accurate empirical formula as

$$\frac{L_i}{D} = 8 \cdot 0(P - 0 \cdot 528)^2 + 2 \cdot 47 \quad (4.23)$$

where D = nozzle diameter
  P = pressure ratio across nozzle.

To obtain the jet balance-force, the pressure is assumed to degrade in the axial direction, z, according to the spread of the jet. Referring to

Fig. 4.17 again, the jet is assumed not to spread in the core region, and then spreads uniformly at an angle ϕ downstream of the core. Pressure variations in the radial direction are also neglected. Summing the total force at the end of the core region and the force at any arbitrary position downstream of this point:

$$P_s \cdot \frac{\pi D^2}{4}\bigg|_{z=L_i} = P_z \cdot \frac{\pi}{4}[D + 2(z - L_i)\tan \phi]^2 \quad \text{for } z \geqslant L_i \quad (4.24)$$

where P_s = supply pressure.

Experimental data gives the jet spread as

$$\tan \phi = 0.009\,(z - L_i)/D \quad (4.25)$$

Substituting (4.25) in (4.24) gives

$$P_z = P_s\left[1 + 0.018\left(\frac{z - L_i}{D}\right)^2\right]^{-2} \quad (4.26)$$

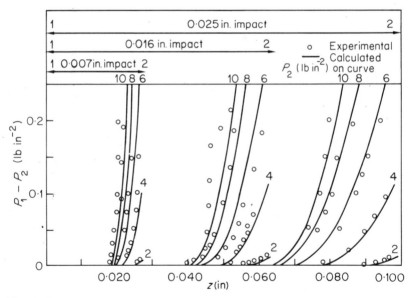

Fig. 4.19 Balance-point positions of two opposed impacting jets (Misevich, Ref. 26, 144)

The impact position can be calculated by applying equation (4.26) to both jets when the core lengths do not overlap. For small gaps, where the cores do overlap, giving rise to a high sensitivity region, the stream pressure of the nozzle nearest the impact is constant. Fig. 4.19 shows the

[4.3] IMPACT MODULATOR 205

correlation obtained between this theory and experiment for three supply nozzle diameters with the opposing nozzles placed 4 nozzle diameters apart. Good agreement is obtained, and it is interesting to note that increased sensitivity is obtained by decreasing the supply pressure levels, even though the jet cores do not overlap. The experimental results quoted earlier for Bjornsen for a practical Impact Modulator are almost certainly for opposed jets set close together (less than 5 nozzle diameters) which should result in the jet cores overlapping. This appears to be the highest sensitivity configuration, although the reported increase in gain with supply pressure level would appear to be at variance with the balance-point sensitivity predictions and data.

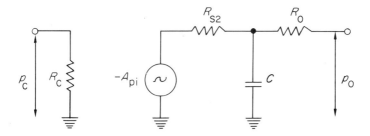

Fig. 4.20 Lumped parameter model of an impact modulator (N.A.S.A. Rep., *Ref. 22*, Fig. 4.3.4-1)

4.3.3 *Dynamic Characteristics*

Only limited information is available on the frequency response of Impact Modulators. Tests on a commercially available unit[22] having a supply nozzle diameter of about 0·016 in. show a phase shift of 45° at a frequency of about 400 Hz under blocked load conditions. The unit shows virtually no attenuation for frequencies up to 30 Hz with blocked output conditions but it exhibits the usual sensitivity to capacitive load changes, the frequency response being reduced to a few cycles per second when loaded with a volume of 5 in³. A smaller device using 0·005 in. diameter nozzles is reported to have an improved frequency response, and also to generate less noise than the larger amplifier. Noise levels are reported to be low under static conditions but to increase under dynamic operating conditions.

A lumped parameter representation of an Impact Modulator dynamic characteristic is shown in Fig. 4.20. The output chamber volume acts as a capacity to ground, sandwiched between the collector chamber orifice R_{s2} and the output channel resistance R_o with in general R_{s2} much larger than R_o. The control nozzle resistor R_c is an orifice and, together with R_{s2} will usually exhibit a non-linear characteristic.

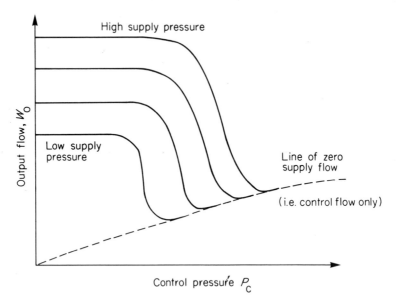

Fig. 4.21 Vortex amplifier transfer characteristics

4.4 Vortex Amplifier

4.4.1 Static Characteristics

A vortex amplifier is a three-dimensional device consisting of a circular chamber which is arranged to admit fluid in a radial direction towards its centre, as shown in Fig. 1.35. An outlet tube, positioned at the chamber centre, allows the flow to pass out in a direction perpendicular to the plane of the chamber. The amplifier has one or more control nozzles positioned so that they are tangential to the periphery of the vortex chamber. In the absence of a control signal, the supply flows radially inwards through the chamber, and passes through the outlet tube with a minimum of losses. When flow is introduced through a control nozzle, it deflects the supply flow by momentum interaction thereby causing a spiral motion to the chamber centre. This effectively increases the pressure drop across the amplifier, thus restricting the supply flow available when operated with a pressure regulated source. The effect is illustrated in Fig. 4.21. The limiting case occurs when the control pressure, and hence control flow, is so large that the supply flow is reduced to zero in order

[4.4] VORTEX AMPLIFIER 207

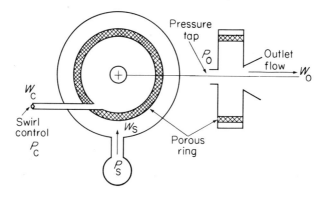

(a) Porous ring type. (Taplin, *Ref* 32, 1, Fig 1).

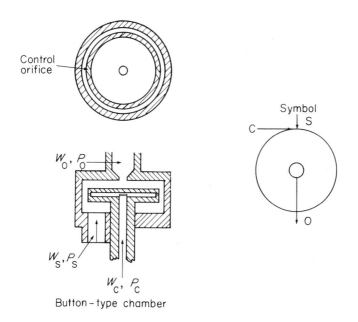

(b) Button-type. (Mayer, *A.F.*, 237)

Fig. 4.22 Practical vortex amplifier designs

to maintain a constant supply pressure. Under such conditions, the output flow is provided entirely by the control flow.

The swirling motion in the chamber has two velocity components, tangential and radial. The conservation of angular momentum requires that the tangential velocity increases as it moves towards the chamber centre and it is this basic mechanism which provides the vortex-valve gain. The action of the vortex amplifier is therefore seen essentially as a flow modulating device with a negative gain characteristic. With a suitable arrangement of output receiver, pressure rather than flow signals may be used with equal effect. However, both Howland[27] and Royle[28] have shown that the design of output receiver is critical, giving a variety of characteristics dependent on the configuration chosen.

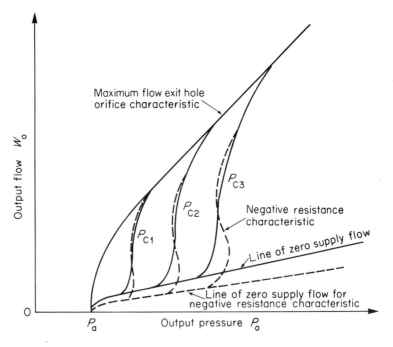

Fig. 4.23 Vortex amplifier output characteristics (Taplin, *Ref. 32*, 4)

Fig. 4.22 shows two practical configurations of a vortex valve. One way of achieving radial supply flow is shown in (a) and consists of an annular chamber, into which the supply flows separated from the circular chamber by porous material. This suppresses any swirling motion which may be present in the annular chamber. The figure also shows the positioning of a receiver to obtain output pressure signal, P_o. A second supply arrangement is shown in (b) and is referred to as a Button-type vortex amplifier. In this case control modulation of the supply flow occurs as it enters the chamber in an axial direction due to tangential control orifices in the

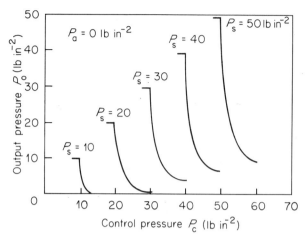

Fig. 4.24 Vortex pressure amplifier transfer characteristics (Taplin, *Ref. 32, 43*)

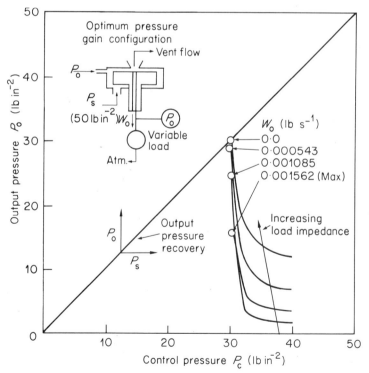

Fig. 4.25 Effect of loading on a vortex pressure amplifier (Taplin, *Ref. 32, 43*)

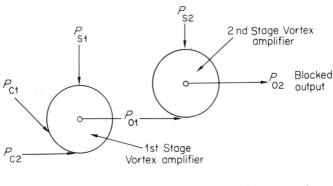

P_C Control pressure (Input signal)
P_O Output pressure (Output signal)
P_S Regulated Vortex amplifier supply pressure

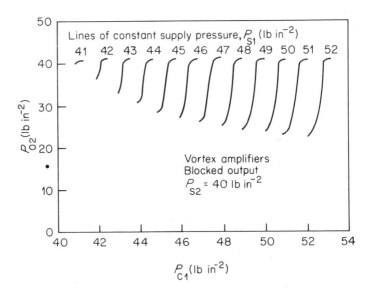

Fig. 4.26 Pressure gain of two vortex amplifiers in cascade (Taplin, Ref. 32, 47)

central 'button'. The flow then impinges on the chamber end wall, and is deflected through a right angle before passing through the chamber into the outlet. The geometry is particularly advantageous as a large, low velocity supply flow is obtained due to the low supply-orifice resistance characteristics. This enables the outlet orifice resistance to be relatively large resulting in a device with a large operating flow range.

An interesting characteristic of vortex valves is that the output may exhibit negative resistance, which implies that multivalued flows can exist at certain supply pressures. Fig. 4.23 shows with full lines a typical output flow-pressure characteristic for various regulated control-pressure levels. P_a is the atmospheric pressure and the curves are confined within an envelope defined by the exit and control nozzle flow characteristics. The slope of the curves is usually positive, signifying a positive resistance, although it is also possible to design an amplifier with characteristics as shown by the hatched lines which exhibit negative gradients over part of the operating range. In general, these negative-resistance characteristics are unstable where supply flow is not independently controlled, and show up when the vortex chamber thickness is relatively large, and the supply-chamber resistance is too low compared to the exit-hole resistance. It is possible to harness negative resistance effects to design an oscillator, an aspect which has been investigated by Sarpkaya[29, 30].

A unique feature of vortex amplifiers is the very wide range of operating pressures which are possible; practical devices working up to 1000 lb in^2 have been reported although 50 to 300 lb in^{-2} is the most common design range. This implies that the flow through the exit hole from the chamber is sonic for a gas under most operating conditions, so that back pressure cannot effect the chamber flow for a given supply and control pressure. The exit hole, which is a vent when the device operates as a pressure amplifier, usually has a diverging conical shape to aid aspiration of the swirling flow. The pressure signal receiver for this configuration is usually situated on the opposite side of the exit hole on the other chamber end wall.

Mayer and Maker[31], Howland[27], Otsap[40] and Taplin[32] have given characteristics of vortex flow and pressure amplifiers. A typical vortex pressure amplifier transfer characteristic is shown in Fig. 4.24 for blocked receiver load conditions giving a negative pressure gain in the order of 10. The linear operating range increases with supply pressure and in all cases the control pressure must exceed the pressure level selected before control flow commences and the output pressure is modulated. Taplin also illustrates that the pressure gain is remarkably unaffected by load variations at the pressure receiver, as shown in Fig. 4.25, and some increase in the operating range as the load is reduced may be expected. This indicates an amplifier which has very small output impedance, which reduces the problem of interconnexion, and causes no degradation in gain when cascaded. This is confirmed in Fig. 4.26 which shows the transfer characteristics for two amplifiers in cascade. The first stage amplifier utilizes two control ports in parallel to give an additive action. Each control has a pressure gain of 5 giving a gain of 10 for the stage, which when combined with the second stage gain of 5, should theoretically give an overall gain of 50. The practical characteristics give values very close to this figure.

The vortex flow amplifier is less useful than the pressure amplifier in control circuits due to the sensitivity to loading effects. One arrangement is shown in Fig. 4.27 in which the exit hole on the pressure receiver are

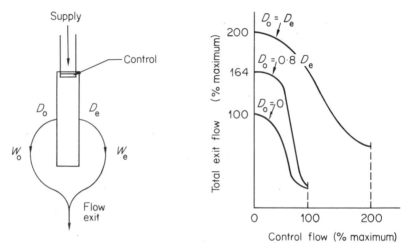

Fig. 4.27 Vortex flow amplifier receiver configuration (Howland, *Ref.* 27, 212)

joined together to provide the output flow. When the diameter of the tubes are made unequal ($D_o = 0.8 D_e$ in this case) higher flow gains appear possible probably due to a lack of axial-flow symmetry in the Button-type amplifier. No-load flow gains as high as 200 have been reported although a more practical figure would be 20. Little general information is available on the geometric shape of vortex amplifiers, but it would appear that fairly short chambers are used, typically with a chamber length/chamber diameter ratio of about 0.1 for two output hole designs. The control nozzle diameter/chamber diameter ratio varies between about 0.05 and 0.15 according to the desired amplifier characteristics. The most usual size of amplifier falls within the range 0.5 to 4.0 in. chamber diameter. Greber and Koerper[33, 34] consider a variety of configuration for a vortex flow

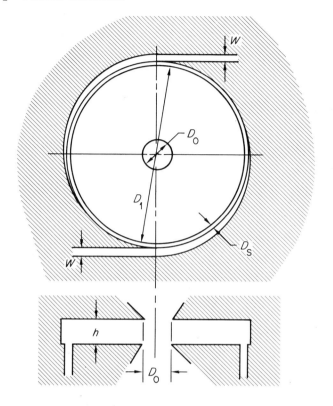

Where $\bar{D} = \dfrac{D_0}{D_1}$, $\bar{D}_s = \dfrac{D_s}{D_1}$, $\bar{H} = \dfrac{h}{D_1}$, $\bar{W} = \dfrac{W}{D_1}$, $\bar{Q} = \dfrac{Q\,\text{max}}{Q\,\text{min}}$

Fig. 4.28 Vortex amplifier design (Greber and Koerper, *Ref.* 33, 234)

amplifier with two control and outlet holes, illustrated in Fig. 4.28, using low pressure air and water sources. Their optimized design for an amplifier with the greatest ratio of maximum to minimum output flow (also known as turndown ratio) $\bar{Q}$ gave a value of 11·5 for this ratio with the following geometry:

Chamber diameter = 1 in. $\bar{H} = 0\cdot 188$ $\bar{D} = 0\cdot 300$
$\bar{D}_s = 0\cdot 045$ $\bar{W} = 0\cdot 010$

where the notation is that of the figure. However, to achieve this, a negative resistance characteristic was introduced into the transfer characteristic as shown in Fig. 4.29. A further optimal design by the same authors was produced to remove this effect at the expense of reducing $\bar{Q}$ to about 7. The following configuration was used:

Chamber diameter = 1 in. $\bar{H} = 0.188$ $\bar{D} = 0.11$ to 0.14

$D_s = 0.045$ $\bar{W} = 0.006$

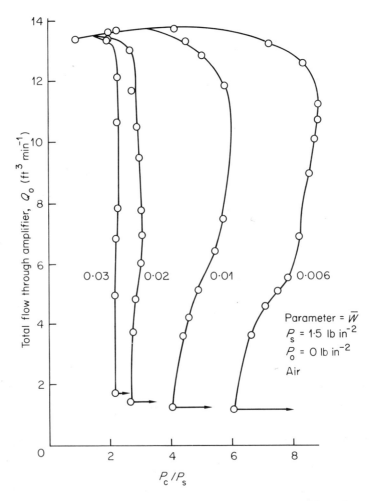

Fig. 4.29 Flow transfer characteristic of a vortex amplifier with maximum turndown (Greber and Koerper, Ref. 33, 238)

[4.4] VORTEX AMPLIFIER 215

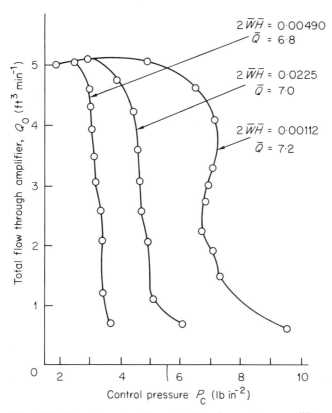

Fig. 4.30 Flow transfer characteristic of a vortex amplifier with no hysteresis (Greber and Koerper, *Ref. 33*, 242)

Fig. 4.30 shows some transfer characteristic data obtained with this design indicating a good, high-gain linear region, although with reduced turndown ratio from the previous design.

The vortex configuration is very versatile; for instance, a summing or subtracting function may be achieved by introducing multiple input nozzles at a tangent to the chamber so as to reinforce or destroy the swirling motion. Alternatively, a device without control tubes, but with an output receiver capable of measuring the tangential velocity component, is an angular velocity sensor if the vortex chamber is rotated. A passive diode is also achieved if flow is possible in either direction[35].

4.4.2 Theoretical Considerations

The swirl motion induced by control flow in a vortex chamber is complex due to viscous effects related to boundary layer formation round the chamber periphery and also along the chamber end walls.

In an efficient vortex valve, the tangential velocity at the chamber periphery varies from stagnation to almost sonic values, as the outlet flow

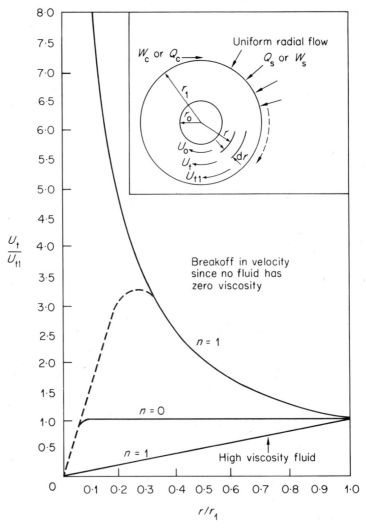

Fig. 4.31 Tangential velocity distribution in a vortex chamber (Taplin, *Ref. 32*, 12)

varies over the operating range. At high tangential velocities, the flow field becomes highly non-uniform in the chamber. Savino and Keshock[36] have shown that under these conditions practically all the flow is carried from the outer wall to the centre by the two boundary layers at the end walls of the vortex chamber. In addition, a strong re-circulation flow occupies the central portion of the chamber outside the boundary layers. Theoretical analyses to date have been confined to a description of average effects within the chamber, no attempt being made to describe the flow field in detail. However, insight into the actual vortex chamber flow field can be gained by consideration of some of the types of flow that can occur.

The distribution of tangential velocity varies with radius and is dependent on the magnitude of swirl, chamber geometry and fluid viscosity. In an ideal lossless or low viscosity vortex field, angular momentum is conserved so that the tangential velocity U_t must increase with decreasing radius and according to the relationship

$$U_t = \frac{U_{t1} r_1}{r} \tag{4.27}$$

where suffix '1' refers to conditions at the chamber periphery. At the other extreme as viscosity effects become very large, the fluid rotates as though it were a solid body. In this case the tangential velocity is given by

$$U_t = \frac{U_{t1} r}{r_1} \tag{4.28}$$

In practice, the tangential velocity distribution may be expressed by the general relationship

$$U_t = U_{t1} \left(\frac{r_1}{r}\right)^n \tag{4.29}$$

where n is an empirical exponent lying between $+1$ (Free vortex) and -1 (solid body rotation).

Fig. 4.31 shows the tangential velocity radial distribution for various values of n. For gases, n is taken as $+1$ until the velocity starts to break down due to viscous effects, as indicated by the hatched line. This breakdown must occur, even for a gas with small viscosity, because the chamber geometry cannot support supersonic flow.

Taplin has given the analysis of both incompressible and compressible flow in the chamber, based on the general tangential velocity distribution. The differential pressure dP developed across a differential radius dr is

$$dP = \frac{\rho U_t^2 \, dr}{r} \tag{4.30}$$

Substituting (4.29) into (4.30) gives

$$dP = \frac{\rho U_{t1}^2}{r}\left(\frac{r_1}{r}\right)^{2n} dr \qquad (4.31)$$

For incompressible flow this expression may now be integrated to give

$$P = -\frac{\rho U_{t1}^2 r_1^{2n} r^{-2n}}{2n} + c \qquad (4.32)$$

where c is a constant. Since P equals the outer wall pressure P_w when $r = r_1$, the pressure drop across the flow field to radius r is given by:

$$P_w - P = \frac{\rho U_{t1}^2}{2n}\left\{\left(\frac{r_1}{r}\right)^{2n} - 1\right\} \qquad (4.33)$$

The pressure drop to the edge of the exit hole, where the radius is given as r_o, is then

$$P_w - P_o = \frac{G^2 \rho U_{t1}^2}{2} \qquad (4.34)$$

where

$$G^2 = \frac{1}{n}\left\{\left(\frac{r_1}{r_o}\right)^{2n} - 1\right\}$$

and is in effect a vortex sensitivity constant. Taplin relates the pressure to the outlet flow Q_o by using the orifice flow equation for the exit hole, given by:

$$Q_o = C_D A_e \sqrt{\left\{\frac{2g}{\rho}(P_o - P_a)\right\}} \qquad (4.35)$$

where C_D = coefficient of discharge
 A_e = exit hole area
 P_a = ambient or back pressure in exit hole.

A similar, but more complex, analysis may be carried out for the compressible flow case.

As the exit hole provides the dominant restriction in most vortex amplifier designs, another approach has been to modify the conventional axial flow exit orifice equations for the effect of swirl. Miller[37] has given some experimental data related to nozzle throttling devices using long, small diameter upstream vortex chambers. A more recent analysis, which appears to have application to a wide range of chamber configurations, has been given by Mayer[38]. He uses the universal orifice equation for either compressible or incompressible flow and modifies the coefficient of

[4.4] VORTEX AMPLIFIER

discharge by an experimentally determined factor W_N which is a function of the swirl. Thus, using the notation of Chapter 2 the output weight flow, W_o is

$$W_o = \frac{W_N K P_s A_e N_{se}}{T_s} \qquad (4.36)$$

where P_s and T_s are the supply pressure and absolute temperature respectively.

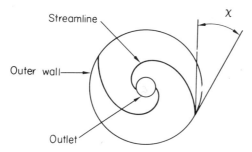

Fig. 4.32 Definition of χ, the flow vector at the chamber wall (Mayer, Ref. 38, 240)

Mayer has found that the coefficient W_N is a function of χ, the direction of the flow vector at the outer wall of the vortex chamber as defined in Fig. 4.32. But, more important, a single universal curve of W_N against χ is valid for a large range of experimental results in which changes of control pressure P_c, supply pressure P_s, ambient pressure P_a and control port diameter D_c, occur. Other changes in the vortex chamber configuration effects the curve. The experiments were carried out on chambers with diameters 0·44 in. to 4·00 in. for a number of outlet orifice sizes and vortex chamber lengths. Fig. 4.33 shows Mayer's experimental curve for $W_N(\chi)$ plotted logarithmically for clarity. For pure radial flow, W_N is expected to be unity and χ is 90° while complete flow shut-off corresponds to $W_N = 0$ and $\chi = 0$. The experimental data approximates to these conditions in the limit but it is interesting to note that for $\chi < 5°$ small changes in χ result in large changes in W_N, i.e. the jet interaction angle has an appreciable effect on the output flow.

The output conditions are correlated to those at the control nozzles in the following way. The value of χ can be calculated from the average radial and tangential velocities, $\bar{U}_r$ and $\bar{U}_t$ respectively, by

$$\cot \chi = \frac{\bar{U}_t}{\bar{U}_r} \qquad (4.37)$$

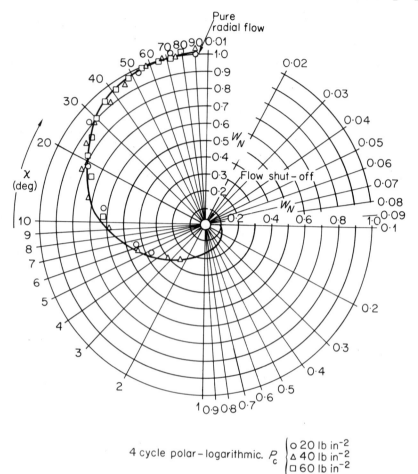

Fig. 4.33 Variation of the outlet discharge coefficient W_N with χ (Mayer, *Ref. 38*, 243)

Also the average velocity is determined by the flow area at the outer wall of the vortex chamber

$$\bar{U}_r = \frac{W_o}{\rho A} = \frac{W_o \mathbf{R} T_s}{\pi D_1 L P_s} \qquad (4.38)$$

where D_1 = chamber diameter
       L = chamber length.

[4.4] VORTEX AMPLIFIER

The average tangential velocity is assumed to be dependent purely on momentum transfer between the control jets and the total flow:

$$\bar{U}_t = \frac{1}{W_o} \sum W_c \bar{U}_c \qquad (4.39)$$

where W_c = control weight flow
$\bar{U}_c$ = control signal average velocity.

The control velocity is usually subsonic so that

$$U_c^2 = g\gamma \cdot R \cdot T_c \cdot \left(\frac{2}{\gamma - 1}\right)\left\{1 - \left(\frac{P_s}{P_c}\right)^{\frac{\gamma-1}{\gamma}}\right\} \qquad (4.40)$$

where γ = ratio of specific heats
R = gas constant.

Thus, for any input conditions, equations (4.37) to (4.40) allow the value of χ to be calculated. From this, the $W_N(\chi)$ curve and (4.36) give the corresponding output flow. The author gives transfer characteristics which show good correlation between the theoretical model and experimental data.

4.4.3 Dynamic Characteristics

Vortex amplifiers, in general, tend to have a lower frequency response than either the Beam-deflexion or Impact Modulator amplifiers. This is mainly due to the relatively large physical dimensions of vortex chambers compared with interaction regions of the other two devices. Taplin[32, 39] regards the delay time for signals travelling at the jet stream velocity to pass through the chamber (transport time) as important. The effect of the time for the tangential momentum to change at the outer wall of the chamber until the effect is sensed at the exit hole is

$$\frac{\Delta p_o}{\Delta P_o} = \frac{1 - e^{-st}}{st} \qquad (4.41)$$

where ΔP_o = steady state change in pressure at the exit hole
Δp_o = corresponding dynamic pressure-change.

For sinusoidal operation this reduces to

$$\frac{\Delta p_o}{\Delta P_o} \simeq \frac{1}{1 + j(\omega t/2)} \quad \text{for } 0 < \omega < \frac{1}{t} \qquad (4.42)$$

At higher frequencies, the fluid inertia within the vortex chamber may become important.

References

1. Simson, A. K., A theoretical study of the design parameters of subsonic pressure controlled, fluid jet amplifiers. *Ph.D. Thesis*, (1963) M.I.T.
2. Reilly, R. J. and Moynihan, F. A., Deflection and relative flow of three interacting jets. *H.D.L.* (2), **I**, 123.
3. Sheeran, W. J. and Dosanjh, D. S., Interaction of transversely impinging jets. *H.D.L.* (1), **I**, 39.
4. Sheeran, W. J. and Dosanjh, D. S., Transversely impinging two dimensional jet flows. *A.F.*, 31.
5. Render, A. B., The design of jet-interaction amplifiers using supercritical pressure ratios. *C.F.C.* (2), Paper C2.
6. Kallevig, J. A., Effect of receiver design on amplifier performance and jet profile of a proportional fluid amplifier. *H.D.L.* (3), **II**, 163.
7. Reid, K. N., An experimental study of the static interaction of an axisymmetric fluid jet and a single receiver-diffuser. *H.D.L.* (3), **IV**, 71.
8. Prosser, D. W. and Fisher, M. J., Some influences of noise of proportional fluid amplifiers. *H.D.L.* (3), **II**, 129.
9. Moynihan, F. A., Jet interaction noise. *H.D.L.* (2), **I**, 111.
10. Weinger, S. A., Effect of aspect ratio on noise in proportional fluid amplifiers. *A.F.*, 94.
11. Dexter, F. M., An analog pure fluid amplifier. *Fluid Jet Control Devices* (*A.S.M.E.*), 41 (1962).
12. Reilly, R. J. and Moynihan, F. A., Notes on a proportional fluid amplifier. *Fluid Jet Control Devices* (*A.S.M.E.*), 51 (1962).
13. Peperone, S. J., Katz, S. and Goto, J. M., Gain analysis of the proportional amplifier. *H.D.L.* (1), **I**, 319.
14. Van Tilburg, R. W. and Cochran, W. L., Development of a proportional fluid amplifier for multi-stage operation. *H.D.L.* (2), **II**, 313.
15. Douglas, J. F. and Neve, R. S., Investigation into the behaviour of a jet interaction proportional amplifier. *C.F.C.* (2), Paper C3.
16. McCabe, A. and Hughes, D. L., Characteristics of proportional fluidic amplifiers. *C.F.C.* (2), Paper C5.
17. Hayes, S. W. F. and Kwok, C., Impedance Matching in bistable and proportional fluid amplifiers through the use of a vortex vent. *H.D.L.* (3), **I**, 331.
18. Camarata, F. J., Analytical procedure for predicting performance of single stage momentum exchange proportional amplifiers. *A.F.*, 87.
19. Belsterling, C. A. and Tsui, K. C., Application techniques for proportional amplifiers. *H.D.L.* (2), **II**, 163.
20. Manion, F. M., Proportional amplifier simulation. *A.F.*, 62.
21. Boothe, W. A., A lumped parameter technique for predicting analog fluid amplifier dynamics. *I.S.A. Trans.*, **4·1**, 84 (1965).
22. Analytical investigation of fluid amplifier dynamic characteristics. *N.A.S.A.* CR–245 (1965).
23. Bjornsen, B. G., The impact modulator. *H.D.L.* (3), **II**, 5.
24. Lechner, T. J. and Sorenson, P. H., Some properties and applications of direct and transverse impact modulators. *H.D.L.* (2), **II**, 33.
25. Katz, S., Pressure gain analysis of an impacting jet amplifier. *A.F.*, 116.
26. Misevich, K. W., The impacting of opposed axially symmetric jets. *A.F.*, 98.
27. Howland, G. R., Performance characteristics of vortex amplifiers. *H.D.L.* (3), **II**, 207.

28. Royle, J. K. and Hassan, M. A., Characteristics of vortex devices. *C.F.C.* (2), Paper D4.
29. Sarpkaya, T., Characteristics of counter-vortex oscillators. *H.D.L.* (2), **II**, 147.
30. Sarpkaya, T., Characteristics of a vortex device and the vortex breakdown phenomenon. *H.D.L.* (3), **II**, 245.
31. Mayer, E. A. and Maker, P., Control characteristics of vortex valves. *H.D.L.* (2), **II**, 61.
32. Taplin, L. B., Phenomenology of vortex flow and its application to signal amplification. Summer Engineering Seminar, Pennsylvania State University (1965).
33. Greber, I. and Koerper, P. E., Fluid vortex amplifier optimization. *H.D.L.* **II**, 223 (1967).
34. Koerper, P. E., Design of an optimized vortex amplifier. *E.D.C. Rep.* 7-65-6 *Case Inst. Techn., U.S.A.*
35. Baker, P. J., A comparison of fluid diodes. *C.F.C.* (2), Paper D6.
36. Savino, J. M. and Keshock, E. G., Experimental profiles of velocity components and radial pressure distribution in a vortex contained in a short cylindrical chamber. *H.D.L.* (3), **II**, 269.
37. Miller, D. P., Characteristics of a vortex fluid throttle. *H.D.L.* (2), **II**, 125.
38. Mayer, E. A., Large signal vortex valve analysis. *A.F.*, 233.
39. Taplin, L. B., Small signal analysis of vortex amplifiers. *Fluid Control-Components and Systems, AGARD lectures* (1966).
40. Otsap, B. A., Experimental stidy of a proportional vortex fluid amplifier. *H.D.L.* (2), **II**, 85.

5
Analog Signal Control Techniques

5.1 High-Gain Amplifiers

5.1.1 General

Due to its relatively good dynamic performance, the Beam-deflexion amplifier has been widely developed for use in analog fluidic-control systems[1,2]. Early designs exhibited poor overall characteristics and had undesirable features, such as instability under blocked load conditions and around the null operationing region. Noise was also a problem. Development over the last few years has minimized the defects of the amplifier, and much improved commercial designs are available which allow efficient cascading of similar elements.

This has led to the development of high gain blocks suitable for use in operational-amplifier configurations[3,4,5]. The static characteristics quoted for these amplifiers are more than adequate for a large variety of fluidic control circuits but there are limitations on their dynamic characteristics as they tend to exhibit large phase lags.

The basic configuration of a high gain amplifier used in the operational mode is shown in Fig. 5.1 in which an input impedance Z_1 is connected to a high-gain amplifier block of gain G and internal input impedance Z_i. Z_f is a feedback impedance across the gain block. Due to the large gain and finite output pressure signal magnitude P_o, the input pressure and flow signals are very small. The relationship between the output signal P_o and the input signal P_i is:

$$\frac{P_o}{P_i} = -\frac{Z_f}{Z_1} \frac{1}{1 + \frac{1}{G}\left[\frac{Z_f}{Z_1} + \frac{Z_f}{Z_i} + 1\right]} \qquad (5.1)$$

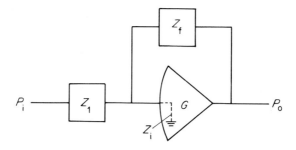

Fig. 5.1 Operational-amplifier (Parker and Addy, Ref. 10)

Thus if

$$G \gg \frac{Z_f}{Z_1} + \frac{Z_f}{Z_i} + 1 \tag{5.2}$$

then

$$\frac{P_o}{P_i} = -\frac{Z_f}{Z_1} \tag{5.3}$$

giving an ideal operational-amplifier.

If the impedances used are linear resistances then summation of analog signals is achieved by modifying the input network as shown in Fig. 5.2. In this case it may be shown that

$$P_o \simeq -R_f \left[\frac{P_1}{R_1} + \frac{P_2}{R_2} + \frac{P_3}{R_3} \right] \tag{5.4}$$

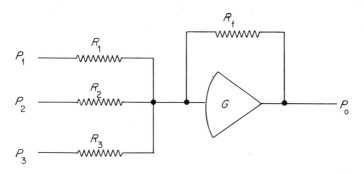

Fig. 5.2 Summation using an operational-amplifier

The precision of this result is dependent on the gain condition of (5.2) being true. In practice, if the gain exceeds approximately 10^5 these inaccuracies are small; present generation fluidic gain-blocks fall well below this figure so the simple relationships of (5.3) and (5.4) must be used with care.

The second term of the product in (5.1) gives the deviation of an intermediate-gain amplifier circuit from the ideal case. Expressed in another way, the percentage deviation may be written as:

Deviation from ideal op. amp.

$$= \frac{100}{1 + G_\text{L}} \text{ per cent} \qquad (5.5)$$

where gain,

$$G_\text{L} = \frac{G}{\dfrac{Z_\text{f}}{Z_1} + \dfrac{Z_\text{f}}{Z_\text{i}} + 1} \qquad (5.6)$$

Wagner and Barrett[5] have given experimental and calculated closed loop gains for various resistive networks coupled to a high-gain amplifier. This information is reproduced in Table 5.1 together with the deviation from the ideal operation based on (5.5). Although the theoretical and experimental correlation is good for the overall gain given by (5.1) for resistive impedances, most of the configurations tested showed serious deviation from the ideal operational-amplifier operation. In this case the summation (5.4) must be modified to account for finite values of block

Table 5.1—Proportional fluidic operational-amplifier data

Experimental data[5]				Calculated data			
R_f/R_1	R_f/R_i	G	P_o/P_i	P_o/P_i	% error of overall gain	G_L	% deviation from op.-amp.
67·3	180	900	53·2	52·7	1·0	3·62	21·7
31·0	83·0	900	28·0	27·3	2·5	7·83	11·3
7·0	32·8	900	6·74	6·68	0·9	22·0	4·35
17·7	83·0	900	15·4	15·7	2·0	8·85	10·2
38·6	180	900	30·9	30·9	0·0	4·10	19·6
17·7	83·0	332	13·1	13·3	2·0	3·26	23·5
17·7	83·0	538	14·1	14·5	3·0	5·29	15·9
17·7	83·0	600	14·7	15·0	2·0	5·90	14·5
17·7	83·0	1850	17·0	16·7	2·0	18·2	5·21
38·6	180	538	27·4	27·2	0·7	2·45	29·0
38·6	180	2240	35·4	35·1	0·9	10·2	8·93

gain G. For the summation of 'n' input resistances (5.6) can be written as:

$$G_L = \frac{G}{1 + \frac{R_f}{R_i} + \sum_1^n \frac{R_f}{R_n}} \qquad (5.7)$$

and substituting in (5.1)

$$\frac{P_o}{P_i} = -\frac{Z_f}{Z_1} \frac{1}{1 + \frac{1}{G_L}}$$

As $G_L \gg 1$, expansion by the Binomial Theorem gives

$$\frac{P_o}{P_i} \simeq -\frac{Z_f}{Z_1}\left(1 - \frac{1}{G_L}\right) \qquad (5.8)$$

so that the more accurate summation of signals P_1, P_2, P_3 etc. is

$$\frac{P_o}{P_i} \simeq \left(\frac{1}{G_L} - 1\right)\left[\frac{R_f}{R_1} P_1 + \frac{R_f}{R_2} P_2 + \cdots + \frac{R_f}{R_n} P_n\right] \qquad (5.9)$$

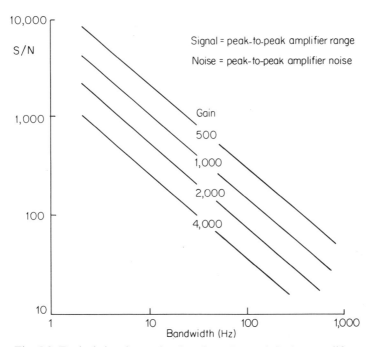

Fig. 5.3 Typical signal-to-noise data for a Beam-deflexion amplifier gain block (Urbanosky, *Ref. 4*)

Noise generated within fluidic-amplifier circuits has been a significant obstacle in the development of high-gain amplifiers. A practical design feature now incorporated in cascaded amplifier circuits to overcome noise is etching out of thin laminations so that each stage in the gain block consists of a number of very low aspect ratio devices operating in parallel. Also etched out of the laminations are supply ducts and laminar flow dropping resistors so that each supply has the correct pressure. Urbanosky[4] gives typical signal-to-noise (S/N) data for a five-stage gain block, as shown in Fig. 5.3, based on peak-to-peak values. The S/N tends to be inversely proportional to the gain for a given configuration and is significantly better than most system requirements.

A more detailed investigation of noise-reduction in amplifiers has been made by Kelley and Shinn[6]. They tested a proportional amplifier in two geometrically similar sizes based on 0·02 in. and 0·04 in. wide power nozzles. The profiles were etched in 0·04 in. thick metal laminates which allowed the effect of size, supply pressure, aspect ratio and the use of parallel paths to be studied in relation to amplifier noise. Although the tests were somewhat artificial in that the differential input pressure was always maintained at zero, the results were highly significant. Their conclusions were:

(a) Output noise occurred below 400 Hz in the nozzle sizes tested.
(b) Decreasing amplifier size can increase output noise.
(c) Providing multiple parallel paths decreases noise amplitude, S/N is approximately proportional to the square root of the number of paths.
(d) Noise amplitude increases as a function of supply pressure, but S/N increases faster than noise amplitude.
(e) Aspect ratio is not important as far as noise is concerned, provided total nozzle area is held constant.
(f) S/N values up to 1500, based on r.m.s. values of noise, can be attained by dividing a single element into several parallel flow paths.

5.1.2 Operational-Amplifier Stability

The operational-amplifier forms a complete closed-loop network internally as an impedance Z_f is fed back around the gain block. Ideally the gain block exhibits a high gain without amplitude-attenuation or phase shift throughout the operating range of the closed loop system. While a fluidic-gain block will usually give acceptable amplitude characteristics the phase shift may be quite significant, leading to closed loop instability unless further compensation is added to the system.

In order to produce a high-gain fluidic-amplifier it is necessary to

[5.1] HIGH-GAIN AMPLIFIERS

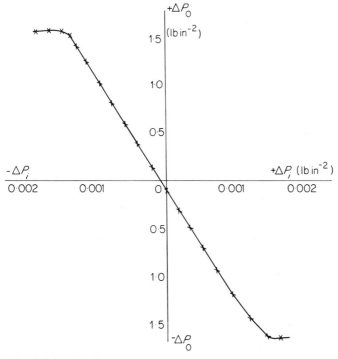

Fig. 5.4 Static gain characteristic of a four-stage amplifier (Parker and Addy, *Ref. 10*)

cascade a number of elements. In the case of a Beam-deflexion amplifier the interstage push–pull gain is typically 6, so that by cascading four elements a forward gain of more than 1000 may be realized. Fig. 5.4 gives the experimental static-gain of such a configuration using elements with

0·010 in. wide by 0·040 in. deep supply nozzles which also incorporated vortex vents. It confirms that a high forward gain is obtained, thus satisfying the static condition for an operational amplifier.

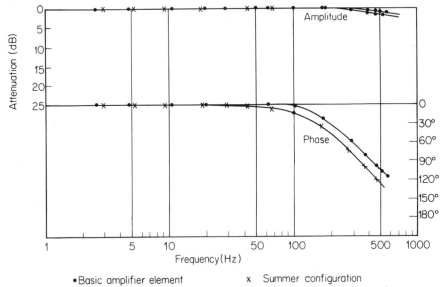

• Basic amplifier element   × Summer configuration
Push–Pull input, sinusoid 0·1 lbin^{-2} each side of null. Supply at 5 lbin^{2}. Load is similar element with resistance at input of 9×10^{4} lbs ft^{-5}.

Fig. 5.5 Dynamics of basic amplifier and Summer (Parker and Addy, *Ref. 10*)

As the amplifier must be dynamically stable when the feedback loop is closed, the phase lag must be less than 180° when the gain exceeds unity. A typical experimental frequency response of the same Beam-deflexion amplifier loaded with a similar element is shown in Fig. 5.5. It exhibits a characteristic phase lag which becomes important at a lower frequency than gain drop-off, due to the finite transport time lag of the supply jet and to vent effects. This is additive for a cascade configuration, giving a large phase lag before gain degradation, resulting in an amplifier which is unstable for closed-loop operation, unless modified. Currently available miniaturized integrated amplifier blocks[3] minimize the phase lag. For a typical five stage block we might have

$$\text{phase lag (degrees)} = 0\cdot 2 \text{ frequency (Hz)} \qquad (5.10)$$

In general, it is not attractive to stabilize the feedback loop by increasing the feedback resistance alone. This is because the loop gain would be

[5.1] HIGH-GAIN AMPLIFIERS   231

reduced to less than unity, implying a large value of feedback resistance, which is incompatible with the conditions associated with (5.2) for an operational amplifier. In fact (5.5) shows that the deviation from ideal operation would be at least 50 per cent.

Another alternative is to use a lag network, such as a volume and resistor combination, to provide the necessary gain reduction at a phase lag of 180°. Using this technique stabilization is achieved at the expense of a severe reduction in frequency bandwidth. The overall gain tends to be low, due to the small values of feedback resistance, although the deviation would also be correspondingly less. For example, using (5.10) the frequency response corresponding to 40° phase margin is 250 Hz, assuming the network has 90° lag at this frequency. If the gain of the amplifier block is 1000, the break frequency for the first order lag would be 0·25 Hz, which is too low for most control systems.

A compromise solution is to use both resistive-feedback and first-order lag compensation for stabilization while retaining reasonable operating bandwidth. This may be achieved if a low overall gain is acceptable together with a significant but not serious deviation from ideal operation. An example shows the relative magnitudes involved.

Assuming the first order lag must allow a 10 Hz bandwidth, then the feedback resistor R_f must provide the remaining attenuation, based on an amplifier block gain of 1000 and a phase margin of 40° at 250 Hz. The lag network gives approximately 28 dB attenuation so the loop gain must be no greater than about 25. In practice it is usual for a fluidic amplifier of this type to have $R_1 \simeq R_i$ and $R_f/R_1 > 1$, so that the loop gain may be written as

$$G_L \simeq \frac{G}{\text{const.} \frac{R_f}{R_1}} \qquad (5.11)$$

where the constant would lie in the range 2–4. This gives $R_f/R_1 \geqslant 13\cdot3$ and a deviation $\geqslant 3\cdot9$ per cent. This illustrates that the bandwidth of a fluidic operational amplifier may be limited by the allowable deviation from the ideal case. Many analog fluidic circuits are characterized by phase lags due to transport time, and tend to be phase-critical for stability. This means that although the amplifier bandwidth of 10 Hz would be compatible with many control systems, any series cascading to produce multi-functional shaping could well interfere with the system stability.

5.1.3 Integration

Integration is one of the two important dynamic requirements for

control system synthesis. In general it must conform to the following specifications:
- (a) The output must increase at a rate proportional to the change of input.
- (b) The frequency response must exhibit a 90° phase lag and associated gain characteristics of -20 dB/decade at all operating frequencies.
- (c) It must be possible to take output flow from the network without altering the dynamic characteristics and also the input impedance must be compatible with the preceding stage and isolated from any feedback paths.

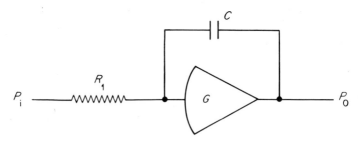

Fig. 5.6 Integration using an operational amplifier

The standard electronic circuit for an integrator based on a high-gain amplifier uses a series capacitor in the feedback circuit, as shown in Fig. 5.6, which leads to a relationship between the output signal P_o and the input signal P_i of the form

$$\frac{P_o}{P_i} = \frac{1}{s\tau} \qquad (5.12)$$

where $\tau = R_1 C$ provided the gain G is high. Unfortunately, a series fluid capacitor without moving parts cannot be realized so that a 'bootstrap' configuration must be used for a fluidic integrator, with the capacitor grounded in the input circuit. Both Urbanosky and Doherty[3] and Wagner and Barrett[5] use a similar circuit in which integration is accomplished by means of a resistance-volume positive-feedback circuit, as shown in Fig. 5.7. Using the notation of Wagner and Barrett it can be shown that:

$$\frac{\Delta P_o}{\Delta P_i} = \frac{K}{\left[1 + \dfrac{R_1}{R_5} + \dfrac{R_1}{R_2}\right] - \dfrac{KR_1}{R_5} + s\tau_1} \qquad (5.13)$$

where

$$K = \frac{(R_3 + R_4)/R_2}{1 + \frac{1}{G}\left[\frac{(R_3 + R_4)}{R_2} + \frac{(R_3 + R_4)}{R_1} + 1\right]} \quad \text{and} \quad \tau_1 = R_1 C$$

The condition for a perfectly balanced integrator circuit is

$$\frac{KR_1}{R_5} = 1 + \frac{R_1}{R_5} + \frac{R_1}{R_2} \tag{5.14}$$

which simplifies (5.13) to the required form

$$\frac{\Delta P_o}{\Delta P_i} = \frac{K}{s\tau_1} \tag{5.15}$$

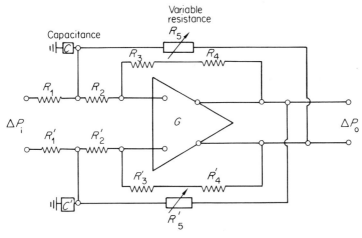

Fig. 5.7 Integrator circuit using an earthed capacitor feedback
(Wagner and Barrett, *Ref. 5*, Fig. 14)

The authors give experimental data showing the operation of this type of integrator. Fig. 5.8 illustrates the frequency response with various feedback capacitors. Although the bias pressure setting of the amplifiers does not appear to be critical, variations in supply pressure directly affect the overall gain K thereby violating the balance condition of (5.14). Under such conditions the circuit acts as a first-order lag as shown in Fig. 5.9 from the same source.

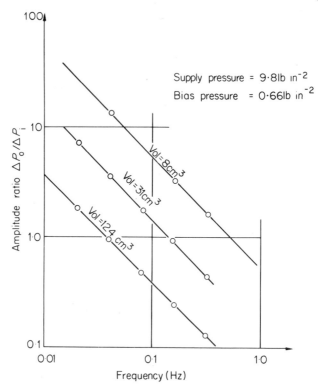

Fig. 5.8 Variation of integrator circuit frequency response with integrator feedback volume (Wagner and Barrett, Ref. 5, Fig. 16)

5.1.4 Differentiation

The general requirements for differentiation are:
(a) A step-input must produce an output pulse.
(b) Input pressure levels of different magnitudes must not alter the magnitude of the output pressure.
(c) The frequency response must produce a 90° phase lead and associated gain characteristic of $+20$ dB/decade over the design frequency range. A lag is usually introduced at a high frequency to eliminate noise problems.
(d) It must be possible to take output flow from the network without altering the network characteristics, and the input impedance must be compatible with the previous stage.

A lead-lag circuit has been described by Wagner and Barrett[5], based

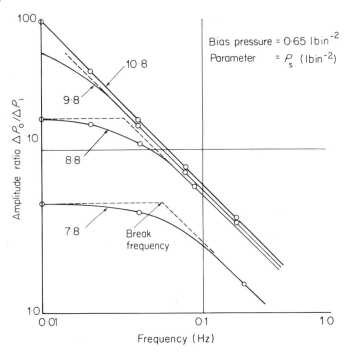

Fig. 5.9 Variation of integrator circuit frequency response with supply pressure (Wagner and Barrett, *Ref. 5*, Fig. 18)

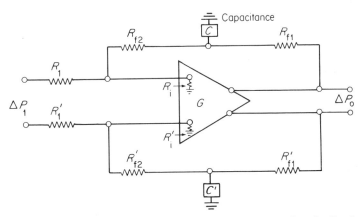

Fig. 5.10 Differentiator circuit using an earthed capacitor feedback (Wagner and Barrett, *Ref. 5*, Fig. 8)

on the pneumatic capacitor to earth, as shown in Fig. 5.10. This leads to the following transfer function:

$$\frac{\Delta P_o}{\Delta P_i} = \frac{K(1 + s\tau_2)}{(1 + s\tau_1)} \qquad (5.16)$$

where $K = \dfrac{R_i}{R_1 + R_i} \dfrac{G}{1 + GK_f}$

$$K_f = \frac{1}{1 + \dfrac{R_{f1}}{R_i} + \dfrac{R_{f2}}{R_i}}$$

$$\tau_1 = \frac{\tau_2}{1 + GK_f}$$

$$\tau_2 = K_f \frac{R_{f1}}{R_i}(R_{f2} + R_i)C$$

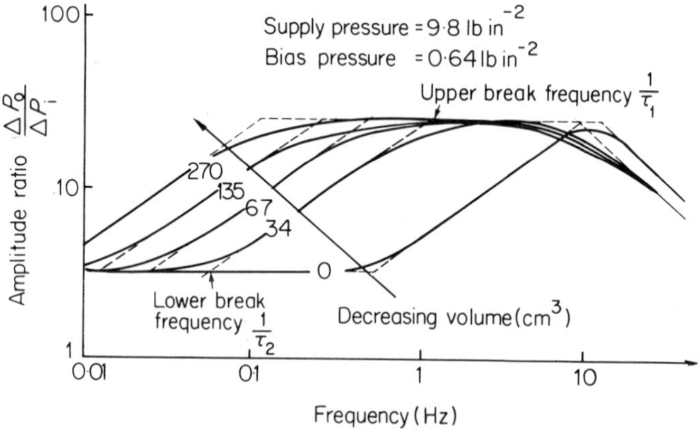

Fig. 5.11 Variation of differentiator/circuit frequency response with differentiator feedback volume (Wagner and Barrett, *Ref. 5*, Fig. 11)

The ratio of time constants τ_2/τ_1 may be adjusted to provide approximately two decades of lead compensation. Experimental results presented by Wagner and Barrett include a frequency response for fixed resistance values and different capacitor volumes, as shown in Fig. 5.11. The cut-off frequency above 10 Hz is due to the amplifier frequency limitations. The authors also found that increasing the supply pressure significantly reduced the operating bandwidth of the circuit but variations in the amplifier bias pressure levels had no appreciable effect.

5.2 Low-Gain Amplifier Signal Shaping

5.2.1 Inversion, Addition and Subtraction

It is apparent that fluidic servo-amplifier design is somewhat more involved than for the conventional electronic equivalent using operational-amplifier microcircuits and may be compared instead with electronic systems of the 1950's. The chief problems arise from static non-linearities, low frequency dynamics in the amplifiers themselves and also the passive matching resistors. The fluidic servo-amplifier must be considered as an inherent system component particularly for a high performance servo requiring shaping networks in which the bandwidth of the overall system could be greatly affected.

The shaping networks described in this section contain few active elements, so that even a complicated servo-amplifier could be accommodated in a single integrated circuit with considerable power and cost advantages. While not exhibiting such good linearity as a fluidic operational amplifier, the shaping networks have smaller phase shifts which make them attractive for complex circuits and, unlike passive networks, produce 'black-box' dynamic shaping. The majority of even high performance control systems operate with a bandwidth of less than 10 Hz and it has been shown that fluidic circuits using low-gain amplifiers are quite capable of operating up to and above these frequencies, without excessive phase lag.

The frequency response of a Beam-deflexion amplifier has a characteristic in which the phase lag becomes important at lower frequencies than the amplitude ratio degradation. Typical results have already been shown in Fig. 5.5. It is convenient to represent the phase ϕ and amplitude ratio M of the basic elements in the derived networks by the relationship:

$$\frac{P_o}{P_i} = f(j\omega, M, \phi) \tag{5.17}$$

as they are not linear transfer functions. For brevity non-linear element characteristics will be denoted by f in the following equations.

With a push–pull amplifier inversion is achieved simply by changing the sense of the output with respect to the input signal, so that

$$\frac{P_o}{P_i} = -f \tag{5.18}$$

as shown in Fig. 5.12.

Again, using a push–pull amplifier, single ended subtraction is easily arranged by inputs P_1, P_2 to each of the control ports when the output

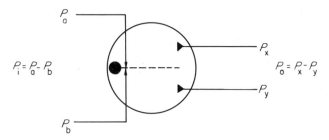

Fig. 5.12 Inversion (Parker and Addy, *Ref. 10*)

P_o will be equal to the product of the forward gain and the input difference. In other words

$$P_o = (P_1 - P_2)\text{f} \qquad (5.19)$$

Addition or subtraction of push–pull inputs is arranged by feeding the outputs of two active elements into respective common outputs, as shown in Fig. 5.13, depending on the input signal sense. The gain of the two amplifiers used in this mode is less than the single element gain and the

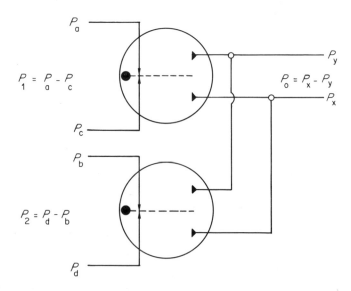

Fig. 5.13 Addition and subtraction (Parker and Addy, *Ref. 10*)

[5.2] LOW-GAIN AMPLIFIER SIGNAL SHAPING

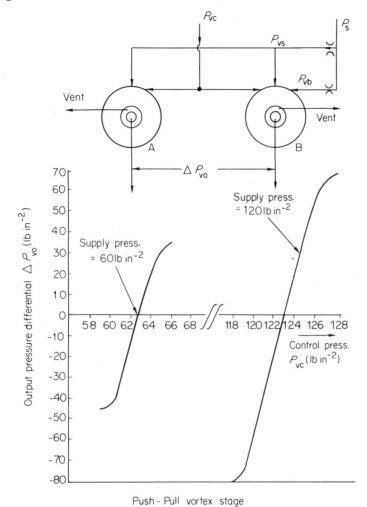

Fig. 5.14 Push-pull operation of vortex amplifiers (Howland, *Ref.* 7, 216)

dynamic response is slightly degraded as shown in Fig. 5.5. For multiple inputs:

$$P_o = [(P_a + P_b) - (P_c + P_d)]f_s \qquad (5.20)$$

where f_s is the modified dynamic characteristic of the summer.

Some of the advantages of using a push–pull circuit include reduced sensitivity to supply pressure changes and noise, insensitivity to both material thermal expansion effects and gas temperature changes, and an improved tolerance to stage supply pressure mismatching.

Although vortex amplifiers are essentially single output devices, they may be constructed with a wide variety of input configurations for different shaping requirements. Fig. 5.14 shows a typical single input vortex amplifier push–pull circuit given by Howland[7]. The single control input P_{vc} is applied to both amplifier A and B so that an increasing control signal causes the impedance of A to increase and B to decrease against the bias signal P_{vb}. If amplifiers A and B are identical and symmetrical the total impedance to supply flow remains constant and independent of control signal level. High amplification can be obtained by cascading pairs of amplifiers in the conventional manner, with the supply pressures to each stage maintained at the design value by restricting the source

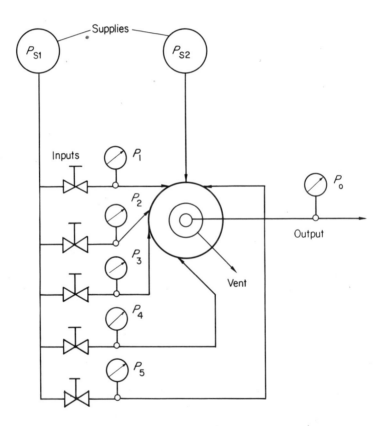

Fig. 5.15 Vortex summing amplifier test configuration (Taplin, Ref. 8, 56)

[5.2] LOW-GAIN AMPLIFIER SIGNAL SHAPING

pressure. The figure also shows the good linearity possible over a large percentage of the useful range, over a wide range of supply pressures.

Taplin[8] describes a proportional vortex amplifier, with five isolated inputs, acting as a summing amplifier. Fig. 5.15 illustrates the input connexions of the amplifier and Fig. 5.16 shows typical summing

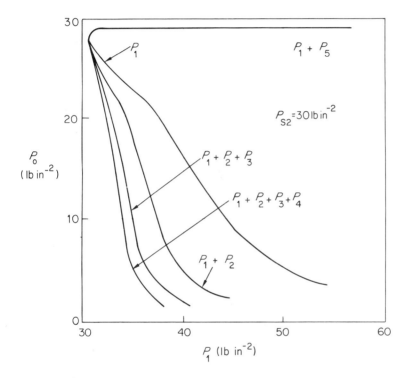

Fig. 5.16 Typical vortex summing amplifier static characteristics (Taplin, Ref. 8, 56)

characteristics obtained from it. All the inputs have the same incremental pressure gain of about 1·5, but the control ports 1 to 4 introduce clockwise chamber swirl, while the remaining port, 5, produces anti-clockwise swirl. This means that the four similarly signed inputs add together, whereas the fifth one subtracts from the output. The characteristics exhibit only approximate static linearity, with marked deterioration with increasing signal pressure. The amplifier would appear to be more useful for dynamic signal shaping as compensation networks can be readily attached to the appropriate control ports.

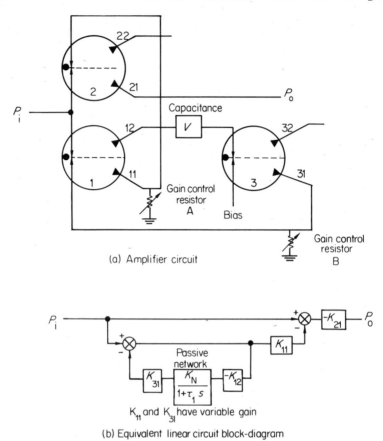

Fig. 5.17 Integrator and lag-lead circuit using low gain amplifiers

5.2.2 Integration and Lag-lead Networks

There have been several references to fluidic integrator circuits in the literature but the only specifically described circuits are those due to Tsui and Stone[9] also Parker and Addy[10]. The former reference gives details of an integrator using a single active Beam-deflexion amplifier with positive feedback and an effective summing junction. Experimental results are given which show that the output increases at a rate proportional to the change of input. However, no information was available on the frequency response and output loading characteristics of the circuit.

Parker and Addy found that a balanced 'bootstrap integrator' configuration based on Beam-deflexion amplifiers as shown in Fig. 5.17, was

[5.2] LOW-GAIN AMPLIFIER SIGNAL SHAPING

compatible with the input and output impedance requirements of fluidic shaping networks. Amplifier 2 in the figure forms the final stage thus eliminating load interactions and the isolated input has an impedance equal to half that of a single active element, a quite acceptable value for loading the preceding stage.

As the transfer characteristics for the amplifiers may be slightly different for each output line, the output connexions are labelled in the figure, firstly, for the amplifier number and, secondly, for the line number. For example, the lower output connexion for amplifier 2 is labelled 21. Also shown in Fig. 5.17 is the approximate linear equivalent circuit with the amplifiers represented as linear elements of a static gain K. Analysis of the circuit gives

$$\frac{P_o}{P_i} = -f_{21}\left[\frac{(1 - f_{12}f_{31}K_N - f_{11}) + (1 - f_{11})\tau_1 s}{(1 - f_{12}f_{31}K_N) + \tau_1 s}\right] \quad (5.21)$$

where $\tau_1 = VR_{o1}R_{i3}/nP_m(R_{o1} + R_{i3})$, the passive network time constant
$K_N = R_{i3}/(R_{o1} + R_{i3})$, the passive network gain
P_m = capacitance mean pressure
R_{o1} = total output resistance for upper line of amplifier 1
R_{i3} = total input resistance for the upper line of amplifier 3.

For low frequency analysis the transfer function f for each amplifier may be replaced by their static gain K. In the circuit f_{11} and f_{31} are considered to have variable gains as the variable shunt resistors provide approximate static impedance matching by attenuating the output signal without significantly altering the dynamic characteristics. Two types of network result depending on the value of the loop gain, $f_{12}f_{31}K_N$, selected.

If $f_{11} = 1$ and $f_{12}f_{31}K_N = 1$ then we have an integrator:

$$\frac{P_o}{P_i} = \frac{f_{21}}{\tau_1 s} \quad (5.22)$$

However, if $f_{11} = 1$ and $f_{12}f_{31}K_N < 1$ then:

$$\frac{P_o}{P_i} = \frac{f_{12}f_{21}f_{11}K_N}{(1 - f_{12}f_{31}K_N)} \frac{1}{(1 + \tau_2 s)} \quad (5.23)$$

which represents a lag network with a time constant τ_2 given by

$$\tau_2 = \frac{\tau_1}{1 - f_{12}f_{31}K_N} \quad (5.24)$$

Although this circuit uses positive feedback, with resultant difficulties in accurate setting to obtain an integrator, it is found that with f_{11} set slightly above unity gain, using control resistor A, the circuit can be adjusted by using control resistor B only. Starting with a loop gain of less than unity is desirable for final adjustment.

The circuit may be analysed by using an equivalent electrical network and neglecting the dynamics of the active elements. This can be an aid to understanding the operation of the circuit although active-element dynamics may have to be considered in a complete system in a similar way to the operational-amplifier analysis. The circuit can be simulated on an analog computer using first order lags to represent the amplifier functions f_{11}, f_{12}, f_{21}, f_{31}. A comparison of the frequency response obtained with that of an experimental integrator circuit is shown in Fig. 5.18.

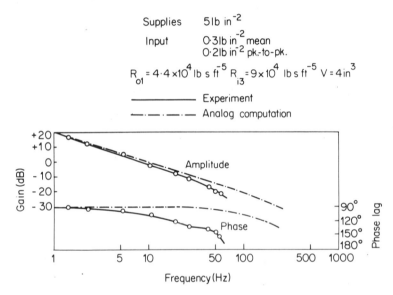

Fig. 5.18 Integrator frequency response (Parker and Addy, *Ref. 10*)

It is evident that the low frequency correlation is reasonable but the phase lag is more severe at higher frequencies for the fluidic circuit. This is due to the circuit being representative of an operating circuit in which no extreme care would be taken to ensure that the gains of the amplifiers 1 and 3 were as close to unity as possible. If the extraneous phase shift of 25° with only 1 dB amplitude degradation at 10 Hz is considered unsuitable for a system, the integrator can be set more accurately or lead compensation introduced.

Boothe[11] has suggested a lag-lead circuit based on positive feedforward

[5.2] LOW-GAIN AMPLIFIER SIGNAL SHAPING

which does not need critical gain adjustment, and can readily be used for single port or push–pull configurations. Fig. 5.19 illustrates the circuit and equivalent linear block diagram. The relationship between the output differential pressure signal ΔP_o and the input differential signal ΔP_i may be expressed as

$$\frac{\Delta P_o}{\Delta P_i} = -f_3(1 + f_1 f_2 K_N)\left[\frac{1 + \tau_2 s}{1 + \tau_1 s}\right] \qquad (5.25)$$

where $\tau_2 = \tau_1/(1 + f_1 f_2 K_N)$, the lead time constant
$\tau_1 = V R_{o1} R_{i2}/n P_m (R_{o1} + R_{i2})$
$K_N = R_{i2}/(R_{o1} + R_{i2})$
R_{o1} = total output resistance for each line of amplifier 1
R_{i2} = total input resistance for each line of amplifier 2.

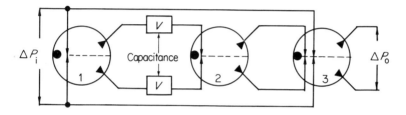

(a) Amplifier circuit

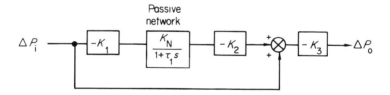

(b) Equivalent linear circuit Block diagram

Fig. 5.19 Lag-lead circuit using positive feedforward

As the interstage lines between amplifiers 1 and 2 should have identical characteristics for differential signal operation, some volume and shunt resistance adjustment may have to be incorporated in the circuit.

Using a vortex amplifier summing amplifier, a lag-lead circuit may be constructed using resistance and volume networks connected to two ports which are additive, as illustrated in Fig. 5.20. One control line contains a resistance and a capacitance while the other contains only a resistance. For any steady input signal the output is at a low level and remains at this value until the break frequency of the resistance-capacity network is exceeded. At this point the phase lag of this control signal becomes significant and the output signal starts to rise by an amount proportional to the time derivative of the input; a characteristic expected of a negative gain amplifier.

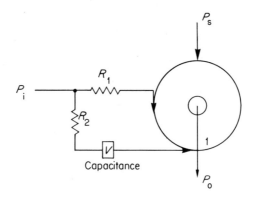

(a) Amplifier circuit

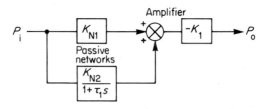

(b) Equivalent linear circuit block-diagram

Fig. 5.20 Vortex amplifier lag network

5.2.3 *Differentiation and Lead Networks*

There have been references to lead-lag networks[11], using low gain amplifiers which produce up to 60 degrees phase lead but there has been little information related to pure D.C. blocking or differentiation networks. This may be an important requirement in a circuit and cannot be even approximately implemented with passive fluidic components, as is the case for lag networks.

[5.2] LOW-GAIN AMPLIFIER SIGNAL SHAPING

A circuit which satisfies the requirements for differentiation using low gain Beam-deflexion amplifiers is shown in Fig. 5.21 and has been described briefly by various authors[2,10,12]. The transfer function for the circuit is:

$$\frac{P_o}{P_i} = f_{32}\left[\frac{f_{11}K_{N1}(1 + \tau_2 s) - f_{22}K_{N2}(1 + \tau_1 s)}{(1 + \tau_1 s)(1 + \tau_2 s)}\right] \quad (5.26)$$

where $\tau_1 = V_1 R_{o1} R_{i3}/nP_m(R_{o1} + R_{i3})$
$\tau_2 = V_2 R_{o2} R_{i3}/nP_m(R_{o2} + R_{i3})$
$K_{N1} = R_{i3}/(R_{o1} + R_{i3})$
$K_{N2} = R_{i3}/(R_{o2} + R_{i3})$

the notation being the same as for (5.21).

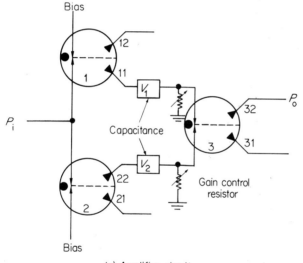

(a) Amplifier circuit

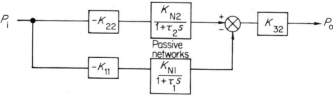

(b) Equivalent linear circuit block-diagram

Fig. 5.21 D.C. Blocking lead-lag network

If the gains through each of two input paths are arranged to be identical so that $f_{11}K_{N1} = f_{22}K_{N2}$ and one of the time constants of the passive networks is significantly greater than the other, say $\tau_2 \gg \tau_1$, then (5.26) becomes

$$\frac{P_o}{P_i} = f_{11}f_{32}K_{N1}\left(\frac{\tau_2 s}{1 + \tau_2 s}\right) \qquad (5.27)$$

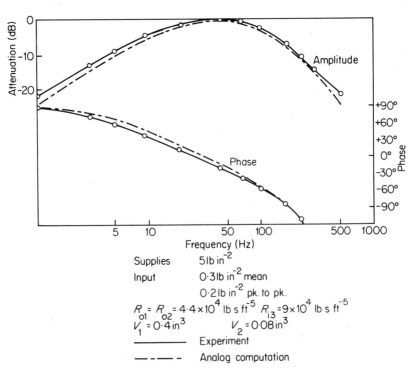

Fig. 5.22 Differentiator frequency response (Parker and Addy, *Ref. 10*)

giving a lead circuit with break frequency $1/\tau_2$. If the break frequency is high the circuit approximates to a pure differentiator with D.C. blocking capability. To realize this circuit fluidically requires the maintenance of equal static restrictive loading on the two inputs of final stage amplifier 3, but also significantly different time constants associated with these inputs. This requires that each input connexion has the same flow cross-sectional area, and time constant variations are made by, say, keeping volume V_1 to a minimum and adjusting volume V_2.

A normal lead-lag circuit may be constructed with any desired low

[5.2] LOW-GAIN AMPLIFIER SIGNAL SHAPING   249

frequency break by altering the interstage gains and volumes. If $f_{11}K_{N1} > f_{22}K_{N2}$ and $\tau_2 \gg \tau_1$ then

$$\frac{P_o}{P_i} = f_{32}(f_{11}K_{N1} - f_{22}K_{N2})\left[\frac{1 + \tau_3 s}{1 + \tau_2 s}\right] \quad (5.28)$$

where $\tau_3 = \tau_2/(f_{11}K_{N1} - f_{22}K_{N2})$.

Usually the lag time constant τ_2 is chosen so that the desirable feature of high frequency noise supression is obtained.

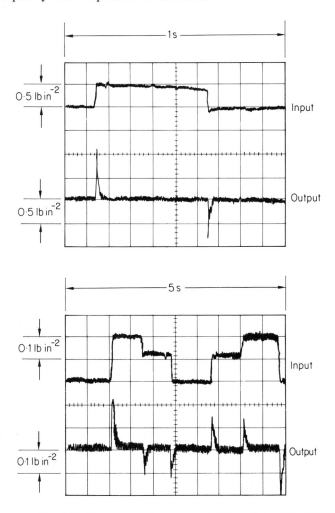

Fig. 5.23 Experimental response of differentiator (Fig. 5.21) to step inputs (Parker and Addy, *Ref. 10*)

The circuit given by (5.27) was simulated by Addy on an analog computer using first order lags for the amplifier functions and the frequency response compared to that obtained from a typical fluidic circuit configuration. This is presented in Fig. 5.22 where the correlation between experiment and computer model is found to be very good. This is fortuitous in that the exponential phase lag caused by the jet transport time in the final amplifier improves the correlation though not taken into account in the model. The circuit is not critical to set up and the low frequency characteristics are very adequate for D.C. blocking or differentiation. The response of the circuit to a step input is shown in Fig. 5.23, and indicates satisfactory performance.

Boothe[11] has described a further circuit for lead-lag shaping using low gain amplifiers, which is particularly useful for circuits requiring differential signals. Fig. 5.24 shows the circuit proposed by Boothe which, in

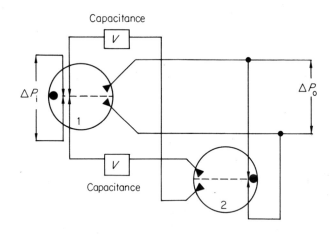

(a) Amplifier circuit

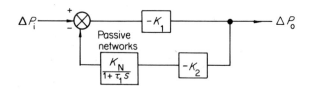

(b) Equivalent linear circuit Block diagram

Fig. 5.24 Lead-lag circuit

[5.2] LOW-GAIN AMPLIFIER SIGNAL SHAPING

practice, may require an addition amplifier at the output in order to effectively isolate the feedback path through amplifier 2 from loading effects. The relationship between the output pressure ΔP_o and the input pressure signal ΔP_i is

$$\frac{\Delta P_o}{\Delta P_i} = -\frac{f_1}{(1 + f_1 f_2 K_N)} \cdot \left[\frac{1 + \tau_1 s}{1 + \tau_2 s}\right] \quad (5.29)$$

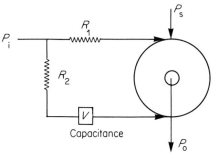

(a) Amplifier circuit

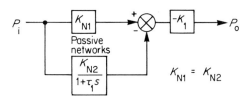

(b) Equivalent linear circuit block-diagram

Fig. 5.25 Vortex amplifier lead network

where $\tau_1 = V R_{o2} R_{i1}/n P_m (R_{o2} + R_{i1})$
$K_N = R_{i1}/(R_{o2} + R_{i1})$
R_{o2} = total output resistance for each line of amplifier 2
R_{i1} = total input resistance for each line of amplifier 1
$\tau_2 = \tau_1/(1 + f_1 f_2 K_N)$

(5.29) shows that the gain is less than unity which indicates the desirability of an output amplifier. As differential signals are used, both interstage passive networks should have an identical time constant τ_1 and gain K_N

which may necessitate the use of adjustable volumes and shunt resistances in the feedback lines.

A vortex amplifier may be constructed with a resistance-volume network to give a low frequency lead network in a rather similar manner to the lag network previously described. In this case the resistance-volume line is connected to provide anticlockwise swirl in opposition to the resistance line as shown in Fig. 5.25. For any steady input signal the two control ports are balanced and the output flow or pressure is independent of the input signal level. For a dynamic input signal above the break frequency of the network the clockwise swirl component starts to dominate as the capacitance causes the pressure in the anticlockwise port to lag. This causes the output pressure to be reduced by an amount proportional to the time derivative of the input signal required by a negative gain amplifier.

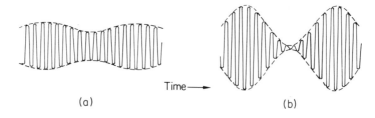

Fig. 5.26 Amplitude modulation for (a) a small-amplitude, and (b) a large-amplitude modulating signal of sinusoidal waveform

5.3 Frequency Modulation and Phase Discrimination

5.3.1 General

A carrier wave has three parameters which can be varied to provide modulation; that is its amplitude, phase and frequency. Although all three types of modulation can occur simultaneously, systems are usually designed to achieve one type and minimize the other two. In amplitude modulation (A.M.) the carrier amplitude varies above and below its average or unmodulated value at the same frequency as the modulating signal, as shown in Fig. 5.26. The variations in carrier amplitude are proportional to the modulating signal-amplitude. The second diagram in the figure shows that the limit of modulation is reached when the carrier amplitude excursion is between twice its average unmodulated value and zero. Apart from this inherent limitation to A.M. systems, they are not attractive for fluidic applications as most fluid networks are very amplitude

[5.3] FREQUENCY MODULATION AND PHASE DISCRIMINATION 253

sensitive, even at modest frequencies, thereby introducing significant signal distortions.

In phase modulation (P.M.) the phase angle varies with signal amplitude, above and below its unmodulated value, at the frequency of the modulating signal. The resulting phase-modulated carrier is similar to a frequency-modulated carrier. Frequency modulation (F.M.) requires the frequency to vary above and below its average value at the frequency of the modulating signal, as shown in Fig. 5.27. The operating range of frequency modulation is usually small compared with the carrier frequency (less than 1/1000th of the carrier frequency) but there is no theoretical limit to this range as in A.M. systems. The amplitude of the F.M. wave is nominally the same as the unmodulated wave although variations can be tolerated provided significant sideband frequencies are not appreciably attenuated—causing harmonic distortion of the detected signal.

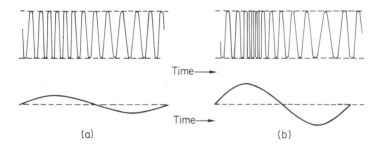

Fig. 5.27 Frequency modulation for (a) a small-amplitude and (b) a large-amplitude modulating signal of sinusoidal waveform

Thus both P.M. and F.M. techniques may be used in fluid circuits provided the fluid system components are capable of responding to the frequencies involved. As the Beam-deflexion amplifier may be operated on air in the frequency range 1 to 4 kHz by using existing amplifier designs, acceptable frequency and phase modulation ranges may be obtained for pneumatic control systems. In addition noise problems are minimized and a variety of control parameters, such as position, temperature, and pressure, can be sensed under a wide variety of environmental conditions.

5.3.2 Temperature Measurement

Two types of carrier signal techniques have been described in the literature for dynamic temperature control systems. Kelley[13] has used F.M. and phase discrimination to measure temperature in a circuit using beat-frequency detection. A hybrid technique based on beat-frequency detection together with a digital frequency converter has been used by Halbach et al.[14] and Waters[15]. In this Section the approach of Kelley will be described while further discussion of the other references is to be found in Section 5.4.4.

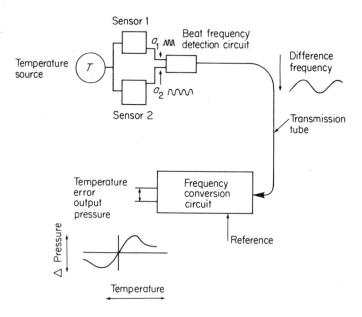

Fig. 5.28 Schematic of a corrected temperature-control system
(Kelley, *Ref. 13*, 123)

Fig. 5.28 shows diagrammatically the temperature-control system described by Kelley for use in a control loop of an aircraft gas-turbine. It uses two temperature-sensitive oscillators, a beat-frequency detector circuit and a phase-discrimination circuit to provide a temperature error signal. Pure fluidic oscillators have a frequency output proportional to the square root of absolute temperature, and may be designed to operate at high ambient temperatures in adverse environments. Several different types of pneumatic oscillators for temperature sensing have been reported. In Refs. 15 and 16 a sonic delay time is utilized in negative feedback loops of bistable amplifiers to generate the frequency signal. Other techniques include acoustic coupling of edgetone frequencies with cavity resonant frequencies[15], and the excitation of two resonant tube frequencies, one open and the other nominally closed[14]. Fig. 5.29 shows a temperature sensitive oscillator operating on the latter principle. Flow is divided by the knife into the feedback and output resonance channels which causes

[5.3] FREQUENCY MODULATION AND PHASE DISCRIMINATION

self-excited oscillations at the natural frequency of the coupled resonance channels. Optimum performance occurs when the feedback-channel length is about twice the output-channel length.

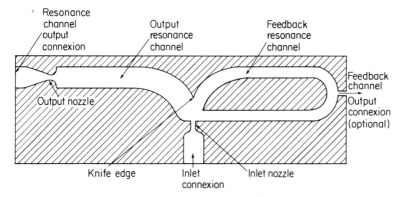

Fig. 5.29 A self-excited pressure insensitive fluidic oscillator (Halbach, *Ref. 14*, 303)

The beat-frequency technique is necessary to obtain satisfactory resolution of a temperature signal. This is because the frequency converter can be utilized over its full range when using a beat-frequency signal, whereas only partial utilization results from using the oscillator frequency directly. For example, if the oscillator frequency varies from 800 to 1000 Hz over the operating temperature range, a direct coupled frequency converter operating in the range 0–1000 Hz is required with operation over only one-fifth of its full range. With beat-frequency detection, only a 0–200 Hz converter is required to operate over its full range giving an improvement of 5:1 in resolution.

Beat-frequency detection is achieved by summing two signals of slightly different frequency and then removing the difference frequency from the resulting signal. Fig. 5.30 shows the detector used by Kelley which consists of a Beam-deflexion amplifier in series with a rectifier element. Inputs a_1 and a_2 receive slightly different frequency signals from oscillator temperature sensors 1 and 2 (see Fig. 5.28), which produces low frequency modulation of the carrier wave as depicted by signal 'c' in Fig. 5.30. The rectifier, which has only one receiver on the jet centre-line, has static characteristics as shown in Fig. 5.31. The element has an average output level which is negatively proportional to the magnitude of the signal for sinusoidal inputs, as illustrated in Fig. 5.32. Returning to Fig. 5.30, the rectified signal 'd' is seen to have a similar form although it has the carrier

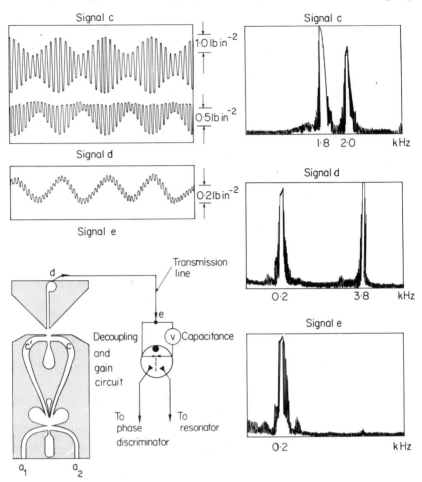

Fig. 5.30 Measured beat frequency detector waveforms (Kelley, *Ref. 13*, 125)

frequency superimposed on it. This is filtered down to an acceptable level in a five foot long transmission line so that the signal '*e*' entering the decoupling circuit is essentially the low frequency beat signal alone. The decoupling circuit provides D.C. blocking of signal '*e*' together with amplification in a further Beam-deflexion amplifier. One output signal from the amplifier passes into a pneumatic resonator circuit while the other is connected directly to a phase discriminator.

[5.3] FREQUENCY MODULATION AND PHASE DISCRIMINATION

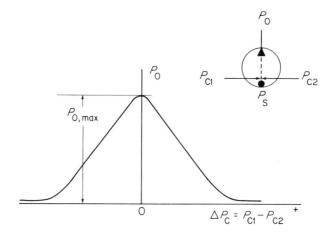

Fig. 5.31 Rectifier input-output static characteristic (Kelley, *Ref. 13*, 124)

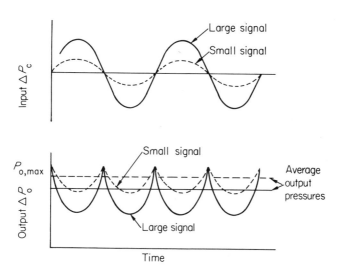

Fig. 5.32 Rectifier input and output waveforms for A.C. operation (Kelley, *Ref. 13*, 124)

The phase discriminator provides an output analog pressure proportional to the difference in phase of two sinusoidal input signals. If the signals are 90° out of phase there is no error and deviations from this phase angle give positive and negative error signals. Fig. 5.33 shows the

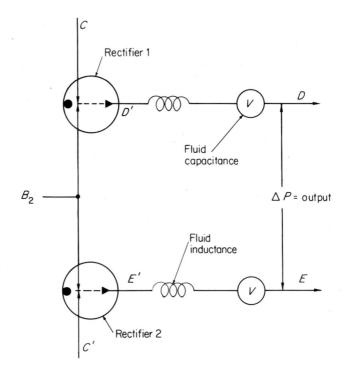

Fig. 5.33 Phase discriminator circuit (Kelley, *Ref. 13*, 126)

phase discriminator circuit used by Kelley which consists of two rectifiers with inductance-capacitance output filters to give a D.C. output pressure. Inputs C and C' in the figure are push–pull signals always 180° out of phase which are compared in phase with signal B_2. If a 90° phase relationship exists between the signals C, C' and B_2, the vector diagram shown in (a) of Fig. 5.34 results. Vectors $(B_2 - C)$ and $(B_2 - C')$ are the rectifier inputs, which are equal in this case. Cases (b) and (c) illustrate when output D is greater and less than output E respectively.

The complete frequency-conversion circuit is shown in Fig. 5.35 in which the push–pull signals C and C' are generated by an adjustable pneumatic resonator, as described in Chapter 2. The initial 90° phase

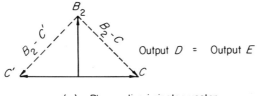

(a) Phase discriminator vector diagram for zero error

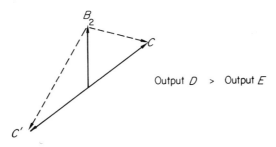

(b) Phase discriminator vector diagram for positive error

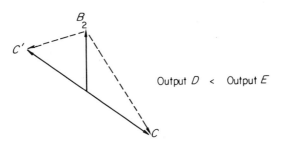

(c) Phase discriminator vector diagram for negative error

Fig. 5.34 Phase discriminator vector diagrams (Kelley, *Ref. 13*, 127)

shift required for zero error is inserted in the amplifier stages, as the resonator has zero phase shift at resonance. When the frequency of the temperature signal varies above or below the resonator natural frequency a phase shift of C and C' occurs relative to B_2, which by-passes this circuit, thereby producing positive and negative analog error-signals.

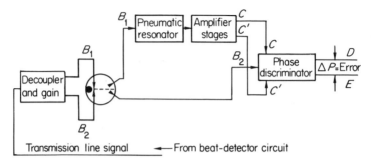

Fig. 5.35 Block diagram of a temperature error circuit (Kelley, *Ref. 13*, 127)

Kelley gives experimental results for the system described above using temperatures up to 2000°F. Fig. 5.36 shows output curves for different reference settings corresponding to resonant frequencies varied by altering the volume. The dynamic response of the system is given in Fig. 5.37 to frequency-modulated input signals. The phase shift is the amount the output lags the input, and the bandwidth is defined for 45° phase lag. For carrier frequencies (f_c) between 270 and 440 Hz the bandwidth varied between 4·5 and 11 Hz. For circuits using 2 kHz carrier frequencies it is anticipated that a bandwidth of 50 Hz is possible.

5.3.3 Speed Measurement

Speed is more easily sensed than temperature in fluidic systems, as a frequency proportional to speed is readily generated from a rotating shaft.

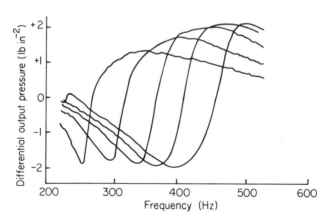

Fig. 5.36 Output curves of high-temperature error circuit (Kelley, *Ref. 13*, 129)

[5.3] FREQUENCY MODULATION AND PHASE DISCRIMINATION

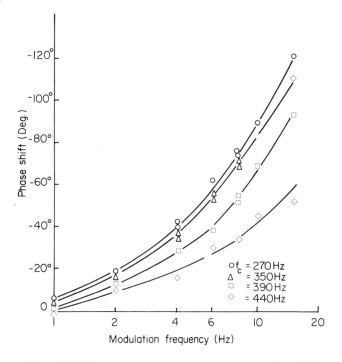

Fig. 5.37 Dynamic response measurements of a high-temperature error circuit (Kelley, *Ref. 13*, 129)

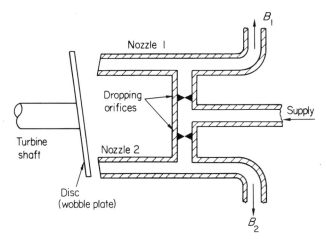

Fig. 5.38 Wobble-plate speed signal generator (Boothe, Ringwall and Shinn, *Ref. 2*, Fig. 1)

Frequency modulation and phase discrimination techniques may be readily used for fluidic speed-control systems for this reason and result in relatively simple circuits.

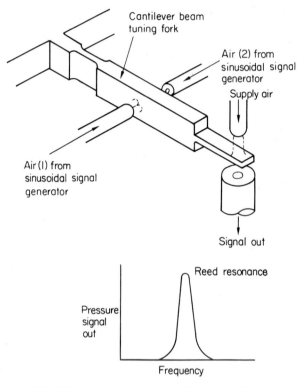

Fig. 5.39 Vibrating reed and pick-off (Boothe, Ringwall and Shinn, *Ref.* 2, Fig. 2)

Boothe, Ringwall and Shinn[2] describe three carrier approaches for control of shaft speed: the reed or tuning fork speed sensor, the pneumatic resonator coupled to a phase discriminator and a pneumatic derivative governor. In all systems the speed signal generator is a 'wobble plate' alternately restricting two back pressure sensing nozzles, as shown in Fig. 5.38. Output pressure signals B_1 and B_2 from the sensors produce sinusoidal waves 180° out of phase, superimposed on a D.C. level. A pneumatic turbine is often used to drive the plate in order to obtain a wide operating frequency range.

The reed sensor concept is appropriate to systems operating at fixed speeds under varying load conditions. Fig. 5.39 shows how the signal

[5.3] FREQUENCY MODULATION AND PHASE DISCRIMINATION 263

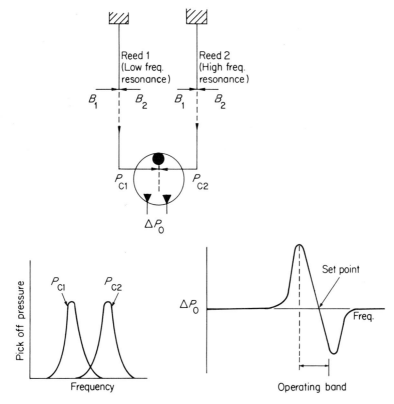

Fig. 5.40 Basic reed speed-governor circuit (Boothe, Ringwall and Shinn, *Ref. 2*, Fig. 3)

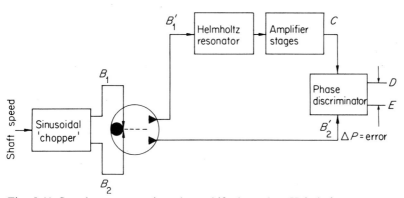

Fig. 5.41 Speed-governor using phase shift through a Helmholtz resonator (Boothe, Ringwall and Shinn, *Ref. 2*, Fig. 8)

generator outputs are applied to the tuning fork to induce vibration. As the speed signal approaches the fork natural frequency, the vibration amplitudes increase thereby intermittently opening a path from a supply nozzle to a receiver, in the other plane. As the fork goes through resonance the pressure signal in the receiver goes through a maximum value. When two forks are tuned to slightly different frequencies and their outputs fed to a Beam-deflexion amplifier, as shown in Fig. 5.40, subtraction of the two signals occurs. The figure also shows the resulting characteristic output differential pressure which is arranged so that the zero output point is half-way between the fork frequencies.

By introducing the speed-signal generator outputs into a resonator and phase discriminator circuit, speed-control can be achieved in a similar fashion to the temperature sensing circuit described earlier. Fig. 5.41 illustrates the circuit configuration. The sharpness of the resonator tuning governs the steepness of the phase-frequency characteristic. A high 'Q' resonator exhibits an abrupt phase change at resonance and so is most useful for narrow range, high accuracy systems. A low 'Q' resonator has the opposite characteristics. The authors give results for a typical circuit using three different resonator volume settings, as shown in Fig. 5.42. Closed loop speed-control to a fraction of 1 per cent is reported using this approach.

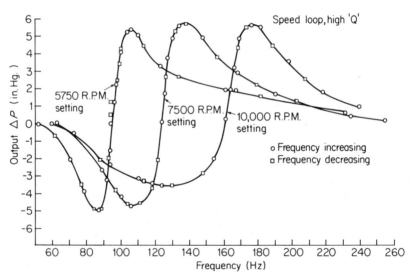

Fig. 5.42 Pneumatic resonator speed-control output characteristics (Boothe, Ringwall and Shinn, *Ref.* 2, Fig. 12)

5.4 Pulse-width Modulation

5.4.1 General

Instead of using proportional amplifiers operating with sinusoidal signals to give carrier wave modulation, digital amplifiers may be used to moderate square wave signals. Although in electronic systems standard techniques exist to transmit very complex information, using coded pattern square wave signals, they may not be used in fluidic systems. This is because fluidic digital switching elements are extremely slow compared with their electronic counterpart, thereby causing the information to be transmitted at an unacceptably slow rate. However, one simple technique used in electronics, known as pulse-width modulation (P.W.M.) has found application in some fluidic circuits.

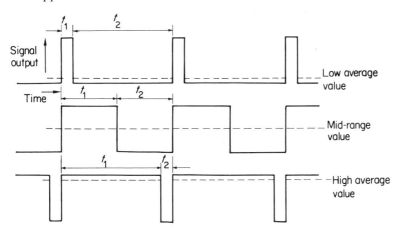

Fig. 5.43 Basic pulse-width modulation (P.W.M.)

Fig. 5.43 shows three different square waveforms in which the positive or 'switched-on' part of each cycle is varied. Starting with a very narrow positive pulse width, the waveform may pass through a symmetrical configuration to a waveform which is positive practically all the cycle. If the waveform is integrated over a long time period it will provide an output signal proportional to the area under the waveform of each cycle. It is possible to sense a physical quantity so that its magnitude is directly proportional to the pulse width of a constant amplitude waveform. The average value of the pulse waveform is the analog output-signal which is proportional to the analog input-signal.

The first and last waveform in the figure show the practical limitations placed on the signal range, which is usually selected to be 10 to 90 per cent of the total maximum cycle area for electronic systems. In a fluidic system using wall re-attachment elements, it is difficult even with highly miniaturized circuits to operate with pulse widths of less than 0·5 ms which places an upper frequency limitation on such systems. The figure shows t_1 as the switched-on time and t_2 as the switched-off time, so that the wave frequency f is

$$f = \frac{1}{t_1 + t_2} \tag{5.29}$$

The maximum frequency occurs when the pulse width is a minimum, i.e.

$$t_1 = t_2 = t_{min}$$

$$f_{max} = \frac{1}{2t_{min}} \tag{5.30}$$

although under these conditions there is no operating range as the pulse width cannot be varied. This means that the upper frequency limitation for present-generation fluidic systems using this technique would be less than 1000 Hz. A more practical frequency is 200 Hz when the corresponding operating range might be 9:1.

Basically two modes of operation are possible. A waveform may be generated in which the positive wave width is held constant in time while the frequency varies. This allows at least a 20:1 operating range at low frequencies provided it is acceptable to work outside the 10 to 90 per cent area range. The other mode is to modulate a fixed frequency by varying the time width of the pulses. For pure fluidic systems it is not convenient, because of the excessive size of the components, to generate pulse widths much outside the range of 0·5 to 4 ms which is equivalent to an 8:1 range.

5.4.2 Jet Reaction Applications

There have been several references to jet reaction stabilization of missiles using P.W.M.[18, 19, 20] Conventional digital on-off control techniques, often known as bang-bang, have been tried for stabilization of missiles using pairs of jet reaction nozzles. This is not particularly successful due to the large overshoots involved resulting in missile oscillation about the desired direction. However, if the reaction jets are supplied from a fixed-frequency oscillator so that they alternately emit from either side, control can be achieved if the oscillating jets are made to pause for a longer or shorter time proportional to the desired correction. If the pulse-width modulated wave has a frequency more than twice the natural frequency of the missile, the inertia effectively smooths out the carrier frequency ripple.

[5.4] PULSE-WIDTH MODULATION

Warren[18] first described fluidic P.W.M. for this type of application and constructed a simple system as shown in Fig. 5.44. A simple negative feedback oscillator supplies the bistable wall re-attachment power amplifier at constant frequency. The output channels of the power

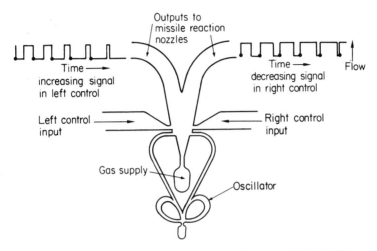

Fig. 5.44 Fluidic pulse-width modulation (Warren, *Ref. 18*, 45)

amplifier lead directly into opposing jet reaction nozzles situated on the missile surface. The analog input error-signals are fed directly into the separation bubble of the bistable amplifier through two additional control ports. This has the effect of slowing down or increasing the rate at which the bubble pressure increases during switching, and effectively varies the output pulse width. The author reports operation up to 250 Hz. Considerable power levels may be required from the reaction nozzles in an actual missile. Campagnuolo and Sieracki[19] have investigated a similar system to Warren but including a three stage digital-amplifier power-stage giving a proportional flow gain of about 500. They give a curve showing the overall performance of the system, as seen in Fig. 5.45, for oscillator frequencies in the range 40 to 55 Hz. A hatched corrected curve for null-balance is shown in the figure which has been achieved by slightly biasing the oscillator output to overcome geometric asymmetry in the amplifier.

A complete anti-tank missile fluidic control system using P.W.M. is described by Eastman[20]. Fig. 5.46 shows the control system. The guidance demand signal is transmitted to the missile via a control wire and converted to a pneumatic low pressure signal. This signal is compared with a gyro pressure signal and the analog error signal is applied differentially

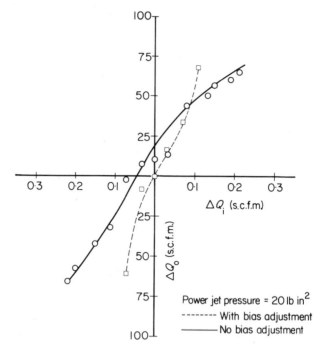

Fig. 5.45 Flow gain characteristics of a P.W.M. missile control system (Compagnuolo and Sieracki, *Ref. 19*, 158)

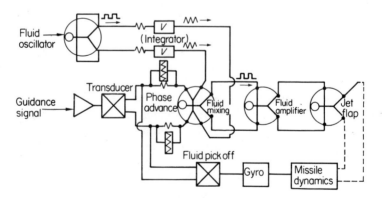

Fig. 5.46 A fluidic P.W.M. missile control system (Eastman, *Ref. 20*, Fig. 1)

[5.4] PULSE-WIDTH MODULATION 269

to the bistable wall re-attachment modulator element after passing through a phase advance network. This network is necessary to provide an adequate stability margin for the missile control loop. The other pair of inputs to the modulator are connected to a fluidic oscillator with a waveform modified to a triangular shape by an integrator network. This provides better control of the pulse-width variation in the modulating element. The P.W.M. signal is amplified by several stages of digital elements and emerges from the missile through jet flap switches situated in the missile wings.

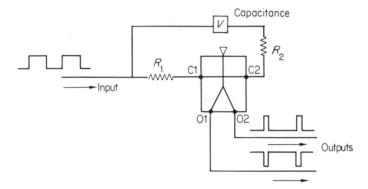

Fig. 5.47 Resistive biased pulse shaper

5.4.3 Speed Measurement

Various other P.W.M. systems have been suggested, including simple speed and torque measuring circuits. In a review of a variety of methods for speed sensing, Foster and Cleife[21] include some simple experiments on a P.W.M. system. A slotted disc is rotated so as to interrupt an air jet between a nozzle and receiver thus providing a rectangular waveform, the frequency and width of which depend on the speed of rotation. The pulse train is fed from the receiver into a pulse shaper which gives an output pulse of constant amplitude and width at the same frequency as the input. A passive resistance-capacitance network averages the pulse train so that the pressure level is proportional to the pulse frequency and hence the rotational speed.

The pulse shaper is shown in Fig. 5.47 and was found to operate reliably at a minimum pulse width of 1·5 ms. The input signal is connected directly to both control ports C1 and C2 of the bistable amplifier but the volume-resistance network delays the signal arrival at C2 relative to C1. This means that if the amplifier output is initially at 01 a positive input

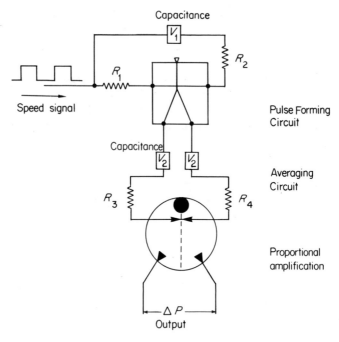

Fig. 5.48 A speed-governor circuit using complementary pulse shapers in push-pull

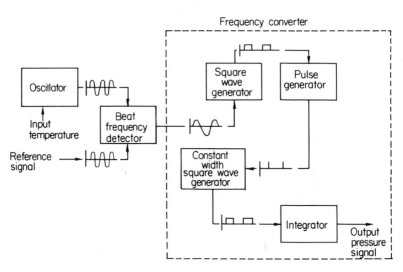

Fig. 5.49 Hybrid fluidic temperature-sensing circuit (Halbach, Otsap and Thomas, *Ref. 14*, 299)

[5.4] PULSE-WIDTH MODULATION 271

signal at C1 will switch the output to 02 for a fixed time period before the same control signal arrives at C2. Normally $R_1 > R_2$ so that the C2 signal overrides the C1 signal thereby switching the output back to 01. As it has two complimentary outputs it would lend itself well to speed governing. A suggested circuit is shown in Fig. 5.48 in which both outputs of the pulse shaper feed into the control ports of a Beam-deflexion amplifier in push–pull. Zero error at a desired frequency occurs when the output from the pulse shaper is a symmetrical square wave. Experiments by the authors using a single ended output showed a linear output pressure range over a frequency range of 50 to 150 Hz. The lower frequency is limited by ripple in the smoothing network.

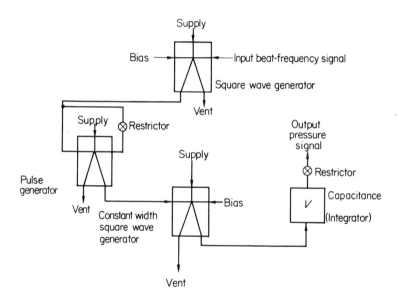

Fig. 5.50 Fluidic frequency converter (Halbach, Otsap and Thomas, *Ref. 14*, 306)

5.4.4 Hybrid Techniques

The work of Halbach, Otsap and Thomas[14] has already been mentioned in connexion with beat frequency detection of a temperature sensitive oscillator. However, instead of using phase discrimination to demodulate the temperature signal, they use digital techniques to produce an analog output pressure signal. Fig. 5.49 shows the circuit used. The beat frequency detector output wave is converted into a variable width square wave using a biased bistable amplifier and the square wave is differentiated to form constant width short duration pulses. A third digital amplifier is used to stretch the pulses to form a constant width longer duration pulse, which is then integrated in a passive volume-resistance network. Fig. 5.50 shows the fluidic wall re-attachment elements used in the frequency converter circuit.

Waters[15] has used a slightly different approach to digital demodulation although still using a beat frequency detector for temperature sensing. Fig. 5.51 shows the demodulation circuit used which consists of a monostable clipping stage and a pulse-width modulator. The clipping stage generates two square waves with approximately 180° phase difference from the beat frequency wave. One of the signals is transmitted directly to a two input OR wall re-attachment element, acting as the pulse-width modulator, while the other signal passes through a delay line before entering the OR element. The delay line creates a constant delay equal

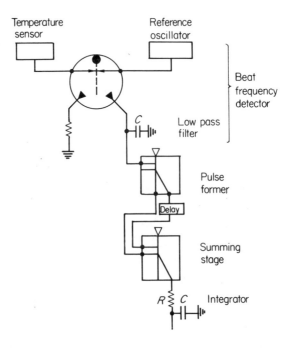

Fig. 5.51 Hybrid temperature-sensing circuit using digital demodulation

to one-half of a cycle at the beat frequency for zero error signal (usually the midpoint of the beat frequency range) in the system described by the author. The pulse-width modulator output is a symmetrical square wave at the beat frequency range midpoint. As the 'on' time is a function of the beat frequency, when the frequency decreases the 'on' time increases while the 'off' time remains constant. This increases the average pressure level produced by the integrator circuit.

References

1. Rose, R. K. and Phipps, W. L., Fluidic control of a J79 turbojet engine. *A.S.M.E. Paper* 67-WA/FE-33.
2. Boothe, W. A., Ringwall, C. G. and Shinn, J. N., New fluid amplifier techniques for speed control. *S.A.E. Aerospace Fluid Power Systems and Equipment Conf.*, **1965**, 5.
3. Urbanosky, T. F. and Doherty, M. C., Fluidic operational amplifiers. A new control tool. *Control Eng.*, **1967**, 57.
4. Urbanosky, T. F., Fluidic operational amplifier survey. *S.A.E. Aerospace Systems Conf.*, **1967**, 670707.
5. Wagner, R. E. and Barrett, J. A., Study of basic analogue amplifier networks using high gain fluid amplifiers. *C.F.C.* (3), Paper K6.
6. Kelley, L. R. and Shinn, J. N., Noise in fluidic proportional amplifiers. *C.F.C.* (3), Paper G1.
7. Howland, G. R., Performance characteristics of vortex amplifiers. *H.D.L.* (3), **II**, 208.
8. Taplin, L. B., Small signal analysis of vortex amplifiers. *AGARD Lecture series 'Fluid Control—Components and systems'* (1966).
9. Tsui, K. C. and Stone, J., Rep AD-634, 527 (Feb. 1966).
10. Parker, G. A. and Addy, M. K., The use of low gain fluidic beam deflection amplifiers in shaping networks. *J.A.C.C., Boulder, U.S.A.*, 559–567 (1969).
11. Boothe, W. A., Feasibility study of the application of fluid amplifiers to reactor-rod control. *N.A.S.A. Rep.* CR-54005.
12. Shinn, J. N., Fluidic compensation for a pneumatic position servo. *S.A.E.*, A-6 Com. (1968).
13. Kelley, L. R., A fluidic temperature control using frequency modulation and phase discrimination. *J.A.C.C.*, **1966**, 123–131.
14. Halbach, C. R., Otsap, B. A. and Thomas, R. A., A pressure insensitive fluidic temperature sensor. *A.F.*, 298.
15. Waters, K. L., Synthesis of a pure-fluidic temperature-control system. *A.S.M.E. Paper* 68-WA/FE-30.
16. Reeves, D., Inglis, M. E. and Airey, L., The fluidic oscillator as a temperature sensor. *C.F.C.* (1), Paper D1.
17. Spyropoulos, C. E., A sonic oscillator. *H.D.L.* (2), **II**, 27.
18. Warren, R. W., Pulse duration modulation. *H.D.L.* (1), **I**, 41.
19. Campagnuolo, C. J. and Sieracki, L. M., A digital-proportional fluid amplifier for a missile control system. *H.D.L.* (3), **III**, 131.
20. Eastman, N., Missile control by fluidics. *C.F.C.* (3), Paper C3.
21. Foster, K. and Cleife, P. J., The selection of a fluidic transducer for sensing rotational speed over a wide range. *C.F.C.* (3), Paper E4.

6
Boolean Algebra Theorems and Some Methods of Manipulation of Switching Functions

6.1 Introduction

In a short chapter it is impossible to convey all the meaning contained in a subject as extensive as switching theory. The best that can be achieved is that a summary may be made of the appropriate theorems and further that an indication may be given of some of the available methods of manipulation of switching functions that have been found to be useful for the design of digital fluid logic systems. For further information the reader must necessarily consult the literature.

First of all, the distinction must be made between analog variables and the digital ones which are the concern of this chapter. An analog signal may take on any positive or negative signal between certain limits and will vary continuously between them. A digital variable will be either ON or OFF and will change suddenly and discontinuously between the two conditions. These two states will be denoted by 0 or 1 for convenience. The 1 level may be chosen quite arbitrarily in any physical units depending upon the system; in some pneumatic diaphragm switching circuits the ON signal level may be 1 atm. whereas in a pure fluid system it may be as low as 0·05 atm. but still be denoted by 1 for the purpose of algebraic manipulation.

6.1.1 The AND Function

Secondly, the idea that logic statements are useful in the design of machines must be introduced. A typical example may be a machine tool where operation C may only be carried out after operation B has been completed and button A has been pressed. The flow of information may be indicated diagrammatically (Fig. 6.1) where the semi-circle denotes

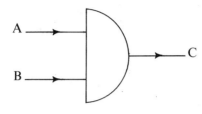

Fig. 6.1

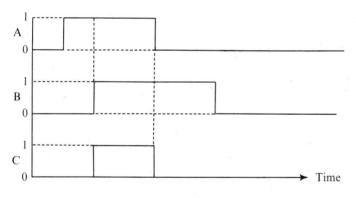

Fig. 6.2

the AND function. The signals may be represented on a time chart (Fig. 6.2) where it may be seen that C has a value 1 only when A AND B are both 1. This could be implemented by using two switches the first of which is closed when A is ON and open when A is OFF, whilst the second is controlled similarly by B.

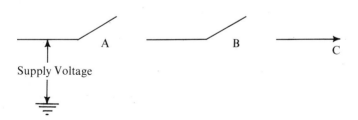

Whilst a time chart is useful in illustrating the sequence of events, the time intervals in operating circuits of this kind are not uniform and the concept of time is not very explicit. A better way is to draw up a *truth table* where all the possible combinations of A and B are tabulated in two columns, with the corresponding values of output C tabulated in a third column.

Table 6.1

A	B	C
0	0	0
0	1	0
1	0	0
1	1	1

Truth tables may be used to list the possible combinations of any number of variables, but the table becomes rather long with numbers greater than 5 or 6.

Lastly, the Boolean algebra notation for the AND function is simply a (.) between two variables, and thus is often omitted, so that

$$C = A.B \quad \text{or simply} \quad C = AB$$

means that C is 1 when, and only when, A AND B are both 1.

6.1.2 The OR Function

Alternatively, the same situation on the hypothetical machine may have been postulated in a different way. The machine could have been required NOT to work if either the guard is NOT down or the button NOT pressed. A bar over the variable denotes a signal which is OFF when that variable is ON, so that $\bar{A}$ denotes NOT A. The circuit is shown diagrammatically in Fig. 6.3.

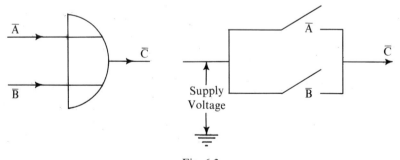

Fig. 6.3

[6.1] INTRODUCTION

The semi-circle with the input lines passing across it denotes the OR logic function. The parallel circuit of switching provides the OR facility, and the NOT function for A and B may be obtained by applying signals A and B to switches which are normally closed.

The Boolean algebra notation for the OR logic function is the (+) sign between the two variables, and the equation representing the output from the circuit becomes

$$\overline{C} = \overline{A} + \overline{B}$$

Combinations of AND, OR and NOT symbols can be used to describe quite complicated logic statements. Complex expressions can be manipulated and often simplified by using certain rules of Boolean algebra which are presented in the following sections.

6.2 Postulates of Boolean Algebra

In order to create the algebra of switching, certain rules must be formulated, from which theorems may be derived. There are alternative sets of independent postulates from which the same algebra is derived, but the most straightforward is perhaps the following.

(a) The operation (.) and (+) are binary. An expression formed with binary variables whose value can only be 0 or 1 must have an output with a value of either 0 or 1.

(b) The use of the AND function gives the following results:

$$1.1 = 1$$
$$1.0 = 0.1 = 0$$
$$0.0 = 0$$

These illustrate the proposition that a statement is true (i.e. has value 1) if two other conditions X, Y are both true, i.e. if X AND Y are both 1. It is false (i.e. has value 0) if either X or Y is false or both are false.

(c) The use of the OR function gives the following result:

$$0 + 0 = 0$$
$$0 + 1 = 1 + 0 = 1$$
$$1 + 1 = 1$$

These illustrate the proposition that a statement is true if either of two other conditions X, Y or both of them are true, but not in any other circumstance.

(d) $\qquad \overline{1} = 0 \qquad \overline{0} = 1$

i.e. the complement or opposite of truth is false and vice versa. All other theorems can be derived from these postulates.

6.3 Definitions in Boolean Algebra

Two Boolean expressions can only be 'equivalent' if they both equal 1, or both equal 0.

We have already defined the 'complement' of a single variable; in the same way one expression is the 'complement' of another if it is 1 only when the other is 0, and 0 only when the other is 1. The *complement* of a Boolean expression is obtained by:

> Changing all .'s to +'s
>
> Changing all +'s to .'s
>
> Changing all 1's to 0's
>
> Changing all 0's to 1's
>
> Complementing all literals

Each occurrence of a variable in an expression is called a literal. Thus the complement of

$$A.\bar{B} + 1.C + 0$$

is

$$(\bar{A} + B).(0 + \bar{C}).1$$

The complement of a switching network produces a network which is open when the original one was closed and closed when the original one was open.

The *dual* of a Boolean expression is obtained by:

> Changing all .'s to +'s
>
> Changing all +'s to .'s
>
> Changing all 1's to 0's
>
> Changing all 0's to 1's
>
> Leaving all literals unchanged

Thus there is a similarity between the obtaining of a dual and the obtaining of a complement. The dual of

$$A.\bar{B} + 1.C + 0$$

is

$$(A + \bar{B}).(0 + C).1$$

A *transmission* function represents the occurrence of 1's from a Boolean expression due to the appearance of 1's in the variables. In a switching

[6.3] DEFINITIONS IN BOOLEAN ALGEBRA

system this means the appearance of a signal at the output due to the closing of the switches represented by the variables. A *hindrance* function, as its name implies, represents the disappearance of a signal at the output due to the opening of the appropriate switches. In general terms this means the occurrence of 0's from a complete Boolean expression due to the appearance of 0's in the variables. One expression is the *dual* of another if the transmission function of one is the same as the hindrance function of the other.

To illustrate the difference between a complemented (or inverse) circuit and a dual circuit, consider the circuit shown in Fig. 6.4(a) whose transmission function is

$$T = A \cdot (B + C)$$

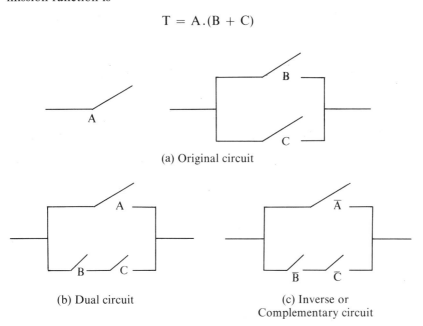

(a) Original circuit

(b) Dual circuit

(c) Inverse or Complementary circuit

Fig. 6.4

The dual of Fig. 6.4(a), shown in Fig. 6.4(b), is given by:

$$T_{dual} = A + B \cdot C$$

The complement, shown in Fig. 6.1(c), is given by

$$T_{comp} = \overline{A} + \overline{B} \cdot \overline{C}$$

The truth table (Table 6.2) shows the different transmission functions of each.

Table 6.2

A	B	C	T	T_{comp}	T_{dual}
0	0	0	0	1	0
0	0	1	0	1	0
0	1	0	0	1	0
0	1	1	0	1	1
1	0	0	0	1	1
1	0	1	1	0	1
1	1	0	1	0	1
1	1	1	1	0	1

Thus the transmission function of the complemented or inverse circuit is the opposite of the original one. At first sight the dual is not connected. However, it will be observed that the pattern of 0's in A, B, C and T_{dual} is the same as the pattern of 1's in A, B, C and T. Thus the hindrance function of the dual circuit is the transmission function of the original one.

Each of the postulates was presented in complementary form where each variable is changed to its opposite value and AND is replaced by OR. Similarly, if an equation is valid or a proposition is true, the complement of the equation or proposition must also be valid or true because of the complementary postulates. Similarly, all variables may be replaced by their complements because their values are merely changed from 0 to 1 and vice versa, so that any arguments used in the proof of the statement remain unchanged. Therefore, if a Boolean statement is true, not only is its complement true but so also is its dual.

6.4 Theorems of Boolean Algebra

(a)
$$0 \cdot x = 0$$
$$1 + x = 1$$
(Th. a)

Multiplication by 0 gives 0 and addition of 1 gives 1†.

(b)
$$1 \cdot x = x$$
$$0 + x = x$$
(Th. b)

---

† Multiplication and addition do not mean the same as in normal algebra but merely indicate the use of (.) and (+).

[6.4] THEOREMS OF BOOLEAN ALGEBRA

Multiplication by 1 or addition of 0 does not change the expression.

(c)
$$x \cdot x = x$$
$$x + x = x$$
(Th. c)

(d)
$$x\bar{x} = 0$$
$$x + \bar{x} = 1$$
(Th. d)

Multiplication of a variable by its complement gives 0, whereas the addition of a complement to a variable gives 1.

(e) Commutativity:
$$xy = yx$$
$$x + y = y + x$$
(Th. e)

(f) Associativity:
$$xyz = x(yz) = (xy)z = (xz)y$$
$$x + y + z = x + (y + z) = (x + y) + z = (x + z) + y$$
(Th. f)

(g) Distributivity:
$$x(y + z) = xy + xz$$
$$x + (y \cdot z) = (x + y) \cdot (x + z)$$
(Th. g)

(h) de Morgan's theorems:
$$\overline{x \cdot y \cdot z \cdots} = \bar{x} + \bar{y} + \bar{z} + \cdots$$
$$\overline{x + y + z + \cdots} = \bar{x} \cdot \bar{y} \cdot \bar{z} \cdots$$
(Th. h)

These two theorems are particularly useful in changing from logic functions using one type of operation, say the AND, to functions using the OR, and vice versa.

Table 6.3

x	y	z	x + y·z	(x + y)·(x + z)
0	0	0	0	0
0	0	1	0	0
0	1	0	0	0
0	1	1	1	1
1	0	0	1	1
1	0	1	1	1
1	1	0	1	1
1	1	1	1	1

The proof of any of the preceding theorems may be obtained by constructing a truth table for the one in question and testing the accuracy of the statement for all the possible combinations of values of the variables. This is possible because each variable has only two values, 0 and 1, and the original postulates may be used for evaluating combinations of the variables.

As an example, consider the second of the distributive laws (g).

The possible combinations of the variables x, y, z are listed in the first three columns of the truth table (Table 6.3). Using the fact that $0.1 = 0$, $1.1 = 1$ and $1 + 0 = 1$, the values for $x + y.z$ may be listed in the fourth column for each combination of x, y, z. Similarly for $(x + y).(x + z)$ in the last column. It may be seen that the fourth and fifth columns are identical, thus proving our proposition.

For the engineer, an alternative way of visualizing the basic rules is to consider each variable to close a switch and by inspection to observe that one circuit gives identical results to another. The circuits for two variable AND and OR functions have already been shown, and using them to illustrate the problem we have the circuits of Fig. 6.5.

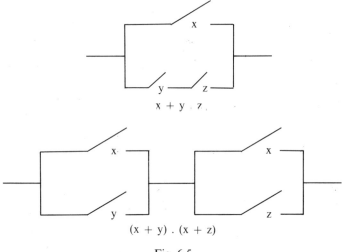

Fig. 6.5

When x is closed there is a path through each circuit. When x is open there is a path through only when y and z are both present. Therefore, the transmission functions are the same.

Neither of these approaches is appropriate for either the proof or the simplification of complicated logic statements, and the theorems already presented must be used to manipulate such expressions.

[6.4] THEOREMS OF BOOLEAN ALGEBRA

The chief purpose of Boolean algebra in engineering is to reduce a switching circuit to an equivalent form in the interests of determining the minimum number of switches required. The following procedure is therefore adopted:

1. Draw the logic diagram of the original circuit from the word statement of the problem.
2. Derive the Boolean expression for the circuit.
3. Reduce the expression to a minimum form.
4. Convert the minimum expression back to a logical diagram.
5. Use a truth table to check the result.

For example, a signal may be required to allow the start of a hypothetical operation in an automatic lathe under the following set of circumstances. The coolant must be on, AND the workpiece must be in position, AND a further signal from the machine must be present. This latter signal may be created in each of three situations; a signal representing tool wear is absent, AND a reject signal from an in-process gauge is absent, AND tool No. 1 is in position; OR tool No. 1 has moved away, AND tool No. 2 is in position, AND the tool wear signal is absent and the reject signal from the gauge is absent; OR tool No. 2 has moved away, AND the workpiece is in position.

Let A represent the coolant ON, B the workpiece in position, C the tool wear signal, D the reject signal from the gauge, E tool No. 1 and F tool No. 2. The circuit representing the word statement is now shown in Fig. 6.6.

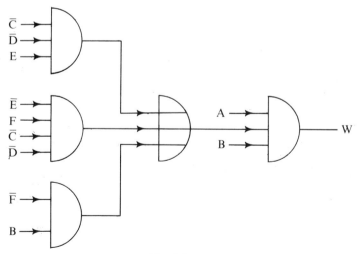

Fig. 6.6

The Boolean expression for this is:

$$(\overline{CD}E + \overline{CD}\overline{E}F + \overline{F}B)AB = W$$

but using (Th. g)

$$\overline{CD}E + \overline{CD}\overline{E}F = \overline{CD}(E + \overline{E}F)$$

Using (Th. g) again,

$$E + \overline{E}F = (E + \overline{E})(E + F)$$

and from (Th. d),

$$E + \overline{E}F = E + F$$

This latter result is a generally useful one in circuit simplification. Therefore

$$W = (\overline{CD}E + \overline{CD}F + \overline{F}B)AB$$

from (Th. g),

$$W = (\overline{CD}EB + \overline{CD}FB + \overline{F}BB)A$$

from (Th. c), $B \cdot B = B$, so that,

$$W = (\overline{CD}EB + \overline{CD}FB + \overline{F}B)A$$
$$= (\overline{CD}E + \overline{CD}F + \overline{F})AB$$

But using the result just obtained that $x + \overline{x}y = x + y$

$$W = (\overline{CD}E + \overline{CD} + \overline{F})AB$$
$$= (\overline{CD}(E + 1) + \overline{F})AB$$
$$= (\overline{CD} + \overline{F})AB$$

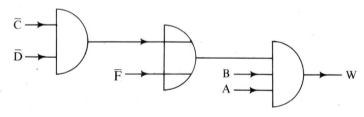

Fig. 6.7

and the circuit is shown in Fig. 6.7 which is a considerable simplification. Of course, this amount of logic will not be sufficient to control the operation of the automatic lathe in the correct sequence, but some of the inter-

[6.4] THEOREMS OF BOOLEAN ALGEBRA

locking conditions for allowing the sequence to proceed have been formulated and simplified.

The truth table for six variables would normally contain 64 rows, but since the circuit can only be ON when A and B are ON, it is only necessary to consider the other four variables. A truth table for these is shown in Table 6.4, and this confirms that the two circuits are equivalent. Checking of more complex circuits in this way can become very tedious.

Table 6.4

C	D	E	F	Original circuit	New circuit
0	0	0	0	1	1
0	0	0	1	1	1
0	0	1	0	1	1
0	0	1	1	1	1
0	1	0	0	1	1
0	1	0	1	0	0
0	1	1	0	1	1
0	1	1	1	0	0
1	0	0	0	1	1
1	0	0	1	0	0
1	0	1	0	1	1
1	0	1	1	0	0
1	1	0	0	1	1
1	1	0	1	0	0
1	1	1	0	1	1
1	1	1	1	0	0

6.5 The Use of Duality in Simplification of Switching Functions

It is sometimes convenient in simplifying switching functions to convert sums to products when the simplifying algebra is more straightforward. This could be done by complementing the original function, simplifying it and then converting back. The use of the dual is easier, however, because each variable is left untouched, and the procedure becomes:

(a) Determine the dual of the original function.
(b) Simplify the dual.
(c) Determine the dual of the latter circuit.
(d) Simplify the result.

For example, the circuit shown in Fig. 6.8 is to be simplified.

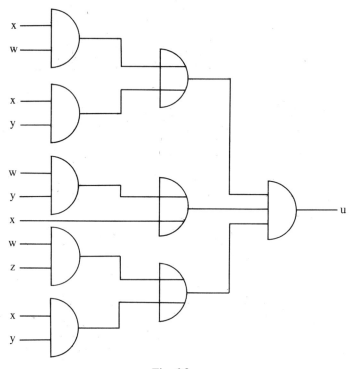

Fig. 6.8

$$u = (xw + xy)(wy + x)(wz + xy)$$
$$= x(w + y)(wy + x)(wz + xy)$$

Take the dual of this (denoted by d)

$$d(u) = x + wy + (w + y)x + (w + z)(x + y)$$

Collecting terms in x together

$$d(u) = x(1 + w + y + x + z) + wy + (w + z)y$$
$$= x + (w + z)y$$

Now find the dual of this, which must be the equivalent of the original function, i.e.

$$u = x(y + wz)$$

and the revised circuit is shown in Fig. 6.9.

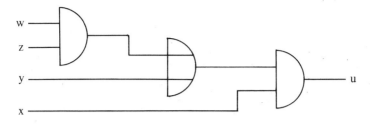

Fig. 6.9

6.6 Canonical Forms of Boolean Functions

If a switching function consists of a number of variables, it is possible to expand the function until the same number of variables is contained in each of the groups of the expression and the expression is homogeneous in appearance. It may seem odd that any attempt should be made to complicate a switching function when always emphasis is placed on minimization in one form or another. The reason is that when dealing with complex expressions there are often various possible minimum solutions, and a straightforward attack on the expression may not result in the best solution. Of the possible methods of manipulation which have been successfully used for this purpose, the tabulation method requires that the expression be in expanded form initially.

Two expanded forms are possible, and these are best illustrated by example.

6.6.1 The Expanded Sum of Products

Consider a switching function

$$A + B\bar{C}$$

By using the fact that $B + \bar{B} = 1$, etc., this may be made homogeneous in the occurrence of the variables, i.e.

$$A + B\bar{C} = A(B + \bar{B})(C + \bar{C}) + (A + \bar{A})B\bar{C}$$
$$= ABC + AB\bar{C} + A\bar{B}C + A\bar{B}\bar{C} + \bar{A}B\bar{C}$$

so that the original function is now represented by 5 out of the possible 8 products involving ABC.

6.6.2 The Expanded Product of Sums

This may be produced using the distributive law and the fact that $C\bar{C} = 0$, etc.

$$\begin{aligned}
A + B\bar{C} &= (A + B)(A + \bar{C}) \\
&= (A + B + C\bar{C})(A + B\bar{B} + \bar{C}) \\
&= \{(A + B) + C\}\{(A + B) + \bar{C}\}\{(A + \bar{C}) + B\} \\
&\quad \{(A + \bar{C}) + \bar{B}\} \\
&= (A + B + C)(A + B + \bar{C})(A + \bar{B} + \bar{C})
\end{aligned}$$

Alternatively, complementing may be used to obtain the product of sums directly from the expanded sum of products, which is often relatively easy to obtain. For example, the three products which are not used in the expanded sum of products must represent the complement of the original function. If the expression consisting of these three terms is complemented, it automatically assumes the form of the product of sums, since all +'s are changed to .'s and vice versa.

The three remaining products give the function

$$\bar{A}\bar{B}\bar{C} + \bar{A}BC + \bar{A}B\bar{C}$$

Interchanging (+)'s and (.)'s gives

$$(\bar{A} + \bar{B} + \bar{C})(\bar{A} + B + C)(\bar{A} + B + C)$$

Complementing all the variables completes the process to give

$$A + B\bar{C} = (A + B + C)(A + \bar{B} + \bar{C})(A + B + \bar{C})$$

6.7 Simplification of Switching Functions: Tabulation Method

In simplifying a complex Boolean expression, the aim is to minimize some parameter in the system. This parameter may, for example, be the number of switches used, the number of logic gates of a particular type, or the overall cost of the circuit. Of the methods available, the most systematic is the tabular method where the original function is arranged as an expanded sum of products. The general expression is then examined for all possible simplifications and the most appropriate one is chosen at the end of the exercise.

This procedure is best illustrated by an example, but first one more concept must be introduced. Consider the four-variable switching function

$$T(w, x, y, z) = \bar{w}\bar{x}y\bar{z} + \bar{w}x\bar{y}\bar{z} + w\bar{y}z + \bar{y}xz$$

[6.7] TABULATION METHOD OF SIMPLIFICATION

which gives the condition that a circuit should be ON. Suppose now that the circuit must be OFF when the following switching function is valid

$$U(w, x, y, z) = \bar{x}y\bar{z} + \bar{w}yz + wxy$$

By multiplying by $x + \bar{x} = 1$, etc., the terms with only three literals may be changed to terms with four literals and the two switching functions in the expanded sum of products form become

$$T(w, x, y, z) = \bar{w}\bar{x}y\bar{z} + \bar{w}x\bar{y}z + \bar{w}xy\bar{z} + w\bar{x}y\bar{z} + wx\bar{y}z$$

$$U(w, x, y, z) = \bar{w}\bar{x}y\bar{z} + \bar{w}\bar{x}yz + \bar{w}xyz + w\bar{x}y\bar{z} + wxy\bar{z} + wxyz$$

Together these two expressions account for 11 out of the 16 possible combinations of four variables. In the particular machine under consideration, the other five combinations probably never occur; for example, if two switches p and q are mechanically connected so that p cannot be on when q is on and vice versa, the two combinations pq and $\bar{p}\bar{q}$ could never occur. Such combinations are called *don't care* conditions or *optional combinations*. By allowing the switching function to cater for some or all of these terms, the expression may be simplified without affecting the machine in which the switching circuit is to be used, since those combinations will never occur. The optional combinations left over in the exercise are $\bar{w}x\bar{y}z$, $\bar{w}xy\bar{z}$, $w\bar{x}\bar{y}\bar{z}$, $wx\bar{y}\bar{z}$ and $w\bar{x}yz$.

All the switching function products and don't care conditions are now tabulated using a binary code where 1 means the presence of a variable and 0 means the presence of the complement of the variable in a given product. For example, 0101 signifies $\bar{w}x\bar{y}z$. The products are tabulated in groups with similar numbers of 1's in ascending order of these numbers. The initial table is shown on the left of Table 6.5 and each row is identified by the decimal number represented by the binary number produced by the particular combination of the variables.

The problem next is to find which products of four variables combine with other products to produce products of only three variables. Using the fact that $x + \bar{x} = 1$, etc., we may observe that terms differing only by one term can be combined together. Therefore, starting with the first row of the first group, it can be tested against any row of the second group where there are possibilities for combination. If any terms can combine, they are tabulated with a (–) indicating the variable which has disappeared. For example, 0001 ($\bar{w}\bar{x}\bar{y}z$) will combine with either 1001 ($w\bar{x}\bar{y}z$) to form –001 ($\bar{x}\bar{y}z$) or with 0101 ($\bar{w}x\bar{y}z$) to form 0–01 ($\bar{w}\bar{y}z$), and these combinations of 1 and 9 or of 1 and 5 head the next table. When all possible mergers of the first group have been obtained, mergers of the second group with the third group are considered. When a row has achieved one merger, a tick is placed by the side of it.

Table 6.5—Possible products of variables

Products of four variables						Products of three variables						Products of two variables					
		w	x	y	z		w	x	y	z			w	x	y	z	
	1	0	0	0	1	1, 9	–	0	0	1		1, 9, 5, 13	–	–	0	1	F
	2	0	0	1	0	1, 5	0	–	0	1		4, 12, 5, 13	–	1	0	–	G
	4	0	1	0	0	2, 6	0	–	1	0	A						
	9	1	0	0	1	2, 10	–	0	1	0	B						
Optional	5	0	1	0	1	4, 5	0	1	0	–							
	6	0	1	1	0	4, 6	0	1	–	0	C						
	10	1	0	1	0	4, 12	–	1	0	0							
	12	1	1	0	0	9, 11	1	0	–	1	D						
Optional	11	1	0	1	1	9, 13	1	–	0	1							
	13	1	1	0	1	5, 13	–	1	0	1							
						10, 11	1	0	1	–	E						
						12, 13	1	1	0	–							

Next, the possible mergers of the products of three variables are considered in the same way in order to produce possible products of two literals, which are then tabulated in the right-hand side of Table 6.5. In the middle table, some rows remain unticked. These rows cannot be merged any further and the products which they represent are called *Prime Implicants*. These prime implicants are allocated letters and so also are the two rows of the third table, which are also unmergeable and are, therefore, prime implicants.

Finally, the choice of the sum of prime implicants which covers all the terms of the original switching function is made by a further tabulation. In this case, all the original switching products are allocated a column and all prime implicants a row of a chart, as shown in Table 6.6, and ticks are placed in the appropriate square whenever a prime implicant includes one of the required terms.

Table 6.6

1	2	4	9	13	
	✓				A
	✓				B
		✓			C
			✓		D
					E
✓			✓	✓	F
		✓		✓	G

[6.7] TABULATION METHOD OF SIMPLIFICATION

The possible minimum switching functions to cover the original variables are listed in this case (Fig. 6.10).

$$T = F + G + A$$
$$= F + G + B$$
$$= F + C + A$$
$$= F + C + B$$

Fig. 6.10

Of these, G is preferable to C because it contains fewer variables, but otherwise the choice is open and will depend upon what connections are most convenient in the switching circuit. As an example, we could choose the first possibility, and the reduced switching function becomes

$$T = \bar{y}z + x\bar{y} + \bar{w}y\bar{z}$$

This method can be used in principle for any number of variables merely by increasing the number of columns of the table. The problem is that the number of rows is 2^n, where n is the number of variables, and the merging procedure becomes very lengthy for relatively small values of n. Nevertheless, by utilizing the capacity of a modern computer, the procedure can be made automatic.

6.8 Simplification of Switching Functions—Karnaugh Maps

6.8.1 General

A Karnaugh map is a diagram on which the values of the various groups of the switching function may be easily displayed for various values of the variables. Again, the best way of describing the method is to give examples, and this will be done for functions of two variables, three variables and four variables in that order. The method becomes unwieldy for more variables than four, and reference should be made to the textbooks on the subject should manipulation of this kind be necessary.

A\B	0	1
0	$\bar{A}\bar{B}$	$A\bar{B}$
1	$\bar{A}B$	AB

Fig. 6.11

A function of two variables is relatively easily displayed because there are only four possible combinations of the variables. Consider two variables A and B, both of which may have the values 0 or 1, and construct a diagram consisting of four squares, as shown in Fig. 6.11.

The two columns represent $\bar{A}$ and A, reading from left to right, whilst the two rows represent $\bar{B}$ and B, reading from top to bottom. Consequently, each square is allocated a particular combination of the variables as shown in the diagram.

By means of a Karnaugh map simplify the switching function

$$T = A + AB + \bar{A}\bar{B}$$

In order to illustrate the method more clearly, we shall consider each term separately:

(1) The first term is A, which gives a value for T of 1, when A is 1 irrespective of state of B. Therefore, '1's are placed in the column which denotes that A is 1. See Fig. 6.12(a).
(11) The second term is AB, which gives a value of 1 only when A and B are both 1. Therefore, a 1 is placed in the appropriate square of the map, as shown in Fig. 6.12(b).
(111) The third term is $\bar{A}\bar{B}$ and a 1 is placed in the appropriate square for that combination as shown in Fig. 6.12(c).

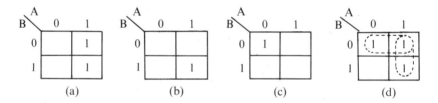

Fig. 6.12

The complete map of the switching function T is obtained by combining the other three maps and the result is shown in Fig. 6.12(d). Although the square AB is filled on account of two of the terms, only a single 1 is entered in the final map. The fact that there is such an overlapping indicates the possibility that the original switching functions contained one or more unnecessary terms.

The final map is now examined to obtain the simplest switching function which describes the entries. First of all, it is examined for adjacent terms, because these indicate that one of the variables can take on either of the value 0 or 1 and may, therefore, be omitted from that particular product

[6.8] SIMPLIFICATION OF SWITCHING FUNCTIONS: KARNAUGH MAPS

in the switching function. In this case, there are two adjacent pairs and these are ringed by a dotted line. The ring occupying the top row gives the term $\bar{B}$ and that occupying the left-hand column gives the term A. Therefore, the reduced switching function is:

$$T = A + \bar{B}$$

This was an elementary example, and the principle can now be extended to examine a function of 3 variables, ABC. In this case, 8 spaces are required and a grid is formed by allocating columns to each of the four possible combinations of A and B and two rows for C, as shown in Fig. 6.13.

C \ AB	00	01	11	10
0	$\bar{A}\bar{B}\bar{C}$	$\bar{A}B\bar{C}$	$AB\bar{C}$	$A\bar{B}\bar{C}$
1	$\bar{A}\bar{B}C$	$\bar{A}BC$	ABC	$A\bar{B}C$

Fig. 6.13

Ex. To illustrate the use of this map, consider the function:

$$T = \bar{A}B + \bar{A}\bar{B}C + AB\bar{C}$$

It is not necessary to draw a map for each separate term, since they can easily be entered one after the other on to the same map, as shown in Fig. 6.14.

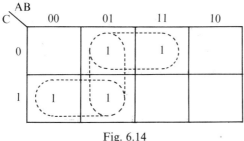

Fig. 6.14

From the diagram, three pairs of adjacent '1's may be found, as indicated by the broken lines around the pairs in Fig. 6.14. These are $\bar{A}C$, $\bar{A}B$ and $B\bar{C}$. However, two of the pairs contain all four of the entries, so that the simplest switching function must consist of these two, i.e.

$$T = \bar{A}C + B\bar{C}$$

294 SIMPLIFICATION OF SWITCHING FUNCTIONS: KARNAUGH MAPS [6.8]

We shall now extend these ideas to a map for four variables, A, B, C, D, which is laid out as a square array of sixteen spaces, as indicated in Fig. 6.15.

CD\AB	00	01	11	10
00	$\bar{A}\bar{B}\bar{C}\bar{D}$	$\bar{A}B\bar{C}\bar{D}$	$AB\bar{C}\bar{D}$	$A\bar{B}\bar{C}\bar{D}$
01	$\bar{A}\bar{B}\bar{C}D$	$\bar{A}B\bar{C}D$	$AB\bar{C}D$	$A\bar{B}\bar{C}D$
11	$\bar{A}\bar{B}CD$	$\bar{A}BCD$	$ABCD$	$A\bar{B}CD$
10	$\bar{A}\bar{B}C\bar{D}$	$\bar{A}BC\bar{D}$	$ABC\bar{D}$	$A\bar{B}C\bar{D}$

Fig. 6.15

Ex. As an example, we shall examine the problem already used to illustrate the tabulation method, again including the concept of 'don't care' conditions. The switching ON function under consideration was

$$T = \bar{w}\bar{x}y\bar{z} + \bar{w}x\bar{y}\bar{z} + w\bar{y}z + \bar{x}yz$$

and the switching OFF function was

$$U = \bar{x}\bar{y}\bar{z} + \bar{w}yz + wxy$$

The switching ON function is given '1' entries in the map of Fig. 6.16, and the switching OFF function is given '0' entries. It will be noticed that the functions may be entered directly on to the map, without any expansion. When these entries have been made, the remaining squares must be

yz\wx	00	01	11	10
00	0	1	X	0
01	1	X	1	1
11	0	0	0	X
10	1	X	0	X

Fig. 6.16

[6.8] SIMPLIFICATION OF SWITCHING FUNCTIONS: KARNAUGH MAPS 295

'don't care' (or 'optional') combinations and X's are used to identify these squares clearly.

The optional conditions may now be used to simplify the switching function as far as possible. This is done by attempting to find groups of either eight, four or two adjacent '1's (in that order), including the X's as '1's wherever this is desirable in order to complete a group. For example, there are no groups of eight 1's (which would yield a term containing only one variable), but by using one of the X squares, the second row is independent of the value of w or x and is described by the two-variable term $\bar{y}z$. One other possible group of two variables is the term $x\bar{y}$ formed by using two of the don't care conditions and two 1's.

The remaining 1 in the bottom left-hand square may be combined with the X next to it, or with the X in the bottom right-hand square (which is also adjacent to it logically), to obtain the switching functions $\bar{w}y\bar{z}$ or $\bar{x}y\bar{z}$. The '1' in the square $w\bar{x}\bar{y}z$ could also be combined with the X below it, to form a three-variable group $w\bar{x}z$, but it already is included in the term $\bar{y}z$, so that the extra grouping is redundant.

The results, therefore, are identical with those produced by the tabulation method, and again the switching function will be

$$T = \bar{y}z + x\bar{y} + \bar{w}y\bar{z}$$

More complex problems may be carried out by using these methods, but the brief survey given here is sufficient for the purposes of this book.

6.8.2 The Design of a Petherick to a Binary Decoder

Ref. 3 gives an excellent example of the use of Karnaugh maps to obtain logic functions in the design of a Petherick to Binary code decoder that is described. If the input signals to the Petherick encoder are the four digits ABCD, and the equivalent binary digits are WXYZ, the respective codes may be listed, i.e.

	Binary Coded Decimal				Petherick Code			
	W	X	Y	Z	A	B	C	D
0	0	0	0	0	0	1	0	1
1	0	0	0	1	0	0	0	1
2	0	0	1	0	0	0	1	1
3	0	0	1	1	0	0	1	0
4	0	1	0	0	0	1	1	0
5	0	1	0	1	1	1	1	0
6	0	1	1	0	1	0	1	0
7	0	1	1	1	1	0	1	1
8	1	0	0	0	1	0	0	1
9	1	0	0	1	1	1	0	1

The Petherick code may be mapped as shown in Fig. 6.17. The unit distance nature of the code is clearly seen from the progressive way the map is traversed in passing from 0 to 9. The unused combinations are 'Don't care' conditions.

CD\AB	00	01	11	10
00	X	X	X	X
01	1	0	9	8
11	2	X	X	7
10	3	4	5	6

Fig. 6.17

The most significant bit of the binary code W is '1' when decimal numbers 8 and 9 occur. This may be related to ABCD through the map of the Petherick code, and by making use of the don't care conditions a minimum function for W may be obtained, Fig. 6.18, i.e. $W = A\bar{C}$.

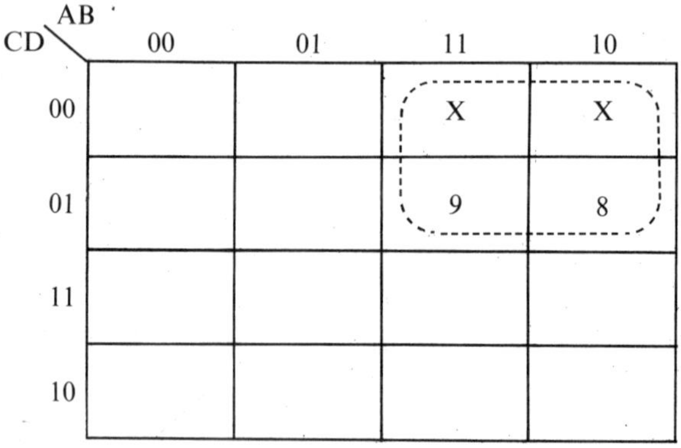

CD\AB	00	01	11	10
00			X	X
01			9	8
11				
10				

Fig. 6.18

Similarly, X occurs for numbers 4, 5, 6 and 7; Y is '1' for decimal numbers 2, 3, 6, 7, whilst Z is '1' for 1, 3, 5, 7 and 9. The appropriate maps are given in Fig. 6.19. Hence

$$X = AC + BC = C(A + B)$$
$$Y = \bar{B}C$$
$$Z = AB + ACD + \overline{ABC} + \overline{ABD}$$

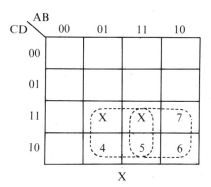
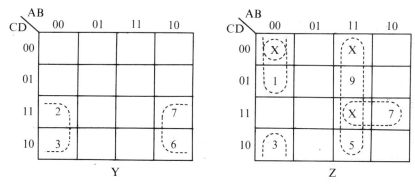

Fig. 6.19

References

1. Torng, H. C., *Introduction to the logical design of switching systems*, Addison-Wesley (1964).
2. Marcus, M. P., *Switching circuits for engineers*, Prentice-Hall (1962).
3. Ramanathan, S., Aviv, I. and Bidgood, R. E., The design of a pure fluid shaft encoder. *C.F.C.*(2), Paper E4.

7
The Design of Pure Fluid Digital Elements

7.1 General

When fluid at a higher pressure is allowed to pass through a nozzle or a slit into a region of uniform lower pressure, a jet is formed. The high velocity fluid in the jet mixes with that of low velocity surrounding it and the velocity of the jet is gradually lowered as the distance from the nozzle is increased. At the same time the velocity profile becomes 'flatter' with less sharp gradients and acquires points of inflexion because the fluid situated at a considerable distance from the jet centre line is given some small velocity. Providing no boundary walls are present other than the cover plates, the pressure downstream of the nozzle will be nearly uniform so that the momentum of the jet at any cross-section will be the same. Because the velocity reduces as the distance from the nozzle increases, the width of the jet must increase in order to maintain the initial value of the momentum. The result is that the mass flow increases, since the momentum is proportional to velocity squared times the width of the jet, whereas mass flow is proportional to velocity times the width. Fluid is therefore drawn into the jet; this is the entrainment process and is the basis for the 'jet pump' in which a low-flow, high-velocity jet gives rise to a high-flow, low-velocity stream, drawing in the extra flow from the surroundings. For a laminar jet, the mixing is relatively weak and is a function of Reynolds Number; the entrainment is largest at very low Reynolds Numbers and, as shown in Chapter 3, may be greater than that for a turbulent jet for Reynolds Numbers below 50. The mixing process, and therefore entrainment, of a turbulent jet is strong and is independent of Reynolds Number.

More information on laminar and turbulent jets and also about the transition from one state to the other is given in Chapter 3. The transition to turbulence does not occur at a clearly defined point, but as a rough guide the critical Reynolds Number is usually taken to be

$$\text{Re}_{\text{crit}} = \frac{U_s d}{v} = 2000$$

where U_s is the velocity of fluid in the nozzle; v is the kinematic viscosity of the fluid and d is the hydraulic diameter of the nozzle or slit from which it emerges.

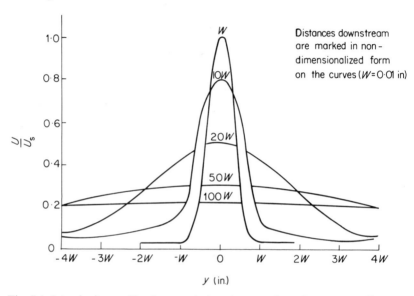

Fig. 7.1 Jet velocity profiles for a turbulent jet at various downstream distances from the nozzle

The velocity profiles for a turbulent jet issuing from a two-dimensional nozzle, 0·010 in. wide are given in Fig. 7.1 for various distances downstream of the nozzle. It is easily seen that the energy available in the jet is very much reduced by the distance from the nozzle, and receivers placed further downstream than $10W$ will not give good pressure recovery. A laminar jet, on the other hand, will spread at a very much lower rate if operated close to the Critical Reynolds Number. In Chapter 3, an expression for the ratio of centre line velocity of the laminar and turbulent jets (3.48) shows that at a distance of $20W$, at Re = 1500, the ratio

$$\frac{U_{\text{lam}}}{U_{\text{turb}}} = 3.56$$

[7.1] GENERAL

Consequently, the pressure recoveries, being proportional to velocity squared, are in the ratio of approximately 12·7:1.

The various theoretical velocity profiles are given in Chapter 3 for both the laminar and turbulent jets. An analysis attempting to describe the performance of amplifiers inevitably must make use of one or the other of them, but the choice depends upon convenience in the particular method of analysis followed rather than on accuracy, since all are approximate and semi-empirical.

7.2 The Turbulence Amplifier

The property provided by the difference in pressure recoveries for a laminar and turbulent jet may be utilized to produce a digital amplifier giving a NOR or NOT function. The sensitivity of a laminar jet to external disturbances is also described in Chapter 3, and it may easily be appreciated that if a jet is close to the normal transition from laminar to turbulent condition but still laminar, any small disturbance will cause the jet to break up into turbulence, with the consequence as we have just seen that the available energy at the receiver will be much reduced. This phenomenon is the basis of the turbulence amplifier first described by Auger[1], and shown in Fig. 7.2.

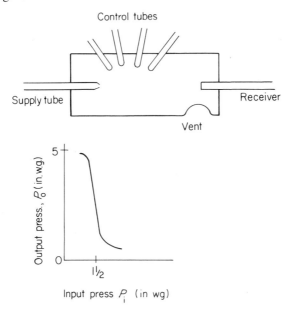

Fig. 7.2. A 4-input turbulence amplifier and a typical input-output characteristic (Auger, *Ref. 1*)

The jet issues from a long supply tube mounted in the end of a larger cylinder, and is collected by a receiver some distance away at the other end of the large cylinder. The jet is caused to become turbulent by a transverse control flow, and in the design shown, four alternative inputs are provided. To the right of the device, the output pressure into a blocked load is indicated, and as may be seen, the initial output pressure of 5 in. w.g. falls

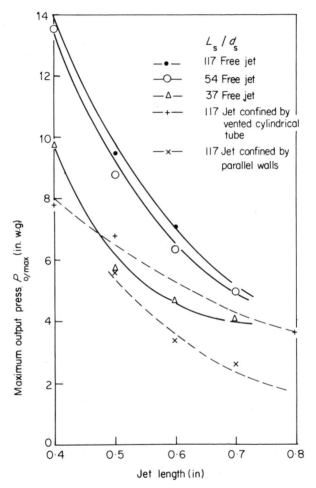

Fig. 7.3 Effect of supply tube length and distance between nozzle and receiver on the pressure recovery of a turbulence amplifier (L_s is the length of the supply nozzle, d_s is the diamater of the supply nozzle) (Oels, Boucher and Markland, *Ref. 2*)

[7.2] THE TURBULENCE AMPLIFIER

abruptly to less than 1 in. w.g. at a relatively low pressure at the control tube. According to Ref. 1, a supply tube diameter of 0·30 in gives a suitable compromise between good power gain and miniaturization. Under these conditions, a nozzle-to-receiver distance of 23 to 33 nozzle diameters is appropriate for logic operation. By increasing the distance to greater than 43 diameters, the jet becomes too sound sensitive, and if it is reduced to below 20 diameters, the discrimination between the logic '1' and '0' signals is not sufficiently good.

It is also suggested[1] that careful control of upstream conditions is necessary to achieve good results. The reason for this is that the laminar jet is more stable if the velocity profile emerging from the supply tube or nozzle is parabolic, or in other words, fully developed. To obtain fully developed profiles in a pipe, long inlet lengths are necessary; opinions differ, but 40 diameters is often quoted as the minimum length for certainty. In Ref. 2, this question is examined and experimental results (reproduced in Fig. 7.3) are given for maximum pressure recovery of an undisturbed laminar jet for various supply conditions, plotted against the distance between nozzle and receiver. The best results are for a supply tube 117 diameters in length; reducing this to 54 diameters has little effect, but on further reduction to 37 diameters, the pressure recovery falls considerably, thus confirming the view that for good results a certain minimum length of supply tube exists. The reason is probably that as the supply tube length is decreased, the velocity profile begins to be sharper and the laminar jet less stable so that is not possible to work at such a high Reynolds Number. Thus the supply pressure must be reduced and the output pressure therefore falls.

The effect of confining the jet is also to reduce the available pressure recovery.

Increasing the receiver distance also reduces pressure recovery, but as already indicated, the receiver must be placed far enough away from the nozzle for the pressure to fall significantly when the jet becomes turbulent, and for the simple design shown, a receiver distance of at least 20 times the supply tube diameter would be required, i.e. 0·54 in.

Verhalst[3] shows the effect of different output tube diameters as in Fig. 7.4. For receivers larger than the supply tube, the blocked output pressure is lowest, but the available flow is highest, whilst for a receiver smaller than the supply tube the converse is true. The shape of the output characteristics as the output flow is changed are typical of pure fluid elements. Other curves are given in the paper in order to illustrate the effect of change of parameters. Fig. 7.5 is reproduced from this source. As may be seen from the P_o–Q_s curves, the effect of increasing the free jet length x is to decrease the maximum possible supply flow Q_s. The reasons

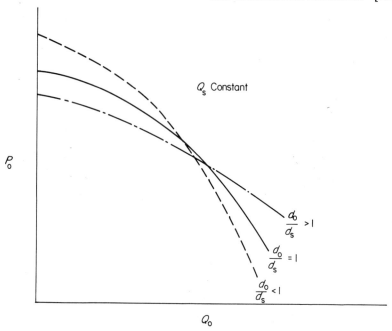

Fig. 7.4 Effect of output-tube diameter on the pressure recovery of a turbulence amplifier (Verhalst, *Ref. 3*, Fig. 4)

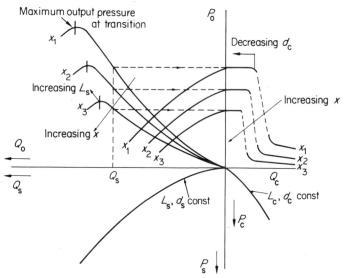

Fig. 7.5 Effect of input-output distance on the characteristics of a turbulence amplifier (Verhalst, *Ref. 3*, Fig. 7)

[7.2] THE TURBULENCE AMPLIFIER

for this are clear from the discussion of Chapter 3 on the transition of a jet, where it is pointed out that transition begins at a low critical Reynolds Number which depends upon the local width of the jet. If the receiver is moved further downstream, the new part of the jet may become turbulent, so that the supply flow has to be reduced to prevent this happening. On the other hand, the supply tube length can be lengthened up to a point in order to stabilize the jet at the higher supply flow. These moves may be necessary in order to decrease the OFF signal at the receiver for good discrimination. The sensitivity may be increased by reducing the control tube diameter d_c and thus increasing the velocity of the control flow. It is also increased by increasing x, but this reduces the output pressure so that the fan-out is little altered. Clearly it is possible to optimize on the characteristics to give optimum values to fan-out or power gain, and Drazan[4] considers the optimization of this latter parameter.

Verhalst goes on to describe the planar turbulence amplifier which is a two-dimensional version of the original cylindrical design. This presents an advance in performance; but its main significance is that it may be produced by injection moulding in a very cheap form—an estimated cost of 3d. is quoted for a production of more than 100,000 per year. The design, shown in Fig. 7.6, may be clamped on to a manifold and has two input channels to provide a two input NOR function. Later designs have four inputs.

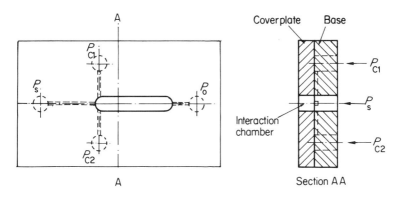

Fig. 7.6 Schematic view of two-dimensional (planar) turbulence amplifier (Verhalst, *Ref. 3*, Fig. 11)

One virtue of the planar design is that a faster switching speed is obtained. In the older, cylindrical design, a switch OFF time of 2 ms was quoted with a switch ON time of 5 ms, giving a cycle time of 7 ms, which was poor compared with equivalent wall attachment elements. The planar design on the other hand, has times of approximately 1 ms OFF and 2 ms ON.

Siwoff[5] describes a modification to the basic turbulence amplifier that both improves the response time and lowers the residual pressure in the switched OFF mode. Beginning with an external feedback back to the input from close to the receiver in order to improve the switching OFF action, Siwoff simplifies the idea till only internal feedback is used as shown in Fig. 7.7.

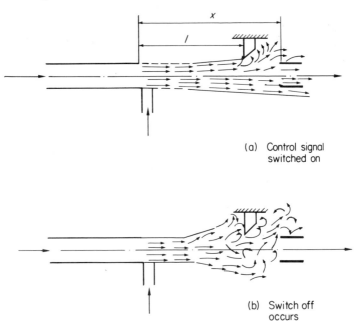

(a) Control signal switched on

(b) Switch off occurs

Fig. 7.7 (Siwoff, *Ref. 5*, Fig. 5)

When a control signal is applied the jet immediately widens and strikes the edge (a). It is claimed that feedback from the edge assists the destabilizing process and fast switching OFF occurs (b); fast switching ON also occurs, but the fast speed may be due as much to the two-dimensional configuration as to the edge since the planar amplifier described by Verhalst also has short switching times. The most important aspect of the design, however, is that the OFF signal is zero, as shown in Fig. 7.8.

The results are given for a device without an edge and also for devices with alternative edge shapes, but essentially any of the shapes gives an excellent cut-off. The reason for the improvement is not explained

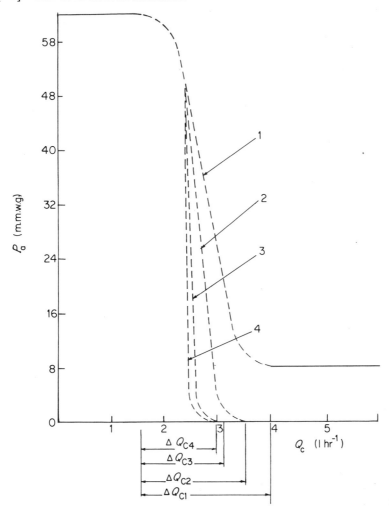

Fig. 7.8 The static characteristics of the turbulence amplifier. 1. Without an inbuilt edge, 2. Flat edge, 3. Semi-circular edge, 4. Circular orifice of the edge the main jet is flowing through (Siwoff, *Ref. 5*, Fig. 8)

clearly, but it is more than likely that the main jet is diverted slightly so that the part of the turbulent jet which enters the receiver is a low velocity one.

O'Keefe[6] uses an alternative turbulent amplifier, the action of which is described in another reference quoted in the paper. Here again, the OFF output signal is very low, but this appears to be due to the design of the interaction region where a negative static pressure is developed when the jet is turbulent. The shape is shown in Fig. 7.9.

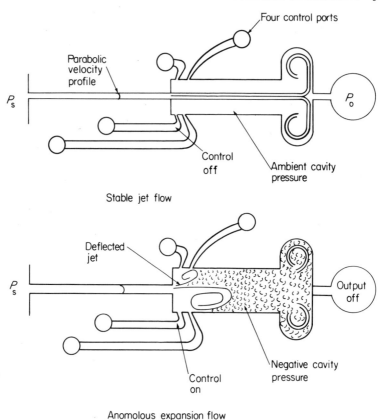

Fig. 7.9. An alternative turbulence amplifier design which gives a low '0' signal (After O'Keefe, *Ref. 6*)

7.3 The Wall Attachment Amplifier

7.3.1 Introduction

Consider a jet emerging from a nozzle into a region where the fluid has lower, or even zero, velocity. Because of the turbulent shearing action in the case of a turbulent jet, the jet carries with it some of the surrounding fluid; it 'spreads' and the mass flow increases with distance downstream from the nozzle. This extra mass has to come from somewhere and generally arrives at the jet from the space around with a low velocity; the pressure throughout the flow field is almost uniform. However, if a wall is placed close to the jet, the fluid drawn in has to pass through the restricted space between the wall and the jet, with a consequent high velocity. The sharp

[7.3] THE WALL ATTACHMENT AMPLIFIER

curvature of flow as it enters the jet creates a low pressure and an embryo vortex. This low pressure 'bends' the jet towards the wall which lowers the pressure still further. The situation is unstable and the jet attaches to the wall some distance downstream from the nozzle and forms a bubble region in which there is a strong vortex flow.

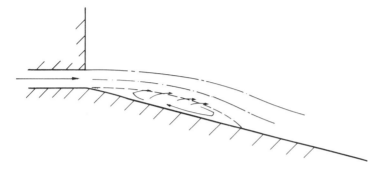

Fig. 7. 10. Formation of a bubble region by jet attachment

Eventually equilibrium is reached, as shown in Fig. 7.10, where the radial acceleration of the fluid in the jet, flowing in a near circular arc, is balanced by the pressure difference across the bubble. Thus the pressure in the bubble is low and its value satisfies a momentum condition in which the entrainment into the jet is supplied by the fluid reversed at the re-attachment point; a situation governed by the angle at which the jet hits the wall and the pressure gradient along the wall. The smaller the angle the jet makes with the wall at re-attachment the more difficult for the flow to be reversed into the bubble and therefore the bubble pressure is low and the bubble short.

To prevent the jet attaching to the wall, it is only necessary to supply enough flow into the re-attachment bubble to satisfy the entrainment demands. It turns out that this flow is relatively small compared to that in the main jet. Thus a digital amplifier may be obtained using this principle, and in Fig. 7.11 a diagram of a typical bistable device is shown.

The main supply flow Q_s issues out of a nozzle of width W and the jet attaches to either of two walls that begin at an offset distance, or 'setback', D from the edge of the nozzle and make an angle α with the nozzle centre line. A control flow Q_{C1} sufficient to cause switching will cause the jet to leave '0' output and pass down channel '1'. This situation may be changed after Q_{C1} has gone off by applying sufficient control flow Q_{C2}, when the jet will switch back to output '0'. The action is a rapid one once the jet has left the wall to which it was attached, and small elements have switching times of less than 1 ms.

If an output is loaded by a restriction or even a blocked load, not all the flow will be able to pass down the output channel and the vents allow this to spill to atmosphere. The position of the vents in relation to the splitter is quite important for good pressure recovery and for sensitive operation,

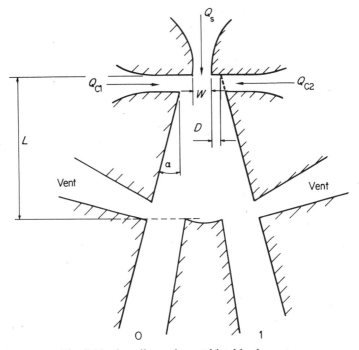

Fig. 7.11. A wall attachment bistable element

and the shape of the splitter strongly affects the stability of the jet in its attachment to the wall over a wide range of load conditions. The pressure recovery also depends very much on the distance L of the splitter from the nozzle.

The fluid mechanics of the attachment process was dealt with in Chapter 3 and will not be discussed here, but the following notes are intended as a guide to the effect of various geometrical parameters on the performance of digital elements of this type.

7.3.2 *The Effect of Wall Length, Wall Angle and Setback*

The first important observation is that the length of the bubble formed on a single wall adjacent to a jet depends upon the 'setback' D, and the angle α the wall makes with the centre line of the nozzle. Fig. 7.12 shows the relationship.

[7.3] THE WALL ATTACHMENT AMPLIFIER

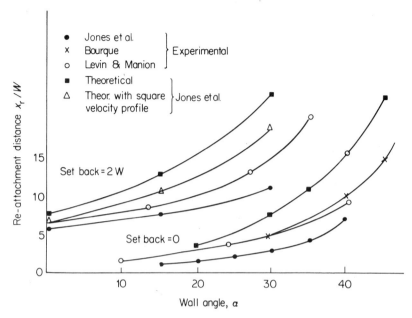

Fig. 7.12. The variation of bubble length with setback (D/W) and wall angle (Foster, Jones and Mitchell C.F.C. (2), paper A3, Fig. 11)

It may be seen that very short bubbles of the order of 3 or 4 nozzle widths may easily be obtained, but that this distance quickly increases with both setback and wall angle. Fig. 3.21 of Chapter 3 shows more clearly that there is a maximum wall angle for attachment of about 60°. Physically, the explanation is that if the wall angle is increased without altering the setback, the angle at which the jet hits the wall is increased so that the required fluid is reversed more easily into the bubbles so that the pressure increases and the bubble grows bigger. At the same time the increase in the wall angle increases the length of the jet around the bubble, which increases the amount of flow entrained into the jet; this only partly offsets the first effect and does not prevent the bubble growth. A similar situation occurs if the setback is increased.

Any receiver to collect the flow must be placed after the re-attachment point, and it has already been shown that the larger the distance downstream of the nozzle, the lower the velocity in the jet and, therefore, the pressure recovery. Therefore, for high pressure recovery it is desirable to work with small wall angles and setback. As will be seen later in the discussion on the splitter shape, some compromise must be reached between the various requirements because it is physically not possible to have a splitter very close to the nozzle with a small wall angle and still have a large enough receiver to pass sufficient output flow.

In order to switch the jet off the wall, a control flow is introduced into the channels which open into the bubble region as shown in Fig. 7.11. The pressure inside the bubble then rises and the bubble lengthens. If the re-attachment point passes into the vent, fluid is drawn into the bubble easily from atmosphere through the vent and the jet no longer attaches to the wall, i.e. it switches off. The amount of flow required to switch the device is, therefore, the quantity necessary to lengthen the bubble by the appropriate distance. This is a function of two parameters; firstly the length of the free surface of the jet inside the bubble increases as the bubble lengthens so that a direct increase in entrainment results; secondly, less is reversed into the bubble as the pressure rises so that a little extra switching flow is consequently required. Whatever the mechanics of the process, it is clear that the closer the vent to the original re-attachment point (and therefore to the nozzle), the smaller the control flow necessary for this action, and therefore the flow gain of the element is increased. This statement is modified in the discussion on the effect of output loading in the next section because there is a considerable interaction of a number of parameters in the operation of a wall attachment device.

The effect of wall angle on the rate of change of bubble length with control flow may be calculated, as reported in Chapter 3, or may easily

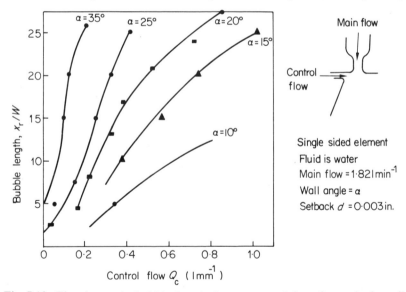

Fig. 7.13. The change in bubble length due to control flow for a single wall (*Ref. 7*, Fig. 7)

[7.3] THE WALL ATTACHMENT AMPLIFIER

be measured; measurements reported in Ref. 7 are given in Fig. 7.13. From these results it may be seen that for a high flow gain a large wall angle is desirable. However, as the wall angle is increased the nominal length of the bubble increases, so that the splitter and vents must be placed further downstream with a consequent loss in pressure recovery. This may not matter if a circuit is designed only on the basis of flow gain, and the successful I.B.M. design of elements with very broad splitters are reported to have been produced with flow gains of between 30 and 40. Certainly, this type of design is easier to manufacture than those with small wall angles and narrow mixing zone and probably is more tolerant of manufacturing inaccuracies. If pressure recovery is important, the splitter may be moved closer to the nozzle by causing the jet to bend through a larger angle more quickly. One way to do this is to effectively have a 'negative' setback. Clearly it is not possible to have the wall interfering with the jet, so that part of the wall which would interfere with the jet is cut away. The result is that the angle at which the jet hits the wall is smaller than it would otherwise have been, resulting in a low bubble pressure and a short bubble. The same may be seen in Fig. 7.14, and clearly curved wall shapes are

Fig. 7.14. A method for obtaining a short bubble length with a large wall angle by using effectively a negative setback with part of the wall cut away

possible. However, it is important to realize that in order to achieve this result, the pressure difference across the jet must be high, so that the bubble pressure on the attached side is necessarily low whatever the shape of the local region where the wall is cut away (providing that the cut away is not too large). Consequently, in order to have a stable jet with open control ports, the impedance of the control channels must be high. The difficulty of this is that to promote switching the pressure at the control inputs must then be high because of the high impedance, and care must be taken to see that this does not reach an unacceptable level.

The sensitivity may be further influenced by the fact that the momentum of the control flow deflects the main jet and effectively increases the wall angle. In addition, a pressure rise is created in the control channel for small setbacks because the small distance between the jet and the wall edge restricts the flow, and this further increases the deflection.

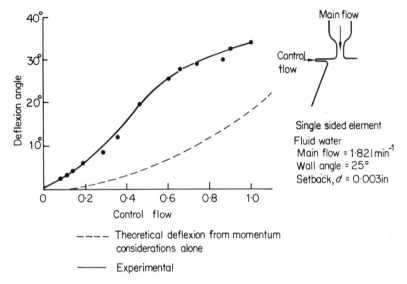

Fig. 7.15 The jet deflexion due to control flow for a single wall (*Ref. 7*, Fig. 6)

Fig. 7.15 shows the deflexion angles obtained in a specific element for a range of control flow in a single wall configuration that is typical of wall attachment element geometry. For such an element a switching flow ratio of 1/6 might be expected which gives a switching flow of about 0·3 l/min^{-1} at which point a jet deflexion angle of 10° is observed. This is considerably more than predicted from momentum considerations and certainly must contribute to the sensitivity of such an element if a second wall is introduced, because the pressure will be locally lowered near the second wall as the jet deflects and will, therefore, assist the lengthening of the bubble. Such an effect will again be increased by having a small setback, and it is reported[7] that for a specific wall angle and wall length, varying the setback showed a maximum sensitivity with $D = (1/4)W$. The problem, however, was that at small setback distances, the sensitivity changed fairly rapidly so that the effect of errors in the manufacture of an element could create great differences in characteristics; this is particularly noticeable in that similar elements from a batch may vary from monostability on one side through bistability to monostability on the other side, even though they appear superficially to be accurately made.

[7.3] THE WALL ATTACHMENT AMPLIFIER

To conclude this section, a note is given on Brown's knife-edge amplifier mentioned in section 3.7.7 and described in Ref. 50 of that chapter. Fig. 3.30 shows the device. The important point is that the 'attachment point' is localized at a knife edge, so that the effect of downstream loading is less complicated, as is the mechanics of switching. The effect of changing the setback is different from in the normal wall effect device; as the setback increases, the radius of curvature must decrease and the bubble pressure reduce. In the normal device, the effect of increasing the setback is to allow the bubble to lengthen, but this is no longer possible in the 'knife-edge' device. Of course, as the setback increases, the jet is turned through a larger angle, and it is relatively simple to design an amplifier intuitively. So far, however, this design does not appear to have been used commercially. Perhaps the edges are sensitive to manufacturing errors.

7.3.3 *The Effect of Loading of the Output Channel*

If the receiver is restricted, that is the element is loaded, the pressure rises in the receiver channel and the flow surplus to requirements passes out of the bleed or across the splitter as indicated in Fig. 7.16.

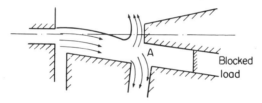

Fig. 7.16. The flow passing out of the vents or across the splitter, in the case when the output port is restricted or blocked

The pressure rises at point A and if reference is made to the theoretical re-attachment model shown in Fig. 3.20, it may be imagined that the pressure rises along the boundary DE. For the momentum equation parallel to the wall now to be satisfied, more flow must reverse back into the bubble. To accept the increased back flow (which has the same effect as control flow) the bubble must lengthen. Hence, the jet may switch because loading makes the bubble lengthen as far as the vent port. Such an element would be *load sensitive*.

Two obvious remedies to load instability are to (*a*) lengthen the wall, and (*b*) open the vents. Method (*a*) also reduces the flow gain, and to be effective, method (*b*) can reduce the pressure recovery under blocked load conditions. Neither of these remedies by themselves leads to satisfactory results, and the influence of the splitter position and shape is found to be critical for load stability and good performance generally.

7.3.4 *The Effect of Splitter Position and Shape*

Consider now a complete element with two walls, two vents and a splitter dividing the output into two receivers, as shown in Fig. 7.17.

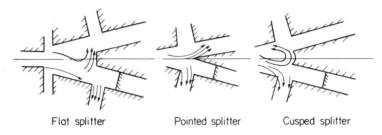

Flat splitter   Pointed splitter   Cusped splitter

Fig. 7.17. The effect of different splitter shapes

If the output is loaded, the surplus flow passes across the splitters. If the splitter is pointed, the tendency is for this flow simply to pass down the other output channel and load stability is often difficult to achieve. If, however, the end of the splitter has a cusped shape, some of the surplus flow tends to reverse back into the jet and raise the pressure on the outer side of the bubble. This counteracts the tendency of the bubble to lengthen. In some designs, the bubble is kept a more or less constant length for all values of the output load, whilst in others, the tendency to lengthen is restricted until so much flow is passing through the vent to atmosphere that the bubble cannot draw in any fluid through the vent and a stable situation results, as shown in Fig. 7.18.

Apart from the shape of the splitter, the position downstream of the nozzle is important relative to the position of the vents, and Rechten[8]

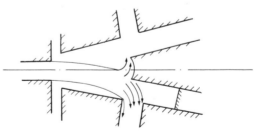

Fig. 7.18. A possible stable jet position when the output is restricted. Here the jet passes straight into the vent and therefore prevents flow being entrained into the bubble from the vents

[7.3] THE WALL ATTACHMENT AMPLIFIER 317

confirms our own experiences at Birmingham in suggesting that there is an optimum position where the change in flow patterns in the main chamber of the element during a full variation of loading is small. One of his element shapes is reproduced in Fig. 7.19 and the results[8] of variation of splitter position for this basic design are shown in Fig. 7.20.

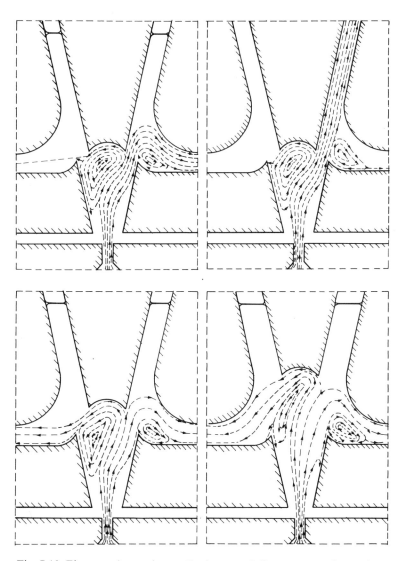

Fig. 7.19 Element shape, due to Rechten, and flow patterns for various downstream positions of the splitter (Rechten, *Ref. 8*, Fig. 8)

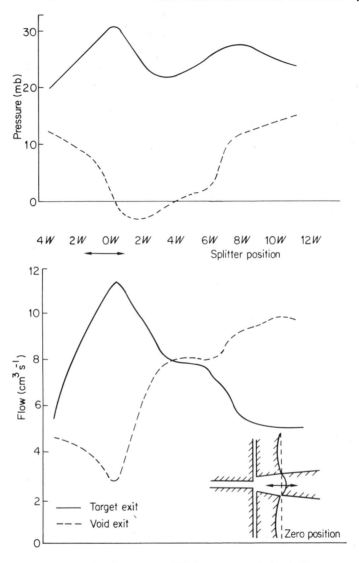

Fig. 7.20. Effect of the position of the splitter on the pressure and flow in the vents of the element (Rechten, *Ref. 8*, Fig. 7)

For small distances from the nozzle, i.e. negative movements according to the horizontal axis of Fig. 7.20, the flows and pressures in both exits are similar. (The 'target exit' is on the side to which the jet should attach,

[7.3] THE WALL ATTACHMENT AMPLIFIER

whereas the 'void exit' refers to the opposite side.) This implies that the jet is not attaching. As the splitter is moved away from the nozzle the difference between the two sides increases to a maximum at a value of around $0.5W$, but then reduces again. The reduction is because the jet width increases and gives a large spillover across the splitter into the 'void' port. Rechten also considers the use of attachment walls that are concave instead of straight and finds some improvement in pressure recovery. As explained in Chapter 3, his reasoning is that there is a better chance of obtaining a 'trapped' single vortex in the attachment bubble which would result in less energy loss; certainly the experimental results quoted appear to bear this out.

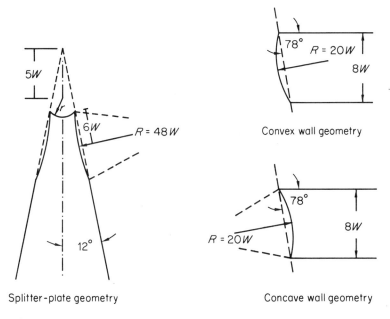

Fig. 7.21. Details of the splitter-plate, convex-wall, and concave-wall geometry (Sarpkaya and Kirschner, *Ref. 9*, Fig. 5)

Sarpkaya and Kirschner[9] on the other hand come to the conclusion that an element with convex walls has the better output characteristics. Fig. 7.21 shows wall geometry that is used in an element that is otherwise conventional. Zero setback is used and the splitter is only $5W$ away from the nozzle so that high pressure recovery is to be expected. Nevertheless, the pressure recoveries given in the output curves of Fig. 7.22 are exceptionally good and the convex wall is shown in a particularly good light. The relatively poor output characteristics of the concave wall suggest

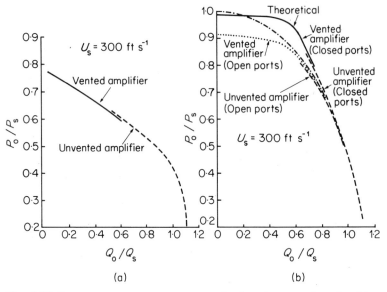

Fig. 7.22 Pressure recovery versus normalized active port flow for (a) a concave-walled amplifier and (b) a convex-walled amplifier (Sarpkaya and Kirschner, *Ref. 9*, Figs. 10 and 11)

that the diffuser action is poor and it may be that the change in geometry upstream of the receiver has had some effect on the flow in the receiver itself; it is almost impossible to cover every possible effect of a design change because so many variables are involved. One rather poor characteristic of the design is that the shape of the control port characteristic changes significantly after switching in this amplifier, but this is an effect observed in any amplifier with very small setback, because, as already discussed, the factor that controls the control port impedance is the gap between the jet and the wall. For very high pressure recovery a small splitter distance and a small setback appear to be necessary, and switching characteristics of this kind must be accepted. The most suitable overall performance must be chosen to suit a given application and it is difficult to arrive at an ideal solution for every case.

The width of the output receiver is also important. If it is wide, a large flow will pass out of the element under no-load conditions, and since it includes that drawn into the jet from the surroundings, an output flow greater than the supply flow may be registered. On the other hand, the pressure recovery is then small because a large proportion of the fluid has low velocity. If the receiver is narrow, the pressure recovery will be high, but the output flow will be reduced. Experience shows that an output

[7.3] THE WALL ATTACHMENT AMPLIFIER 321

no-load flow of about the same, or slightly less than the supply flow is desirable. If it is larger, the change of flow patterns upstream of the splitter between no-load and full load is too great for stability of the element.

7.3.5 The Effect of Vent Shape

Sarpkaya and Kirschner[9] also point out that the output characteristics of unvented and vented amplifiers are continuations of each other, or to put it another way are two different parts of the same curve, and this point is illustrated in Fig. 7.22. With no vents the amplifier load switches when the output pressure rises above a certain level and so the curve for that case ends abruptly. The vented amplifier will operate with the outputs fully loaded, but at the lower output pressure range may not pass so much flow since some is being dumped in the vent (although this is not necessarily so).

The shape of the vents are important and in the early designs of wall attachment elements, it was thought that vents should be angled as far back as possible in order to provide good pressure recovery, as in Fig. 7.23.

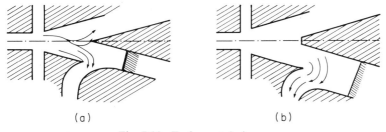

Fig. 7.23. Early vent designs

The idea behind this was that under blocked load conditions, the pressure recovered would be due to the momentum change due to turning the jet through a large angle. The idea is good, but the jet takes the line of least resistance, and there are easier passages out through the element than turning through a larger angle. In Fig. 7.23(a), the fluid will only cross the vent boundary at about 90° to the wall before the rest of the flow passes down the inactive channel, because there is separation from the sharp edge as the flow turns round into the vent. This is physically reasonable because the pressure rise calculated from momentum change should agree with that available from the energy left in the jet at that point, and the flow patterns will adjust themselves accordingly. In Fig. 7.23(b), the extreme of this design philosophy is shown in which it is hoped that the flow will turn through nearly 180°, but the momentum change acts over double the receiver area so that the pressure recovery is no better—in fact designs of this type have proved to be poor.

This is borne out by the results shown in Fig. 7.24 where a number of designs of vents were tested. The worst by far was that of 7.24(g), which corresponds to the sketch 7.23(b). However, it was shown that some advantage could be gained by angling the vent forward a little from the

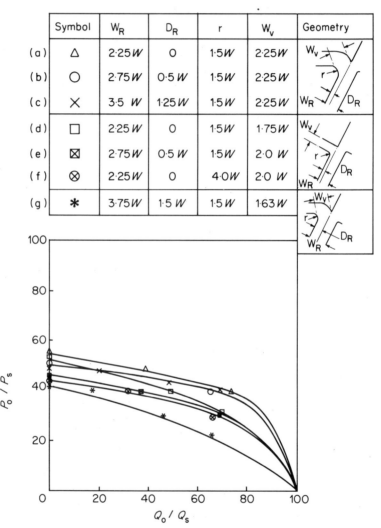

Fig. 7.24. The effect of the vent receiver configurations on the output pressure-flow characteristics for a particular element (Misra, A. K., University of Birmingham, Ph D. Thesis)

[7.3] THE WALL ATTACHMENT AMPLIFIER

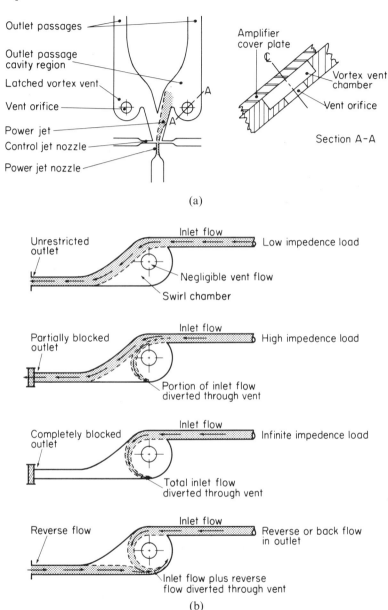

Fig. 7.25. Vortex vent operating principal. The inlet to the vortex vent is the receiver of the fluid element. In practice, the vortex is very close to the receiver so that the inlet length to the vent is short (Hayes and Kwok, *Ref. 10*, Figs. 1 and 2)

90° angle and this may be seen by comparing the results of (a), (b) and (c) with those of (d), (e) and (f) in the figure. The characteristics given by shape (a) was the best that was achieved in a wide range of experiments and it was clear that more acute angles for the vent did not improve the pressure recovery.

The sketches of Fig. 7.23 show rather more flow going out of the vent in the attachment side than going past the splitter. If this is so, then by looking at a momentum balance perpendicular to the wall, it may be deduced that the pressure on the other side of the jet from the vent must

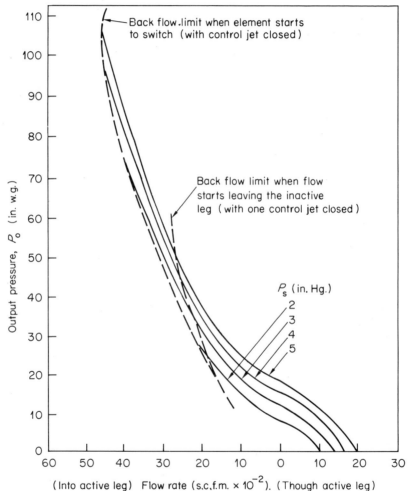

Fig. 7.26. Bistable fluid amplifier with vortex vent back flow characteristics (Hayes and Kwok, *Ref. 10*, Fig. 9)

rise in order to keep the jet on the wall. This may happen because of the splitter if the vent is placed downstream of the splitter end, but if this is not so, the element will load switch. The point about the cusped splitter is that it causes some flow to reverse round into the interaction region and raise the pressure there when the output port is restricted. Any flow surplus will pass out of the vent on the unattached side, and the momentum across the element will be balanced.

In practice, the design shown by Rechten[8] is about as good as one can get, and it is of interest to note that with the best splitter position, the width of the bleeds is not critical.

An alternative design which has something to recommend it is a design of element using the vortex vent described in Ref. 10 and shown in Fig. 7.25(a); the operating principle is shown in Fig. 7.25(b). Here, on loading the element, the surplus fluid is caused to form a vortex around the vent and a pressure difference between the atmospheric pressure at the vent and the higher pressure of the receiver is created. In practice the pressure recovery is little better than that of other elements and the shape of the output flow curves are not too good. On the other hand this design will tolerate a high reverse flow down the active limb, since this will pass straight through the vent, which may be designed to have low impedance under these conditions.

The output characteristic is given in Fig. 7.26. This type of vent is good, for example, in situations where an actuator (or piston valve) is being retarded and reversed by the element. It is also good for pulsed operation at high speed because reflected waves will pass directly out of the vent.

7.3.6 The Effect of the Diffuser

For a good pressure recovery under all flow conditions, the high velocity of the jet must be slowed down smoothly so that kinetic energy is transferred into potential energy, i.e. pressure. This requires a good diffuser. The best shape for a diffuser is now well known, and is a diverging channel with 6° between the walls. It must be long enough to ensure a good retardation of the flow but not too long so that there is appreciable pressure drop due to friction. Fig. 7.24(a) shows a good output pressure-flow characteristic where the pressure is maintained over a wide flow range. It is interesting to compare this with the characteristic of the vortex vent given in Fig. 7.26 which is not quite so good, because the original work on trapped vortices was, as reported in Chapter 3, started with the intention of using vortices trapped in cusps in order to diffuse flow in wind tunnels.

For the element of Fig. 7.24(a), in order to provide a reasonably easy passage to reverse flow, the vent was widened, and in the case of the optimum splitter position, the element performance was not changed appreciably. The width of the receiver downstream of the vent was then increased till the size of the vent again limited the reverse flow; thus the minimum impedance to reverse flow was achieved. Small increases of channel width do not affect the element too much, as in Fig. 7.24(b), but larger increases begin to affect the blocked load pressure recovery, as shown by Fig. 7.24(c). Curve (b) therefore, presents a reasonable compromise in which good forward flow characteristics are achieved together with reasonable tolerance of reverse flow. This design of element is used for the pilot operation of piston valves, where some reverse flow down the inactive leg of the element is to be expected.

Downstream of the actual vents of Fig. 7.24, a 6° diffuser was used so that the point of the exercise was to show the effect of the geometry local to the vent. In the passage of the fluid across the vent, the sudden gap may give rise to some vortex flow and it does appear to be an advantage to have a slight enlargement of the output channel at that point so that some diffusing action takes place as the flow actually crosses the vent.

7.3.7 Combinations of Wall Attachment and Vortex Element

Results using a wall attachment device above the critical pressure ratio are generally poor and some means are required to overcome this. If, as

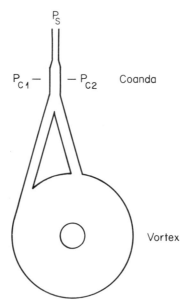

Fig. 7.27. The use of a vortex on the output of a wall attachment device (Royle, *Ref. 11*, Fig. 9)

[7.3] THE WALL ATTACHMENT AMPLIFIER

suggested by Royle[11] a vortex is added to the output of a wall attachment device as shown in Fig. 7.27, the jet will flow out straight through these to the output of the vortex when directed along one wall, but when switched

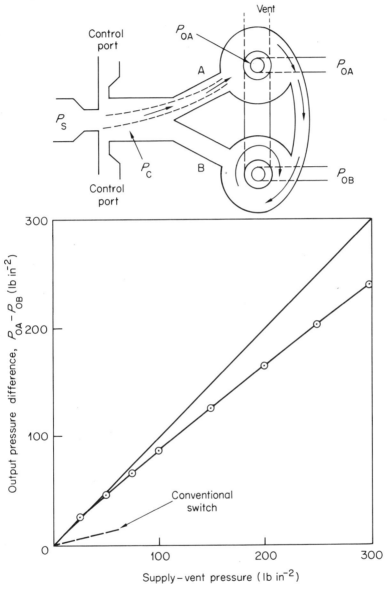

Fig. 7.28. Use of a vortex on each output of a wall attachment device (Rimmer, *Ref. 13*, Fig. 6)

to the other will cause the vortex to operate. In the latter case, the impedance goes up, the pressures rise throughout the system, and the main flow is restricted. The element is stable without vents because the pressure levels rise uniformly in the attachment region. This device has only a single ended output which may be taken either at the periphery of the vortex or preferably from a pitot tube at the vortex output. A use of this type of design is reported by Hart[12].

A further elaboration of this idea is given by Rimmer[13] where a vortex chamber is added to each output leg to make a symmetrical, bistable device that will operate at high pressure ratios (Fig. 7.28). When the element is switched into one channel, some surplus flow passes into the tangential channel connected between the vortex chambers, so that the other vortex is operated. This raises the pressure on the side of the jet opposite the reattachment and so stabilizes the bubble. Very high output pressures may be used.

7.4 Monostable Devices

Most monostable wall attachment devices are derived from the bistable elements because it is relatively easy to make a bistable element monostable by small changes in the design. For example, it is relatively easy to make a bistable element monostable by opening up one control port; enough fluid is then drawn in from the atmosphere to allow the device to switch back when the control signal on the input is removed. However, this is not necessarily the best design, because it relies on weak attachment to the OR wall in order to be able to pass down the OR output channel and a balance of the impedances of the control port and what is now a vent port in the bubble region of the OR wall is required. Some designs have a different setback on the two walls, but little work has been done on testing various asymmetrical configurations. Hart[12] reports on results of changing the angle at which the main jet enters the element and for a particular application finds some improvement in performance. However, his design was the opposite of what one would normally require since it was necessary in his case to have an exceptionally strong attachment on the NOR wall, so that the main jet was angled towards that wall.

In a study of a new laminar NOR device, the shape of which is shown in Fig. 1.28, Walker and Trask[14] did not start with any preconceived idea; the jet does not attach strongly to the NOR wall and therefore the NOR wall is placed parallel to the nozzle centre line. The OR output is achieved by momentum interaction and consequent jet deflexion. The reference optimizes the design for maximum gain.

To follow this type of thinking to a conclusion, a wall attachment NOR

[7.4] MONOSTABLE DEVICES

element could have the OR wall parallel to the nozzle centre line and could be vented so that no attachment took place on that wall. However, some compromise is probably necessary, as usual, and more work could be done in experimenting with asymmetrical geometries in the wall attachment OR/NOR element.

7.5 The Switching Action of a Bistable Wall Attachment Element

If a bistable element is used in a real circuit, the output loading condition can vary widely from complete blocking down to even a suction pressure in the active output channel. This can lead, in turn, to variations in the switching characteristics, but the magnitude and sign of the variations will depend on the shape of the splitter and the position of the vent. These effects are summarized as follows:

(a) Element with a Pointed Splitter

With no output load, the effect of control flow is to increase the bubble length, i.e. move the re-attachment point downstream, and depending on which begins first, switching is either caused when the bubble opens into the vent, thus drawing fluid in from the vent to assist switching, or the jet interferes with the splitter and switching takes place. With an output load, the bubble is caused to lengthen and if the bubble interferes with the splitter before the re-attachment point reaches the vent, switching is assisted by the output loading, so that the switching flow is reduced. If, however, the re-attachment point reaches the vent first, a stable situation can occur where the jet continues on a curved path straight out of the vent. Under these conditions, no air can be drawn in through the vent and the amount of control flow necessary for switching is actually increased.

These variations of flow required to switch may be only of the order of 20 per cent, but the shape of the control port impedance characteristic changes, because the suction pressure in the control port with no control flow is reduced by loading.

(b) Element with a Cusped Splitter

The cusp causes a reversal of flow into the mixing region which creates a vortex on the outer edge of the jet, and for stability the centre of the vortex is positioned nearly opposite the re-attachment point. As the load is applied, the strength of this vortex increases and the low pressure opposite the reattachment tends to reduce, and so counterbalance the rise in pressure along the wall near the re-attachment point, so that the bubble does not elongate. The introduction of control flow in either the loaded or unloaded state does not cause appreciable increase in the bubble length;

probably because the cusp reverses still more flow into the mixing region, causing the pressure to rise there, so that although the bubble pressure rises, the radius of curvature does not change appreciably. This, however, is a critical situation and at some point the vortex at the splitter is shed and the re-attachment vortex moves over to take its place and switching occurs.

In both cases, for small setbacks, a control flow may cause the jet to deflect so much that it touches the opposite wall. A small bubble forms on that wall and assists the main switching process (photographs, Ref. 15). In both cases, if the splitters are far enough downstream of the reattachment point, the pressure in the receiver cannot be transmitted upstream to the bubble, because any small curvature of the jet away from the wall causes the jet to pass down the inactive output channel, even though still attached to the original wall, as indicated in Fig. 7.29.

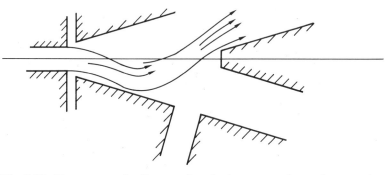

Fig. 7.29. Downstream loading causing the jet to pass down the opposite output channel

There are many variants on these particular effects, since there are a large number of possible variations of geometry in an element, and in particular, alterations in the setback and wall shape can have a large effect. Fig. 7.30, from Ref. 16, shows only isolated, but instructive, examples of the phenomena, where the control port characteristics of an element with a cusped splitter are compared with those for a pointed splitter which are shown for two different vent positions, in the loaded (full line) and unloaded (broken line) conditions.

Considering the sharp splitter first; when the vent position is nearer to the nozzle, the control pressure levels are generally lower than those for the vent a greater distance from the nozzle. Since the setback is almost the same in each case, this is probably because the impedance of the output channels is sufficient to cause a higher pressure in the interaction region. With a shorter vent distance, the switching flow is generally lower than in the other case, but in particular, the switching flow for a loaded output is

[7.5] SWITCHING ACTION OF BISTABLE WALL ATTACHMENT ELEMENT 331

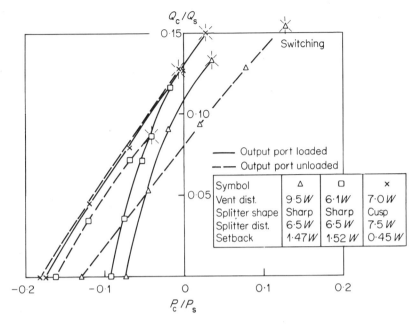

Fig. 7.30. Control port characteristics for an element with a pointed splitter and two vent points, under both output loaded and unloaded conditions. Also shown are the characteristics for a cusped splitter (*Ref. 16*, Fig. 10)

greater than that for an unloaded output. For a greater vent distance, the reverse is true, thus supporting the general remarks made above.

A further important difference is that the slope of the control port characteristics (and, therefore, control port impedance) is less in the loaded condition than in the unloaded condition, and that the slopes do not vary much with vent position. From Schlieren observations, a possible explanation is that the impedance, even with this relatively large setback, is dependent mainly on the gap between the jet and the wall. The effect of loading in both cases is to widen this gap; for the shorter vent, a stable position is reached under load with the jet going straight out of the vent, and this appears to result in a rather larger bubble volume than might be otherwise expected; for the longer vent, the effect of loading is to move the jet off the wall so that it attaches to the splitter, again giving a larger bubble height.

For comparison, the results are given for a cusped splitter, which show no change of impedance with loading, an effect attributed to the fact that the vortex created by the cusp keeps the length and height of the bubble more or less constant. The switching flow is greater in the loaded

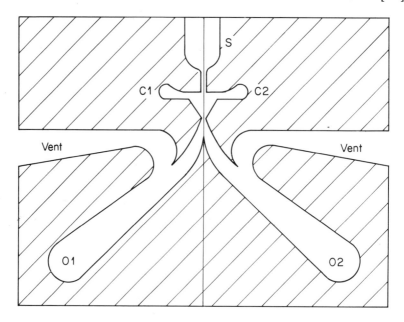

(a) Bistable digital amplifier with converging walls.

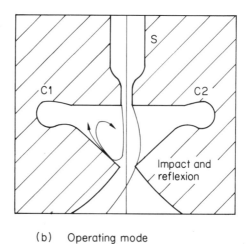

(b) Operating mode

Fig. 7.31. An alternative design of bistable amplifier
(Facon, *Ref. 17*, Figs. 1 and 2)

condition, and again the vent distance is relatively short, so that it would suggest that the distance of the vent from the nozzle is still important in determining the difference in switching flow between the loaded and the unloaded condition.

Note that the switching pressure varies from negative through zero to positive. For good digital operation, the switching pressure should be half that available from the output of an element—and the output characteristics should show little variation of pressure with flow, i.e. a good diffuser action. To obtain a large fan-out, good sensitivity is required, which usually results in a sub-atmospheric switching pressure. This switching pressure may be increased by increasing the control port impedance, which in turn may be realized by having long narrow control channels or by having a very small setback which restricts the flow between the wall and the jet.

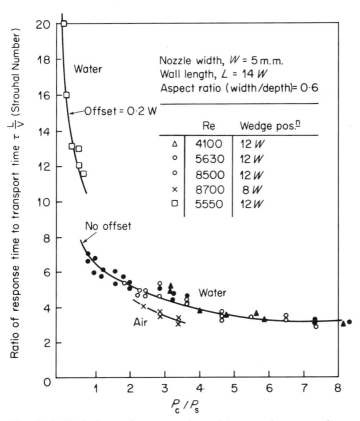

Fig. 7.32. Variations of response time with control pressure for a particular element (*Ref. 18*, Fig. 7)

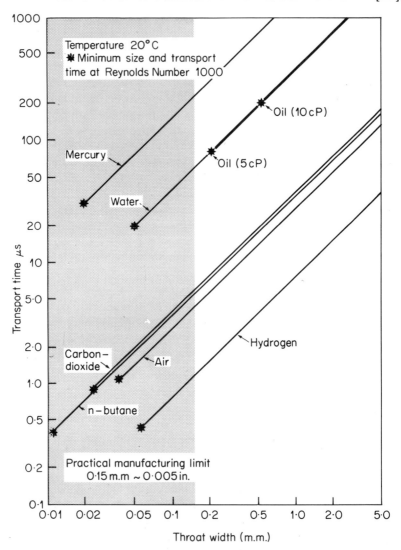

Fig. 7.33. Variation of transport time with linear size for an element length of 10 nozzle widths using a variety of fluids (Comparin, Mitchell, Mueller and Glaettli, *Ref. 18*, Fig. 8)

An alternative design of element was recently described[17] in which the interaction region is ingeniously isolated from the attachment region (Fig. 7.31(a)). The jet flows through a control area, as shown in Fig. 7.31(b), and passes through a narrow gap formed between the two walls

[7.5] SWITCHING ACTION OF BISTABLE WALL ATTACHMENT ELEMENT   335

which are inclined at an opposite angle to the nozzle centre line from that normally accepted. The control pressure forces the jet into an arc of a circle which then directs the jet on to the wall on the *same* side as the control signal, and it subsequently remains attached. The advantage of this geometry is that the element operates on control pressure rather than flow and changes in output loading are less likely to affect the pressure field within the interaction region.

Whatever the switching action, the switching speed appears to depend largely on transport time, that is the time taken for a particle leaving the nozzle to travel as far as the output receiver. Therefore, the higher the operating pressure and the shorter the element, the faster the switching.

Fig. 7.32[18] shows the ratio of response time to transport time (a non-dimensional number known as the Strouhal number), as a function of the ratio of control pressure to supply pressure (and therefore, of control flow to supply flow), for a particular element. Johnson[19] shows that Strouhal number for any element is more or less independent of Reynolds Number. Lush[20] has calculated switching 'off' and re-attaching 'on' times for a turbulent jet on to one adjacent wall, and good correlation between theory and experiment is obtained as explained in Chapter 3. The Strouhal numbers quoted are considerably larger than those of Ref. 18, but the definition of Strouhal number is different, so that the two results are

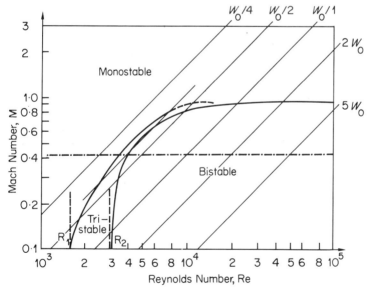

Fig. 7.34. Ranges of stability for one particular element operating on a gas. W_o is the width of the supply nozzle of an element taken as a nominal size (Glaettli, *Ref. 21*, Fig. 2)

similar. In any case the switching time must be a function of element geometry and therefore, the results need not agree exactly.

Fig. 7.33[18] gives the transport time for an element length of 10 nozzle widths as a function of linear size for various fluids. The lines are drawn for maximum jet velocities which are limited by sonic conditions in the gases and cavitation in the liquids. The starred points are given for Reynolds Numbers of 1000 and represent the limit of possible bistable performance where the limiting Reynolds Numbers and Mach or Cavitation Numbers are approached simultaneously. In practice, the manufacturing problems limit nozzle size to a larger value than those given by the starred points. Knowing the transport time, the actual switching time may be obtained by multiplying by the Strouhal number which depends on the control pressure applied.

Glaettli[21] gives more information about the limiting effects of Reynolds Number and Mach and Cavitation Numbers. The graph showing regions of bistable, monostable and tristable performance is reproduced in Fig. 7.34 for one particular element operating on a gas. The region of tristable performance is interesting, but in this case is rather a small area (I.B.M. use tristable elements so that it is possible to design for this mode). The upper limit of Mach number is shown as being asymptotic to unity, but this is a coincidence, and the actual limiting value depends upon the element design. The experiments were carried out on a series of geometrically similar elements and lines of constant size are superimposed across the curves.

The effect of Cavitation and Mach Numbers are similar: as Mach Number increases, the density changes become important and the bubble pressure does not fall as low as it ought, thus the bubble lengthens; in the case of a liquid, as the Cavitation Number increases, air comes out of solution or vapour is formed in the bubble so that the pressure cannot fall enough, and again the bubble lengthens. As Reynolds Number is reduced, flow into the bubble through the boundary layer on the cover plates becomes important with the result, once again, that the bubble lengthens. For a given element, a longer bubble generally means some sort of undesirable operation.

References

1. Auger, R. N., Turbulence amplifier design and application, *H.D.L.* (1), **I**, 357–366.
2. Oels, R. A., Boucher, R. F. and Markland, E., Experiments on turbulence amplifiers, *C.F.C.* (1), Paper D3.
3. Verhalst, H. A. M., On the design, characteristics and production of turbulence amplifiers, *C.F.C.* (2), Paper F2.

REFERENCES

4. Drazan, P., Optimal design of the control jet of a fluid amplifier, *C.F.C.* (2), Paper B2.
5. Siwoff, F., Improvement of the static and dynamic behaviour of the turbulence amplifier by inbuilding of an edge over the distance between the emitter and the collector, *C.F.C.* (3), Paper H2.
6. O'Keefe, R. F., Fluidic decimal counter for digital control applications, *Adv. Instr.*, **23**, 11; *I.S.A. Conf.* (23), Paper 93.
7. Foster, K. and Jones, N. S., An examination of the effect of geometry on the characteristics of a turbulent re-attachment device, *C.F.C.* (1), Paper B1.
8. Rechten, A. W., Flow stability in bistable fluid elements, *C.F.C.* (2), Paper B6.
9. Sarpkaya, T. and Kirschner, J. M., The comparative performance characteristics of vented, unvented, cusped, and straight and curved-walled bistable amplifiers, *C.F.C.* (3), Paper F3.
10. Hayes, W. F. and Kwok, C., Impedance matching in bistable and proportional fluid amplifiers through the use of a vortex vent, *H.D.L.* (3), **I**, 331–368.
11. Royle, J. K., Operational characteristics of a vortex amplifier, *C.F.C.* (2), Paper D4.
12. Hart, R. R., Performance of a coanda device as a pressure maintainer and as a switch selector, *C.F.C.* (2), Paper B4.
13. Rimmer, R., A fluidic pressure ratio sensor for gas turbine control, *I.F.A.C. Symp. on Fluidics* (London), Paper B4.
14. Walker, M. L. and Trask, R. P., Feasibility study of a laminar NOR Unit, *A.F.*, **1967**, 162.
15. Toda, K., High speed Schlieren photography, *H.D.L.* (2), **IV**, 147–166.
16. Foster, K., Misra, A. K. and Mitchell, D. G., A note on the effect of a downstream pressure rise on a reattached jet, *C.F.C.* (3), Paper F9.
17. Facon, P., Fluid logic devices; elementary circuits and applications, *C.F.C.* (3), Paper K7.
18. Comparin, R. A., Mitchell, A. E., Mueller, H. R. and Glaettli, H. M., On the Limitations and special effects in fluid jet amplifiers, *Symp. on Fluid Jet Control Devices*, *A.S.M.E.* (Winter Meeting 1962).
19. Johnson, R. F., Dynamic studies of turbulent reattachment fluid amplifiers, *M.Sc Thesis* (Pittsburgh), 1963.
20. Lush, P. A., Investigation of the switching mechanism in a large scale model of a turbulent reattachment amplifier, *C.F.C.* (2), Paper A1.
21. Glaettli, H. H., Mach and Cavitation Number effects in fluid dynamic elements and circuits, *C.F.C.* (2), Paper B8.

8
Moving-Part Logic Devices

8.1 Introduction

Standard pneumatic valves have been used for many years to provide some degree of logic function in pneumatic systems. The introduction of devices without moving parts has stimulated the development of more complex logic arrangements of pneumatic components, and in particular has stimulated the development of miniature moving-part devices which can be interconnected readily. As the size of the elements has come down, the speed of operation has improved so that they are no longer so inferior in this respect from their pure fluid counterparts. Although reliability may still be a problem, the fatigue life of the moving parts of the newer components is probably better than that of their larger predecessors because the movements required are not so great. On the other hand, the smaller flow passages are probably susceptible to contamination and no clear picture has yet emerged.

The elements fall into two main categories:

(*a*) **those devices that act as diverter valves and**
(*b*) **those devices that act by creating a back pressure in the circuit.**

The former have the distinct advantage that the only air flow required in the system is that necessary to switch the appropriate elements. A system using back pressure devices, on the other hand, has a continual passage of air through a proportion of the elements, and though this will not be as serious as for a pure fluid system, it may nevertheless create a demand on the air supply. It may be, however, that the back pressure elements are somewhat simpler in construction and interconnexion, although these points are by no means clear.

Of the large number of moving-part devices which are available, the

[8.1] INTRODUCTION 339

following miniature ones have received most attention recently:

Diverter Valves

(1) Spool valves used either singly or in pairs
(2) A single latching spool valve
(3) A diaphragm-lever valve
(4) A double-diaphragm valve
(5) A triple-diaphragm valve

Back Pressure Devices

(1) A diaphragm and diaphragm with foil valve
(2) A planar diaphragm element
(3) A spring NOR unit
(4) Ball-valves
(5) The free-foil valve.

In addition, an interesting idea for a threshold logic device has appeared from Czechoslovakia.

Not all of the elements to be described are in large scale production, and it may well be that some good valves not described here are in production.

8.2 Diverter Valves

8.2.1 The Spool Valve

The basic piston type fluid logic element is shown in Fig. 8.1.

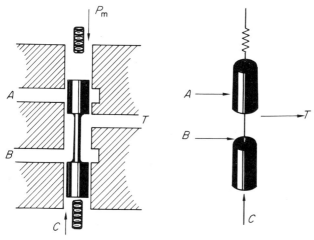

Fig. 8.1. The basic single spool element (Glaettli, *Ref. 1*, Fig. 1 and Togino and Inoue, *Ref. 2*, Fig. 1)

It is described by Glaettli[1] and by Togino and Inoue[2]. The use of this element requires three pressure levels in the system: high pressure, a medium pressure and a low return pressure. In the diagram, P_m signifies the medium pressure; C represents a signal which is at the high pressure of the system when present and at the low return pressure when absent. High pressure is signified by '1' and low pressure by '0' in the diagrams.

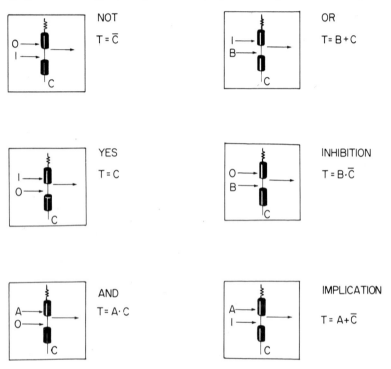

Fig. 8.2. Logic functions using a single spool valve (Togino and Inoue, *Ref. 2*, Fig. 3)

The diagrams of Fig. 8.2 show the logic functions which can be achieved with a single spool device. Notice that the valve diverts one or other of two input signals into a single output channel; this is more powerful logically than using the valve to divert a single input into one of two possible outputs. These functions are useful in themselves, but the single piston unit cannot be used for the EXCLUSIVE-OR, NOR, NAND, or EQUIVALENCE functions, which require the use of two cascaded pistons. There are several ways in which two pistons can be cascaded together to achieve these extra logic functions, and two are shown in Fig. 8.3.

[8.2] DIVERTER VALVES 341

Fig. 8.3. Two ways of cascading spool valves (Togino and Inoue, *Ref. 2*, Fig. 4)

The second of the two arrangements is not so useful because it cannot produce the EXCLUSIVE-OR or EQUIVALENCE functions, and the first one provides the more useful logic combination. Fig. 8.4 lists the functions which can be achieved by different interconnexions between the two cascaded spool valves.

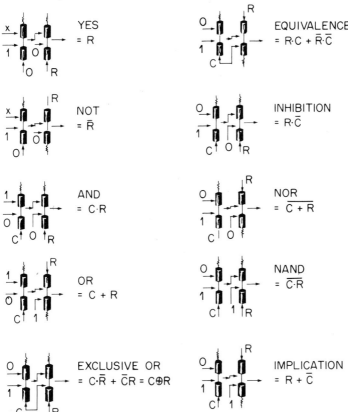

Fig. 8.4. Table of logic functions using cascaded spool valves (Togino and Inoue, *Ref. 2*, Fig. 5)

One example of the use of the valves is the EXCLUSIVE-OR function. Here, in the state shown, a signal appears at the output when R is present but not C. Alternatively, a signal appears when C is present, in which case spool '1' moves up, cutting off the unit supply, and when R is not present, when the C signal passes through to the output. When C and R are both present, the output is zero. An additional use of the two spool valve arrangement is in providing a fluid Flip-Flop. There are several ways of doing this, and three of them are shown in Fig. 8.5.

In this figure, R stands for 're-set' and S stands for 'set'. These are the two signals which change the Flip-Flop from one state to another. For example, in the second of the two illustrations, the 'set' signal will pass

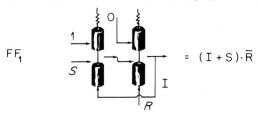

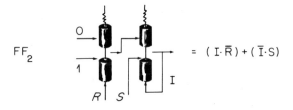

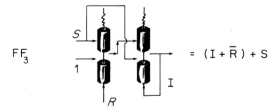

Fig. 8.5. Methods of creating fluid Flip-Flops using cascaded spool valves (Togino and Inoue, *Ref. 2*, Fig. 7)

through the second spool and, on appearing at the output, the feedback will push the spool upwards and so allow the unit signal to pass through to the output, holding the second spool up and cutting off the 'set' signal. On applying the 're-set' signal when the 'set' has disappeared, the first spool cuts off the high pressure to the second spool which is pushed downwards again by the middle pressure on the top of the spool, and the output now goes to the '0' value.

These tabulations of the logic functions obtainable from a piston valve are extremely useful for reference because they apply to any type of diverter device, whether piston or diaphragm.

Switching speeds of spool valves can be quite high because their physical size may be small, i.e. for a diameter of 4 mm, a time of response of 1 ms may be achieved when operated on water at 10 atm.

8.2.2 Integrated Spool Valve Circuits: Scale-of-Two Counter

The I.B.M. Company have done a considerable amount of development work on spool valve combinations in order to integrate them into sandwich-layer constructions[3]. One example of this work is the Scale-of-Two Counter which is shown in Fig. 8.6.

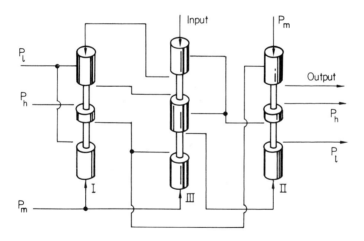

Fig. 8.6. Circuit diagram for binary counter stage using cascaded spool valves (Glaettli, *Ref. 1*, Fig. 3D)

This is a special case of a hydraulic shift register. The central valve acts as a buffer amplifier while the other two are comparable to triodes in an electronic Flip-Flop. Table 8.1 shows a complete cycle including the input and output binary pressure signals and the positions of the three valves: U for up and D for down:

TABLE 8.1

Input	III	I	II	Output
P_h	D	D	D	P_h
P_l	U	U	D	P_h
P_h	D	U	U	P_h
P_l	U	D	U	P_l

Every two complete cycles of the input signal yields one complete cycle of the output signal, thus the scale-of-two counting action. The pressure signals in the output channel indicate that a pulse-shaping action takes place so that several stages can be cascaded without differentiating elements. A multistage binary counter can be formed in this manner.

It is just possible to identify the function of the various parts of this arrangement, although they have been assembled with such skill that

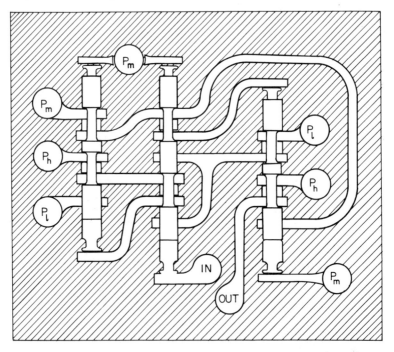

Fig. 8.7. Arrangement of connecting passages for one layer of the binary counter (Mitchell, Glaettli and Mueller, *Ref. 3*, Fig. 9)

[8.2] DIVERTER VALVES 345

analysis is not too easy. The two outer pistons act as a Flip-Flop for which the feedback passes through the central piston. The central piston therefore acts as an AND gate, or steering stage, whilst the two outer ones provide the memory. In Chapter 12, pure fluid devices of this kind are described and in particular two of them have an arrangement to 'inhibit' the feedback whilst a pulse is ON so that the counter cannot change state —there is then no need for a pulse shaper. It may be seen that the central valve also performs this function because when the input is ON and the valve down, both feedback signals to the bases of the outer valves are cut off.

Fig. 8.7 shows the layout of one layer of the combination of spool valves used for the binary counting stage. This is produced by the lost wax casting process in an epoxy resin. The inter-connecting passages are formed in wax first of all and the epoxy resin cast around them. The cylindrical holes for the spool valves are drilled and reamed after the casting process. Ref. 1 shows a collection of the sandwich layers in a block which produces four stages of binary counting.

8.2.3 The Latching Spool Valve

This design of spool valve was produced by the British Telecommunication Research[4]. Fig. 8.8 shows the device.

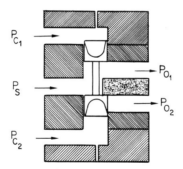

Fig. 8.8. The latching spool valve (Bell, *Ref. 4*, Fig. 7)

When a control signal is present at the lower end of the spool, the spool rises and an output appears at OP_1. The high pressure of the output causes the spool to remain in the 'up' position because of the force differential on the internal faces of the spool, which is created by the extension of one of the partitions of the cylinder into the bore of the valve. A signal on the top of the valve causes it to switch down.

Thus the device provides the Flip-Flop or memory function that requires two ordinary spool valves for its implementation. However, this advantage is probably offset by the more difficult construction. So far the problem of manufacture appears to be the dominating one in determining the commercial success of miniature piston valves in general, and the only ones on the market are still relatively large in size (compared with the experimental ones quoted in this sub-section and the previous one). Certainly no commercial integrated circuit has yet appeared on the market. An example of the use of this valve is the production of a shift register[4], which is slightly simpler than the I.B.M. design, but will shift the information in one direction only.

8.2.4 The Diaphragm-Lever Valve

This valve, shown in Fig. 8.9, is a simple arrangement where an input signal at port 1 depresses a diaphragm against a spring. A lever attached to the spring closes port 5. An input signal to 2, or merely the removal of the signal from 1, causes 5 to open and 4 to close. The three-way operation of this particular design of valve gives it the same limited logic possibilities as the single spool valve, and apart from acting as a diverter valve, it functions as a single element only according to the list of possibilities given in Fig. 8.12. In order to provide an active (that is with some power amplification) device, two elements must be used in combination. Simple

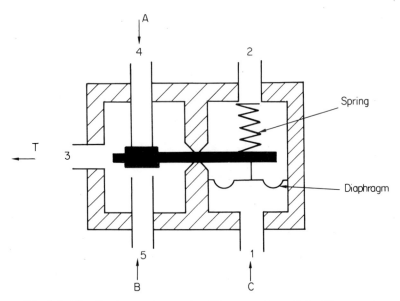

Fig. 8.9. The diaphragm-lever valve (Desoutter Lang Sales Literature)

[8.2] DIVERTER VALVES 347

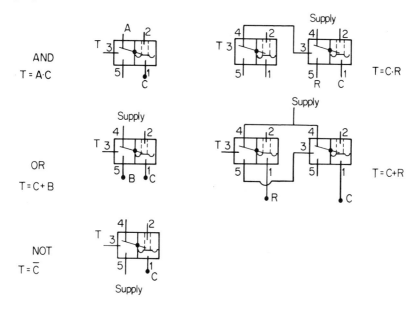

Fig. 8.10 Some logic functions using diaphragm-lever valves (Desoutter Lang Sales Literature)

examples of the method of connexion are included in Fig. 8.10, but more ideas can be obtained from the diagrams of possible connexions of spool valves.

The switching speed of this valve depends on the control pressure applied and may be as fast as 4 ms. Rack mounting facilities are available, and whilst the element is not as small as the double-diaphragm valve, to be described next, compact logic systems using these valves are possible.

8.2.5 The Double-Diaphragm Valve

This valve, shown in Fig. 8.11, is the basic element of a highly sophisticated digital system developed in East Germany[5]. The physical size is very small, the switching time is fast at 1 ms, and even at an operating pressure of 1 atm. the air consumption is low because the valve is essentially 'closed-centre' in operation. Having a working pressure which is relatively high for logic operation has the advantage that little pressure amplification is required for power functions.

The device consists of two diaphragms, connected by a spindle, the result of which is that the working area on the insides of the two diaphragms is less than that on the outside. The two outside channels are used only for switching or bias signals. The input channels are A, B, C, D, and it is clear

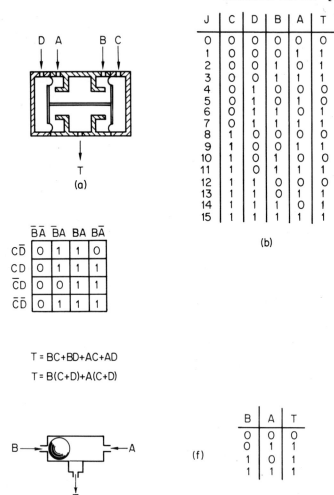

Fig. 8.11. The double-diphragm valve and passive OR element

that the diaphragms close off the channels A or B, depending on the direction of movement. The device is therefore a diverter valve which may be used in a similar way to the previous biased valves, or may be used without bias so that logically it is slightly superior to them. Biasing is slightly more complex, however, and relies upon the differential area across one diaphragm, i.e. in the IDENTITY function in Fig. 8.12, for example, it is achieved by applying supply pressure to both A and D with the result that a small resulting force pushes the diaphragm to the right.

[8.2] DIVERTER VALVES

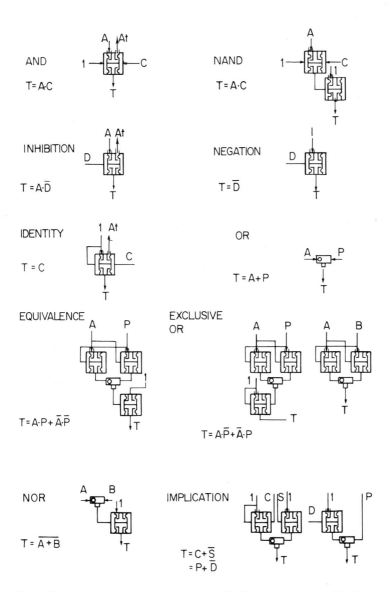

Fig. 8.12. Some logic functions using double-diaphragm valves (Töpfer, Schrepel and Schwarz, *Ref.* 5, Fig. 1)

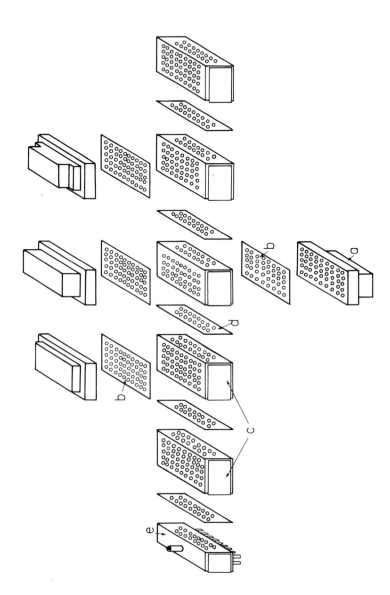

Fig. 8.13 Manifold arrangements for the double-diaphragm valves (Töpfer, Schrepel and Schwarz, *Ref. 5*, Fig. 2)

[8.2] DIVERTER VALVES 351

Any signal in C overrides the bias and allows the pressure signal in A to pass to the output channel T.

To make the operation of the device clear, a truth table for the output T is shown in Fig. 8.11(b) and a Karnaugh map of the output T is given in (c). The resulting Boolean algebra expression for T is thus easily deduced and is given in (d). Again, more complicated functions may be created by connecting two devices together as shown in Fig. 8.12 (the inputs to the second device are P, Q, R, S).

A passive OR element is also used in conjunction with the diaphragm valves in order to provide a means for allowing several inputs into a logic device. This is simply a ball shuttle valve as shown in Fig. 8.11(e)—the ball can be replaced by a floating disc as described later.

The greatest advantage of the use of this valve is the mechanical design of the system components (Fig. 8.13). Universal manifold blocks (c) are available on which several alternative groups of elements may be mounted. For example, five OR elements on one block provides a six-input passive OR function; three relays mounted together may be obtained for general logic use, and there are other groups also. The manifold blocks contain drilled passages passing in several layers and directions through the block with the holes not completely drilled through. The logic blocks are sealed on to the manifold with gaskets (b) which arrive with the holes indicated but not punched out. Several blocks may be connected together through gaskets (d) and an air supply introduced via block (e). By completing the appropriate holes in the gaskets and manifolds, complicated, overlapping connexions can be arranged internally within a unit assembly. An exploded view of the manifolds is shown in Fig. 8.13.

This method of assembly is excellent for a system which is destined for a machine for which the circuit is already developed, since it is ideal for factory assembly. However, the concept is still the subject of debate when breadboard circuits are required for development purposes. Here, a

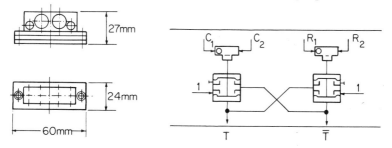

Fig. 8.14. A Flip-Flop formed from two double-diaphragm valves and two passive OR elements mounted in one block (Töpfer, Schrepel and Schwarz, *Ref. 5*, Figs. 5 and 14)

simpler form of connexion is preferable because it is often necessary to make quick changes as experience demands it.

A typical set of connexions carried out by the single block on which two OR elements and two relays are mounted to form a two input Flip-Flop is shown in Fig. 8.14.

8.2.6 Triple-diaphragm Devices

The triple-diaphragm element has been used extensively in Russia, and although it has a lower operating speed than the East German double diaphragm device, and is relatively bulky, it has a very high logic power.

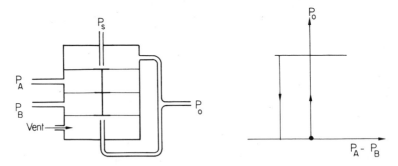

Fig. 8.15. The triple-diaphragm device (Gancikov, *Ref.* 7, Fig. 2)

The construction is shown diagrammatically in Fig. 8.15, and consists of four chambers separated by three diaphragms whose centres are rigidly linked together. If two control pressures are applied at the inputs, the diaphragm assembly distorts and shuts off either one of the two supply inputs. The area of the centre diaphragm is larger than the areas of the two outer ones, so that a pressure in one or the other of two control ports will bias the valve one way or the other.

By applying the correct bias to either of the inputs A or B, the logic functions of Fig. 8.16 may be achieved using a single device. Two devices may be coupled in a variety of ways, some of which are shown in Fig. 8.17.

For more details see Refs. 6 and 7.

8.3 Devices Working on a Back-Pressure Principle

8.3.1 Single-diaphragm and Diaphragm-foil Elements

A system of diaphragm elements has been developed in Czechoslovakia which uses two basic types of devices shown in Fig. 8.18.

The single diaphragm of the first element operates as a back-pressure device to give a straight amplification of the pressure signal in the ratio

[8.3] DEVICES WORKING ON A BACK-PRESSURE PRINCIPLE

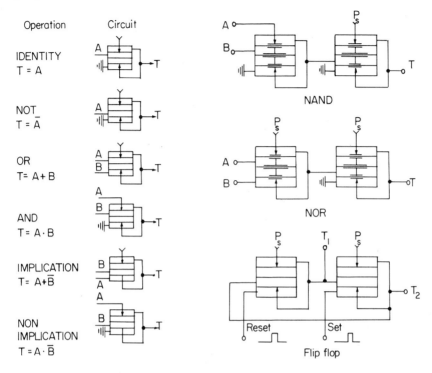

Fig. 8.16 Some logic functions using a single element (Gancikov, *Ref. 7*, Table I)

Fig. 8.17. Some logic functions using two elements (Gancikov, *Ref. 7*, Fig. 4)

of 2:1. The logic of a single element is thus a YES function. Two elements together give an active AND or an active OR function, depending on how they are connected, as shown in the first two arrangements of Fig. 8.18. In the AND function it may be seen that with no inputs on, the air supply passes through the first restrictor, through the two elements in parallel and out through the second restrictor to atmosphere. Thus, there is a continual air flow in this condition, as there would be for a single element performing a YES function with no input on. When both elements have an input signal, the leakage to atmosphere is stopped and a signal out is achieved. Two elements are generally included in one package, the dimensions of which are relatively small, as shown by the dimensioned drawings accompanying the logic diagram.

To produce an inverter with a simple, single diaphragm is not possible because the diaphragm would have to close an output channel which has the full system pressure applied to it. In other words, the control signal would also have to be at full system pressure, which is not practicable. To

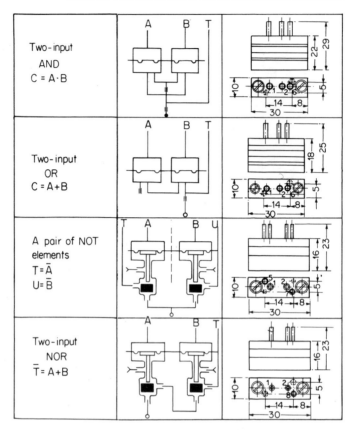

Fig. 8.18 Examples of the single diaphragm and diaphragm-foil valves (Sales Literature from ZPA Czechoslovakia)

produce an inverter function, therefore, the single diaphragm is coupled by a small rod to a foil element (a floating disc as described in 8.3.5) which closes off the main pressure on the receipt of a signal. The large area of the diaphragm compared with that of the foil provides the necessary force amplification. Again, two elements are provided in one miniature package, as shown in the two lower diagrams of Fig. 8.18.

Operating speeds of the elements are a little lower than the double-diaphragm element, being only about 200 Hz maximum, and clearly the inverter element must be the limitation in the circuit from this point of view. The system is easy to interconnect, since 30 pairs of diaphragm devices can be screwed on to a single supply manifold which slides into a rack. Small diameter tubing is then used to make external connexions.

8.3.2 A Planar Diaphragm Element

A recent extension of the ideas of diaphragm logic has been developed by the I.B.M. Company[8], in which all the elements of a circuit are machined or etched into two layers with a single thin membrane, acting as the diaphragm for all the elements in the layer, sandwiched between them. The diameter of an individual device is approximately $\frac{3}{16}$ in., and in cross-section it appears as shown diagrammatically in Fig. 8.19.

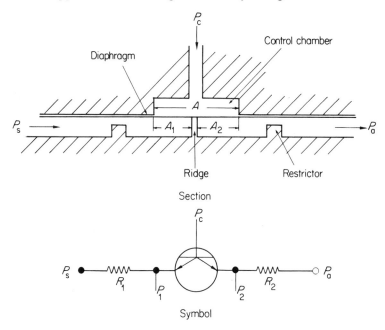

Fig. 8.19. The planar diaphragm element (I.B.M.) (Jensen, Mueller and Schaffer, *Ref. 8*, Fig. 1)

The schematic symbol below the latter diagram indicates the method of working. The restrictors R_1 and R_2 are provided by projections in the channels of the element, and in operation the diaphragm closes off the flow through the element by seating on the ridge. The ratio of area A_1 to A and the relative size of restrictors R_1 and R_2 determine the switching points and the amount of hysteresis in the element. If the output is taken from points labelled P_1, the element is an amplifier, whereas if taken from P_2 it becomes an inverter. Little flow may be taken from a single device, and if a higher power output is required the push-pull circuit of Fig. 8.20 must be used.

An example, given in Ref. 8 of the inverter design, states that when used in this mode the downstream restrictor R_2 is usually made larger than R_1 so that the output P_o is greater than $P_{s/2}$. For switching, the force due to P_c on the diaphragm must become equal to the upward force of P_1 and P_o on A_1 and A_2 respectively. P_1 and P_o are almost equal before switching

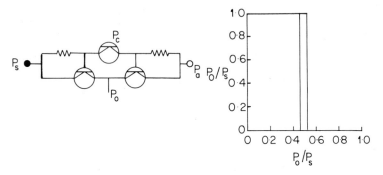

Fig. 8.20. A push-pull inverter (Jensen, Mueller and Schaffer, Ref. 8, Fig. 4)

because the airflow distorts the diaphragm upwards, allowing a clear passage over the ridge. Thus:

$$P_c = P_o = \frac{R_2}{R_1 + R_2} P_s$$

Hence the basic inverter has no pressure gain, as explained in the previous description of the Czechoslovakian designs. Once the diaphragm is seated against the ridge, there is no flow, so that the switch open point depends only on the relative areas A_1 and A_2. The control pressure must decrease until,

$$P_c A = P_s A_1$$

$$P_c = P_s \frac{A_1}{A}$$

There is normally a hysteresis in the switching curve, and for this to be present the switch closed point must be greater than the switch open point, or:

$$\frac{R_2}{R_1 + R_2} > \frac{A_1}{A}$$

The effective area ratio A_1/A for the element described was 0·3 and the restrictor ratio $(R_2/R_1 + R_2)$ was 0·8. It is clear that considerable varia-

[8.3] DEVICES WORKING ON A BACK-PRESSURE PRINCIPLE

tion in the switching characteristic is possible by adjustment of restrictions and areas.

An amplifier is necessary to amplify the inverter output, and in this mode R_2 is small compared with R_1 so that the pressure drop occurs primarily across R_1. When pressure is applied, the diaphragm will deflect to maintain a balance between the forces $P_c A$ and $P_o A_1$. Thus:

$$P_o = P_c \frac{A}{A_1}$$

The output is nearly linear with a gain of A/A_1.

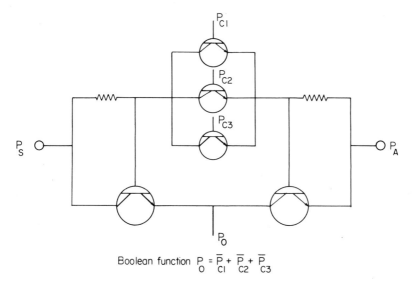

Boolean function $P_o = \bar{P}_{C1} + \bar{P}_{C2} + \bar{P}_{C3}$

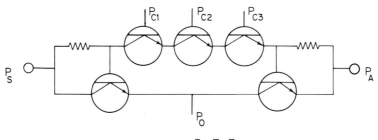

Boolean function $P_o = \bar{P}_{C1} \cdot \bar{P}_{C2} \cdot \bar{P}_{C3}$

Fig. 8.21. A three-input NAND (*top*); A three-input NOR (*lower*) (Jensen, Mueller and Schaffer, *Ref. 8*, Figs. 5 and 6)

Multi-input functions are achieved by interconnecting several devices, as shown in Fig. 8.21 for the circuits for the 3-input NOR and NAND functions. Thus, although the elements are simple, the system is not very economical on the number of components because of their individual low logic power. Nevertheless, the packing density is high and the overall performance appears attractive. In particular, the circuits can be integrated in a similar way to that possible with pure fluid elements.

The switching times are reported to be 1·3 ms for control going on, and 1·0 ms for the control going off, with an average power consumption per element of 0·75 watt at 1 lb. in^{-2}.

For convenience, Table 8.2 at the end of the chapter gives comparative operating figures for the various elements.

8.3.3 The Spring NOR Unit

This ingenious device[9] basically provides a NOT or NOR function. High-pressure air is fed through a restrictor to a close-coiled helical spring. When the spring is deflected by a diaphragm, the pressure is released and a 'zero' appears at the output. Combinations of NOR functions may provide a multi-input OR device, an OR gate, a bistable Flip-Flop, and an AND gate. The device is shown in Fig. 8.22, together with the input/output pressure characteristics. From the characteristics it may be seen that the element is a high gain proportional amplifier and could be used as such.

Fig. 8.23 shows the various methods of interconnecting the NOR units to provide an OR function, an AND gate and a Flip-Flop.

The element is somewhat larger than the other ones described, and the speed of switching possibly lower, but it is robust and operates at pressures which require little amplification for power functions. Some systems have been built for machine tool control using these elements in conjunction with a pneumatic drum card reader described in Chapter 11.

8.3.4 Ball Valves

Two types of ball valves have been suggested, the Kearfott valve shown in Fig. 8.24 and the Stanford variation on this idea, shown in Fig. 8.25[10,11]. Similar devices to the latter one are available commercially.

The Kearfott valve consists of a single ball moving in a short cylinder at each end of which is a seat. Supply pressure is applied through a restrictor to each of the holes through the ball seats. A stable position exists with the ball at either end of the chamber when the high pressure across the small area of the seat is more than balanced by the pressure in the rest of the cylinder, the value of which depends on the restrictor in the supply line and on the impedance to flow out through the control line at

[8.3] DEVICES WORKING ON A BACK-PRESSURE PRINCIPLE

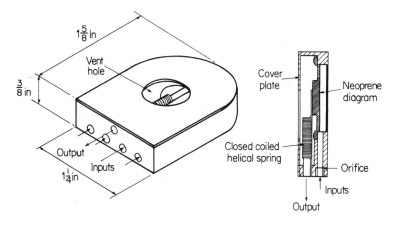

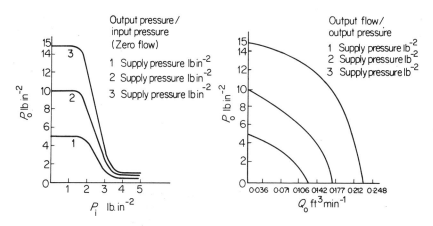

Fig. 8.22. The spring NOR unit and its characteristics (De Bryne, *Ref. 9*, Fig. 3)

the end away from the ball. Any leakage past the seat in which the ball rests passes out through the control port adjacent to it. It is only necessary for this control port to be closed for the pressure across the whole area of the ball next to the seat to approach the supply pressure, so that the ball will switch to the opposite end of the cylinder. Output signal pressure may be taken from the lines to the ball seats, downstream of the restrictors. This is basically a bistable element, but by choice of line impedances it may be made monostable.

With a 0·020 in. diameter and an aluminium ball travelling 0·008 in., a switching time of 100 μs has been reported, but little commercial use has been made of the idea since it was first reported in Ref. 10. The problems

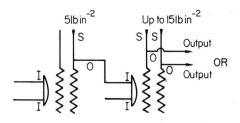

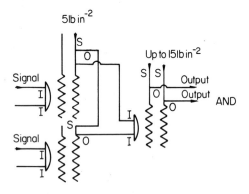

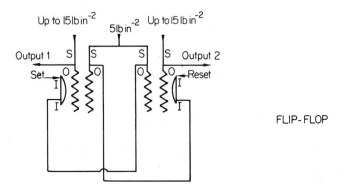

Fig. 8.23 Three functions obtained by interconnecting spring NOR units (Techne sales literature)

[8.3] DEVICES WORKING ON A BACK-PRESSURE PRINCIPLE

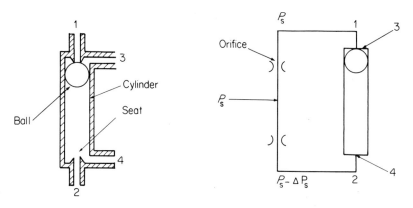

Fig. 8.24. The Kearfott ball valve—a bistable device (Riodan, *Ref. 10*, Fig. 1)

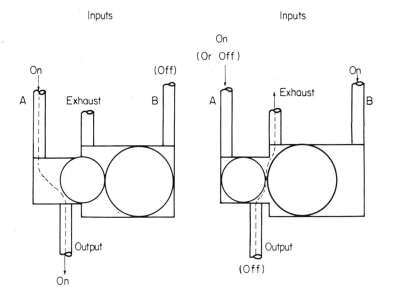

Fig. 8.25. The basic Stanford ball valve—a NOT device (Eige, *Ref. 11*, Fig. 1)

lie partly in good manufacturing to ensure low leakage past the balls and good seating of the ball on the supply channels at the ends of the chambers.

There are also difficulties in using the device in complicated circuits because of its low logic power. A means of improving this is the design of the Stanford valve, in which balls of two sizes are used—always one small one, but there may be more than one larger one.

Fig. 8.26 shows a two-input NOR valve. The larger and more complicated valve must clearly sacrifice speed of response, and a switching time of only 1 ms is claimed. Again, the disadvantage of the device is the relatively large leakage unless the manufacturing is very good.

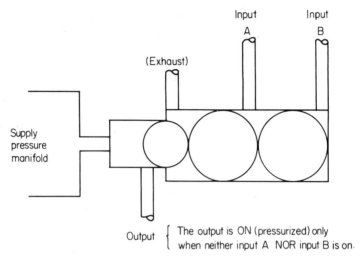

Fig. 8.26. A two-input NOR Stanford ball element (Eige, *Ref. 11*, Fig. 2)

8.3.5 *The Free-Foil Element*

The basis of the element is a free moving foil contained in a chamber of a shape to constrain the foil so that it has only one translational degree of freedom, in a direction normal to the plane of the foil[12, 13] (Fig. 8.27). The parts of the chamber to which the supply, input and output connexions are made determine the logic function of the element. For example, an inverter or NOT gate, an OR function, and a diode arrangement are shown in Fig. 8.28.

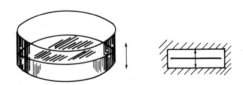

Fig. 8.27. Foil valve basic shape (Foster and Retallick, *Ref. 13*, Fig. 2.1)

[8.3] DEVICES WORKING ON A BACK-PRESSURE PRINCIPLE

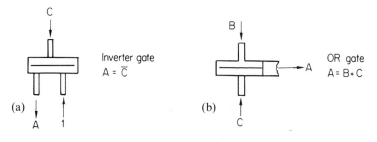

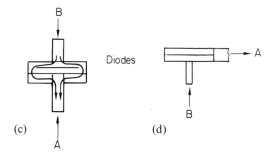

Fig. 8.28. Alternative connections to the foil chamber (Foster and Retallick, *Ref. 13*, Figs. 2.2, 2.3, 2.4, and 2.8)

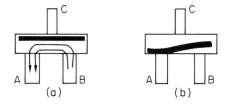

Fig. 8.29. Switching action of the foil inverter (Foster and Retallick, *Ref. 13*, Fig. 2.5)

8.3.5.1 Inverter gate. The switching action is basically the same for all the elements, so it will be explained by referring to the inverter gate in Fig. 8.29.

'B' is the power supply, taken preferably from a constant pressure source
'A' is the output, having an on-off nature represented by 0 or 1
'C' is the control signal which switches A.

The action is as follows:

(a) Assume that B = 1, C = 0, then as A = B.C̄, A = 1.

In this mode, the foil is held at the top of the chamber, as in Fig. 8.29, sealing the control tube 'C'. Thus fluid has a free passage through the chamber from 'B' to 'A' beneath the foil.

(b) If control signal C = 1 is applied, 'C' must be at pressure equal to or greater than 'B', but assume equality for ease of explanation. For flow to occur through the chamber, there must be a pressure drop through it which means that the net pressure on the underside of the foil must be less than supply pressure. So now there is, upon the application of C, an unbalance of pressure on either side of the foil, which causes it to move downwards. Leakage around the edges of the foil is not great enough to impair the action.

(c) The foil continues to move downwards, and at the completion of the switching action the pressure inside the chamber is equal to the supply pressure and the output tube A is covered and sealed by the foil (Fig. 8.29).

A foil element relies for its switching process on a flow of fluid through the chamber and so it cannot be connected into a blocked load. If this is unavoidable, then bleeds must be provided to ensure that leakage flow when the element is switched off can escape freely to atmosphere and yet they must not be so large as to adversely affect the pressure recovery capabilities of the foil element when switched on.

It is essential that all active foil elements are fed from a constant pressure source with little impedance between the source and the element. The reason is that when the element is switched off and flow ceases, the chamber pressure rises to equal that of the source. Thus the required control pressure is equal to that of the source and not that at which the element is normally running. A high line impedance effectively reduces the pressure gain. This is especially important when running a number of foils from one supply. The pressure at which the foil elements are run is not critical, typical extremes for a $\frac{1}{4}$ in. diameter foil being $\frac{1}{2}$ to 40 lb. in^{-2}.

The small physical size of the foil element is an advantage in itself, but it also means that the device has a fast switching speed. It also has a very low quiescent power consumption when switched off, the only flow through the element being due to leakage. A disadvantage is that the pressure gain is less than unity.

8.3.5.2 Foil OR gate. This consists of the same type of chamber as the Inverter Gate but with the tube connections arranged differently as in Fig. 8.28(b).

'A' and 'B' are the two OR inputs and 'C' is the output.

If A = 1 and B = 0, there is a pressure difference across the foil, which thus moves downwards and isolates input 'B', the output appearing as C = 1.

8.3.5.3 The diode. A foil diode is described in Ref. 12, a diagram of which is reproduced in Fig. 8.28(c), but a simpler diode may be made using a configuration more like the inverter and OR gates. If one input is removed from an OR element, the result is a two connexion device, one connexion being normal to the foil and one being in the plane of the foil, as in Fig. 8.28(d). If the element is supplied in the forward direction, i.e. B = 1, then the foil moves to the opposite side of the chamber and the input appears at 'A' with very little impedance. This action resembles that of the OR element. Now if the diode is supplied in the backward direction, i.e. A = 1, then the input is transmitted to both sides of the foil and the switching action is then identical with that of the inverter gate. The foil moves across and seals off the tube 'B'.

The diode is simple to construct and is small, a $\frac{1}{4}$ in. diameter foil adequately passing flows of 0·2 s.c.f.m. In the form of Fig. 8.28(c), the foil diode can be conveniently combined with a pure fluid element to provide input isolation, as shown in Fig. 8.30.

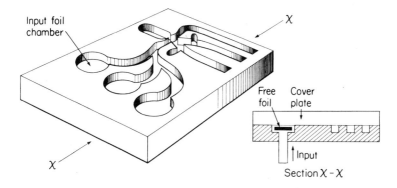

Fig. 8.30 A combination of foil diodes and a wall attachment amplifier to provide a multi-input OR function (Foster and Retallick, *Ref. 13*, Fig. 3.3)

8.4 A Diaphragm 'Threshold Logic' Device

An ingenious device from Czechoslovakia by Stivin[14] is a diaphragm arrangement which acts as a 'threshold' logic device, i.e. that several input units are connected to supply in such a way that a given number of other inputs must be present before the original 'threshold' is overcome.

Such devices have greater logic power, and indeed are more complicated devices than the triple diaphragm elements described earlier. It is apparently possible to obtain simplification of circuitry by use of these units, but it is clear that this is at the expense of relatively complicated construction.

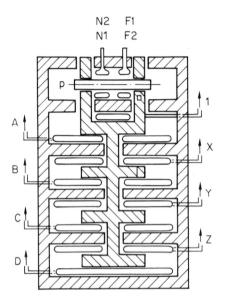

Fig. 8.31. Threshold logic device from Czechoslovakia (Stivin, *Ref. 14*, Fig. 1)

Fig. 8.31 shows the device in cross-section. Flat, rectangular bags with a hole in the centre are stacked around a central spindle which has projections on it. Inputs A, B, C, D push the spindle up and 1, X, Y, Z push it down. The bag 1 is permanently filled with air and is half the surface area of the other bags. Hence if X and Y are connected to supply, any three of A, B, C, D must be ON before the spindle will move up. The complicated bag design at the top of the device ensures that complementary outputs are available from F_1 and F_2. The whole device measures about 3·5 cm cube, the operating pressure is 0·5 to 1·5 atm. and the switching time 5 ms.

In Ref. 15, a list of logic functions is given for the device, and this is a useful catalogue of the possible logic functions for a four-input threshold arrangement.

8.5 Speed of Operation of Elements

8.5.1 Devices Operating under Incompressible Flow Conditions

When the fluid may be regarded as incompressible, the inertia of the fluid and moving pistons, together with friction, are the two factors which determine performance. The first depends on the length of channel over which the fluid is accelerated, whilst dry friction and viscous forces between the piston and bore create an overall frictional force.

The initial acceleration of the fluid is given by

$$a = \frac{p}{\rho l}$$

where l is the overall channel length, p the absolute fluid pressure and ρ is the mean density of fluid and pistons over length l. Inserting the numerical values for a typical unit with a 20 cm channel length, 150 lb. in^{-2} fluid pressure, and mean density of 0·001 kg cm.$^{-3}$ gives a fluid acceleration of 5000 m s^{-1}. System response time can then be obtained from

$$t = \frac{2s}{a}$$

where s is the small valve displacement required to switch fluid flow from one channel to another. For a displacement of about 2 mm, the response time is 1 ms, and the piston reaches a velocity of 5 m s^{-1}.

The loss of pressure due to gain in kinetic energy of the fluid is $pv^2/2g$ and is about 2 lb in^{-2}. This is negligible and indicates that the estimate of switching speed will be reasonable.

The energy required for each actuation of these hydraulic units is

$$E = ps\frac{\pi d^2}{4}$$

When d, the bore diameter, is 4 mm, this works out to be 25 mW.

8.5.2 Device Operating under Compressible Flow Conditions

When the flow is compressible, the response of a valve to a sudden input is not immediate because the fluid in the control input chamber must be brought up to a pressure high enough to overcome frictional and inertia effects. This takes time and may be expressed mathematically for a hydraulic fluid as follows:

Considering Fig. 8.32, for flow into the control chamber:

$$q_1 = A\frac{dx}{dt} + \frac{V_1}{\beta}\frac{dp_1}{dt} \tag{8.1}$$

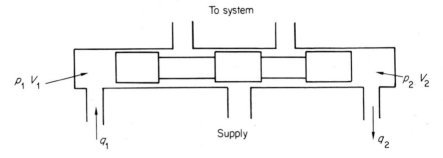

Fig. 8.32. Diagram of a piston valve showing the flows and pressures during switching

For flow q_2 out of the valve:

$$q_2 = A\frac{dx}{dt} - \frac{V_2}{\beta}\frac{dp_2}{dt} \tag{8.2}$$

A is the cross sectional area of the valve and β is the bulk modulus of the fluid, which for isothermal compression of air $= P$, the static pressure in the chamber.

The equation of motion for the load gives

$$p_1 - p_2 = \frac{M}{A}\frac{dv}{dt} + \frac{F}{A} \tag{8.3}$$

where F is the force due to friction and M is the mass of the piston.

Assuming a symmetrical system where $V_1 = V_2 = V$, adding (8.1) and (8.2) and using (8.3) gives:

$$\frac{MV}{A\beta}\frac{d^2v}{dt^2} + \frac{F}{A} + 2Av = q_1 + q_2$$

Assuming orific flow into and out of the control chambers and a symmetrical system:

$$q_1 + q_2 = 2C_q A_o \sqrt{\frac{P_s}{\rho}} \cdot \sqrt{\left[1 - \frac{(P_1 - P_2)}{2P_s}\right]}$$

$$= C\sqrt{\left[1 - \frac{M}{2AP_s}\frac{dv}{dt} - \frac{F}{2AP_s}\right]}$$

and

$$\frac{MV}{2A^2\beta}\frac{d^2v}{dt^2} + \frac{F}{A^2} + v = C'\sqrt{\left(1 - \frac{M}{2AP_s}\frac{dv}{dt} - \frac{F}{2AP_s}\right)}$$

[8.5] SPEED OF OPERATION OF ELEMENTS

where C_q is a coefficient of discharge and A_o an orifice and C and C' are constants.

The L.H.S. is the equation of a spring-mass system with a natural frequency

$$\omega_n = \frac{2A^2\beta}{MV}$$

It is possible to use this to make a dimensionless time scale, i.e. $t = \omega_n \tau$. Also, v may be made dimensionless by dividing by Q_{max}/A, where Q_{max} is the maximum available control flow. If the term $(M/2AP_s)(dv/dt)$ is non-dimensionalized in this way, the coefficient becomes:

$$\alpha = \frac{MQ_{max}\omega_n}{2A^2P_s}$$

The response of the system to a step input, or to any other input for that matter, is decided first of all by the natural frequency and then by the modifying influence of α. From the natural frequency it is easy to estimate approximately the time to the first peak of the overshoot in the response,

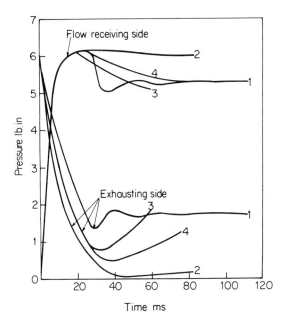

Fig. 8.33. Effect of inertia load parameter α on the response of a spool valve to a step control flow in the case where compressibility is important (Foster and Misra, *Ref. 15*, Fig. 9)

since this is nearly a one-half cycle of the oscillatory motion determined by the spring mass system. The parameter α contributes largely to the damping, and for very large values may ultimately determine the motion. Essentially this means that for small α the inertia effects of the valve are small and the valve could move quickly if enough flow is available. On the other hand, if α is large, the motion of the valve will be dominated by inertia effects, and to have extra flow available will not be of any use.

Fig. 8.33 taken from Foster and Misra[15] shows the response of a large spool valve to a step control flow input for various values of α. There is some static friction present on this valve also, and the first 20 ms is taken up by the pressure on the inlet side rising and the pressure on the exhaust side falling until the pressure difference is enough to overcome dry friction, after which an oscillatory movement occurs.

For a large diameter valve, α is small and the response (case 1) is slow because a lot of flow is required to move the valve. For small diameter valves, α is large and the inertia of the valve dominates the motion (case 2). An optimum was arrived at in case 3.

8.6 Valves for Power Amplification

The general tendency is for logic functions to be performed at relatively low pressures either by moving-part valves or by devices without moving parts. It is therefore necessary to amplify both the pressure level and the flow of the signals obtained from the logic system. The aim of the theory of the previous section was to help in choosing the best layout for a power amplifying valve. If the pressure amplification required is relatively low, say 5:1 or lower, the logic signal may be taken straight to a spool valve. If this is done, the following points should be noted:

(1) The higher the natural frequency ω_n, the faster the response. This means that all volumes of air in the control lines of the power valve should be kept small. This is self-evident, but more important than apparent at first sight.
(2) The travel of the valve should be kept short.
(3) Having taken care of the first two points, the diameter of the valve should be chosen to make the loading parameter α about 0·3 to 0·5.

In Ref. 14 a conventional, large-flow spool valve was switched in 60 ms by a wall attachment element. Since that was written, responses of 40 ms or better have been obtained, and with a smaller power valve designed specifically for the purpose, significantly faster times are possible. One valve which is available commercially is operated direct from a wall

[8.6] VALVES FOR POWER AMPLIFICATION 371

attachment device. This valve is approaching the ideal in layout and will switch in about 30 ms with a relatively low switching flow.

If much more pressure amplification is required, for example when amplifying the signal available from a turbulence amplifier, diaphragm-valve designs are more attractive. In these cases, α is necessarily very low, and great care must be taken to ensure that the enclosed volumes of air are as small as possible and that the travel of the diaphragm is severely limited. Ref. 16 describes one variety of such a valve, and a drawing is given in Fig. 8.34.

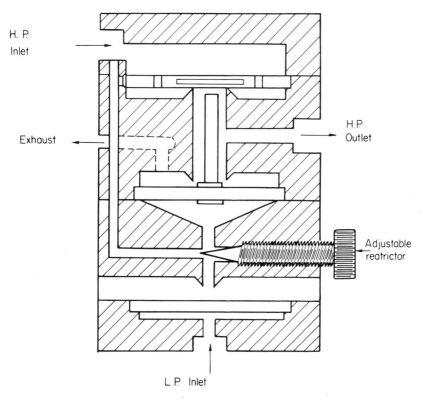

Fig. 8.34. A diaphragm power valve to give high pressure amplification from the signal of a turbulence amplifier (Hodge and Hutchinson, *Ref. 16*, Fig. 3)

In designs of this kind the switching flow requirements are entirely determined by the volume displaced by the motion of the diaphragm, and although the diaphragm movement may appear small at first sight, the volume displacement may be high in relation to the flow rate available from the pure fluid switching element.

8.7 Conclusions

A variety of different valves has been described briefly. Some are easy to interconnect, some have more logic power, and some are faster than others. As yet there is insufficient information available from which to predict fatigue life and reliability, so that it is impossible to compare their complete performance with that of pure fluid devices. One advantage is that they operate at a higher pressure than pure fluid devices so that less pressure amplification is required. From a capital cost point of view they are therefore attractive for the moment, and much will depend in the future on how much the price of pure fluid devices is reduced.

In the end, the choice of element will probably be decided by the application in which they are to be used, and much more experience is required in a variety of areas before a best choice will emerge. Table 8.2 presents a

TABLE 8.2

Element	Volume Occupied by Element	Minimum Switching Speed or Maximum Frequency	Operating Press (atm.)	Remarks
Piston (I.B.M.)		1 ms	10	Difficult to integrate
Diaphragm-lever (West Germany)	16 cm^3	2–5 ms	1	
Double-diaphragm (East Germany)	4·3 cm^3	1 ms	1·4	Interconnexions well engineered with no external pipes
Triple-diaphragm (Russia)	32 cm^3	200 Hz	1·4	High logic power, but bulky
Single-diaphragm and diaphragm-foil (Czechoslovakia)	2 cm^3	200 Hz	0·5	Good packing density on to standard racks
Planar diaphragm (I.B.M.)	0·1 cm^3	1·2 ms	0·1	High packing density—very interesting development
Spring NOR element	15 cm^3	5–10 ms	up to 2·0	
Kearfott ball element	150 elements per cm^3	100 μs	7	High packing density—does not appear to be used much
Foil element	0·1 cm^3	1 ms	up to 0·5	High packing density, but pressure gain poor

brief summary of the characteristics of the devices described in this chapter.

References

1. Glaettli, H. H., Hydraulic logic—what is its potential, *Control Eng.*, **8**, 83–86 (1961).
2. Togino, K. and Inoue, K., Universal fluid logic element, *Control Eng.*, 78–87 (May 1965).
3. Mitchell, A. E., Glaettli, H. H. and Mueller, H. H., Fluid logic devices and circuits, *Trans. Soc. Inst. Technology*, 1963.
4. Bell, R. W., An application of fluid logic to telegraphy as a research vehicle, *C.F.C.* (1), Paper E1.
5. Töpfer, H., Schrepel, D. and Schwarz, A., Universelles Baukastensystem für pneumatische Steuerungen, *Sonderdruck aus der Zeitschrift für Messen, Steuern, Regeln*, **7**, Pt. 2, 63–72 (1964).
6. Berends, T. K. and Tal, A. A., Pneumatic relay circuits, *Proc. 1st I.F.A.C. Congress*, Moscow, 1960.
7. Gancikov, A., Pneumatic diaphragm logic controls material distribution, *Control Eng.*, May 1967.
8. Jensen, D. F., Mueller, H. R. and Schaffer, R. R., Pneumatic diaphragm logic, *A.F.*, 1967.
9. De Bruyne, N. A., Pneumatic NOR blocks, *C.F.C.* (1), Paper C3.
10. Riodan, H. E., High speed pneumatic digital operations with moving elements, *H.D.L.* (1), **I**.
11. Eige, J. J., Multiple-ball pneumatic amplifiers, *H.D.L.* (2), **II**.
12. Bahr, J., The foil element, a new fluid logic element, *Electronische Rechenanlagen*, vol. 7, No. 2, 69–78 (1965).
13. Foster, K. and Retallick, D. A., Some experiments on a free foil switching device, *C.F.C.* (2), Paper D1.
14. Stivin, J., Contribution to the development of pneumatic logic elements and their applications to machine tools, *Proc. 9th Int. M.T.D.R. Conf.*, Manchester, 1968, Pergamon Press.
15. Foster, K. and Misra, A. K., The turbulent reattachment amplifier in a conventional pneumatic circuit, *C.F.C.* (1), Paper F1.
16. Hodge, J. and Hutchinson, J. G., Turbulence amplifiers—principles and applications, *C.F.C.* (1), Paper F2.

9

The Design of Sequential Systems and Counters, with a Note on Number Codes

9.1 Introduction

A sequential system is a system in which the outputs are required to change in a specific order. A counter is an easily recognized example of a sequence circuit because the output patterns must change in an order which is predetermined by the number code allocated to the counter. In this case, the output states will only change when an input pulse is fed into the circuit, and the method of operation is said to be *synchronous*. Since the next output state depends only on the present output state, the counter logic circuit must simply make a choice which depends on the input information consisting of the present output signals. Such a logic circuit is an example of a combinational circuit which may be designed using the methods presented in the chapter on Boolean Algebra, once the order of the change of states is established. The problem of formalizing the change of states will be considered later in this chapter, but first a somewhat more general situation will be examined.

A machine in a manufacturing industry may have a number of operations to perform, each one of which must be controlled to operate at specific points during the work cycle. Here, any one operation can only begin when the previous operation has been completed, a mode of working which is called *asynchronous*. The information on which to base a logic decision to move any particular cylinder may depend, firstly on the state of all the cylinders at that instant of time, and secondly on the operations already completed. In order to illustrate this point let us consider the sequential operation of two cylinders A and B. Suppose, first of all, that A

[9.1] INTRODUCTION 375

must go out, then B must extend, followed by A retracting and finally by B returning to its original position. This is most easily written as $A+\ B+\ A-\ B-$, a notation which we shall continue to use. The state of the cylinders will be given by sensors attached to the cylinder and arranged to give signals at the extremities of the strokes; these signals will be indicated by $x_{A+}, x_{A-}, x_{B+}, x_{B-}$, whilst the outputs are indicated by z_A and z_B.* In order to derive the logic circuit some method of visualization of the sequence is desirable; several methods are available and will be used later, but for the moment a *timing diagram* as shown in Fig. 9.1(a) will be sufficient.

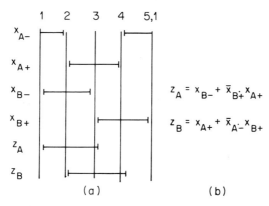

Fig. 9.1 Timing chart for cylinder sequence
$A+\ B+\ A-\ B-$

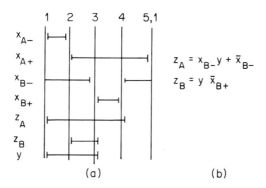

Fig. 9.2 Timing chart for cylinder sequence
$A+\ B+\ B-\ A-$

* Z_A, Z_B are outputs from the logic circuit that cause the cylinders A and B to move.

The sequence is divided up into intervals which indicate the change in the variables, but do not indicate equal intervals of time. Considering the operation of the circuit; at the beginning of the cycle, when x_{A-} is ON and x_{B-} is ON, the output to cylinder A must go ON so that it will start moving, and z_A is indicated as being ON. x_{A-} will go OFF and when cylinder A reaches the end of its stroke at 2, x_{A+} is turned ON. The line representing x_{A-} is terminated a little short of the vertical line representing state 2 and the line for x_{A+} starts a little after state 2 in order to indicate that during the movement of the cylinder both sensor signals disappear. Cylinder B must start to move when x_{A+} goes ON and reaches the end of its stroke at 3; x_{B-} is indicated to go OFF just before state 3, whilst x_{B+} goes ON just after this point and causes cylinder A to begin to retract. At point 4, A has retracted, and then B starts to retract, completing the sequence at point 5, which in reality is the same as the starting point 1.

In this case, the logic equations governing the outputs of the system are straightforward because each output is uniquely associated with a certain set of input combinations which are not repeated. Care must be taken to make certain that signals are not lost when the input signals disappear during the motion of a cylinder, and the negated input variables in the output equations given in Fig. 9.1(b) appear for this reason. Making provision for these extra terms, however, increases the complexity of the logic circuit.

Now consider a different sequence of the same cylinder, i.e. A+ B+ B− A−. The timing chart appears in Fig. 9.2(a). Now, both z_A and z_B are required to be ON during the interval 2 to 5 but OFF during the interval 4 to 5, when the input combinations are exactly the same. There is now too little information available on which to make a decision; something extra must be added that differentiates between the two appearances of the ambiguous input combination. A memory, or *secondary*, circuit is used for this purpose and one possible assignment of its output is indicated by y on the timing diagram. The output logic is given in 9.2(b) and now depends upon the value of y. Some means of obtaining the signal y must be arrived at and this is discussed in the next section.

The latter problem, in which insufficient information originally exists, is called *under-specified*, whereas the first problem, in which no ambiguity occurred, is called *fully-specified*.

9.2 State Diagrams and Flow Tables

A flow diagram gives a useful visualization of the way in which a circuit changes state and can be an aid in formulating a problem. Two flow diagrams in general use are those due to Mealy and Moore, and as an

[9.2] STATE DIAGRAMS AND FLOW TABLES 377

illustration the Mealy flow diagram is given in Fig. 9.3 for the second of our two sequencing problems.

The first of the figures in the circles represents the different internal states of the system, where now a different numbering system is used from

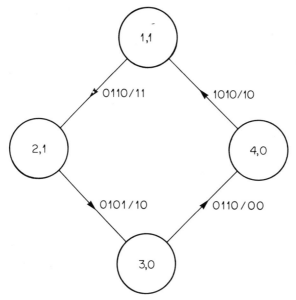

Fig. 9.3 Mealy flow diagram for sequence
A+ B+ B− A−

that of Fig. 9.2. In the latter figure the numbers 1, 2, ... represented the points where the internal states changed, whereas now they represent the intervals between the changes. The second figure in the circle represents the state of the secondary. Transitions of state are indicated by the arrowed lines, whilst the figures on the lines represent the input combinations causing the transition and the output combination resulting from it respectively.

An alternative method of formulating the problem is to use the Huffman flow table shown in Fig. 9.4. Here the possible input combinations form the columns of a table with the secondaries listed to the left and the output combinations to right of the table. Stable states are those where the output is that which is required from the input combinations, and those states are ringed. Possible transitions are then indicated by numbers without rings; for example, from ② in the second row the possible input transitions are indicated by 1 and 3, but the outputs corresponding to these are in the first and third rows respectively, so that after the change of

$x_A x_B$ / y	1010	0110	0101	1001	
1	①	2	–	–	10
1	–	②	3	–	11
0	–	4	③	–	10
0	1	④	–	–	00

Fig. 9.4 Huffman 'Primitive' flow table for cylinder sequence A+ B+ B− A− (Based on an example given in Retallick, *Ref. 3*)

input the outputs must change to take the circuit either to stable states ① or ③.

The next step is to eliminate any redundant or equivalent rows, that is to say, when the input states and output states are the same and when changes of input combinations result in the same transitions. If a row is similar to another in all respects except that an optional transition corresponds to a definite transition in the other one, use may be made of the optional transition to combine the two rows and such rows are called pseudo-equivalent. In Fig. 9.4 this procedure is not necessary because there are no redundant lines.

Up to this point the general procedure for synchronous and asynchronous circuits is the same, since both the flow diagram and the state diagram are merely graphical means of representing the changes of input, secondary and output states of the system. From now on, however, there is a basic difference between the two that is due to the nature of the inputs; in an asynchronous circuit all the inputs are present all the time, merely changing their states at different points in the time cycle, whereas in a synchronous circuit the inputs appear as discontinuous signals so that the circuit has to 'remember' which state it is in until the next pulse arrives. The consequence of this is that the next output state of an asynchronous circuit is determined largely by the input combination (and to a lesser extent by addition of secondary memories) at any one time, whereas in a synchronous circuit the next output state is determined by the previous state of all the memories (i.e. secondary Flip-Flops). Because of this difference, the subsequent methods of secondary and output assignments are somewhat different.

For the moment we shall concentrate on the asynchronous situation exemplified by the sequence of two pneumatic cylinders. If the operation of the circuit is to be controlled primarily by the input combinations, then it is necessary to know whether these by themselves provide sufficient information on which to base a logic decision. Clearly, if two stable states occupy the same input combination column, for example, states ② and ④, then that input combination is insufficient information to distinguish between them, and a secondary memory must be included to provide the discrimination. The assignment of the secondary state may be done by merging the rows of the table wherever possible, or by mapping the states on a map of the input combinations. The process of merging consists of combining rows where a stable state in one row corresponds either to an unstable state or an optional state in the other; the output states are for the moment ignored. Thus rows (1) and (2) combine together and rows (3) and (4) also combine to give the merged flow table of Fig. 9.5.

$x_A x_B$ / y	1010	0110	0101	1001
1	①	②	3	—
1	1	④	③	—

Fig. 9.5 Merged flow table for sequence A + B + B − A − (Based on an example given in Retallick, *Ref. 3*)

This new table merely indicates that the changes from state ① to ② to 3 can be achieved only by changes of input combinations, and similarly for changes from ③ to ④ to 1. In changing from 1 to ① or 3 to ③, the secondary must change as indicated, but this change automatically distinguishes between the two stable states ② and ④.

The alternative is to map the states as shown in Fig. 9.6. It is easy to see that a path can be followed from ① to ② to ③ without duplicating an input combination, and similarly from ③ to ④ to ①. The merged flow table and secondary assignment could easily be determined from this.

9.3 Secondary and Output Excitation

The next step is to determine the switching circuits necessary to give the correct secondary and output states, i.e. to determine the secondary and output excitation functions. In both cases, Karnaugh maps are drawn, but the actual use of the maps depends upon whether or not Flip-Flops are to be used for the memory and output. In electrical circuits, Flip-Flops are generally relatively expensive components and are less used so that memory circuits must be designed from combinational logic; if turbulence amplifiers are used the same is true; but if the circuit is to be implemented

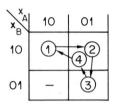

Fig. 9.6 Map of states for cylinder sequence A+ B+ B− A− (Based on an example given in Retallick, Ref. 3)

using wall attachment elements or pneumatic spool valves, then Flip-Flops are the most economical way of achieving a memory.

A Karnaugh map is drawn first of all showing the required state of the secondary (or secondaries) for various existing states of the input and of the secondary. Fig. 9.7 shows such a map for the two cylinder sequence, with X denoting optional combinations.

$x_A x_B$ y	1010	0110	0101	1001
1	1	1	0	x
0	1	0	0	x

Fig. 9.7 Secondary excitation map for cylinder sequence A+ B+ B− A− using combination logic (Based on an example given in Retallick, Ref. 3)

Hence the combinational circuit for the secondary excitation is determined by the logic function

$$y = x_{A-} + x_{B-} y$$

the circuit for which is shown in Fig. 9.8, where the memory is achieved by the use of the feedback of the output y back to the input.

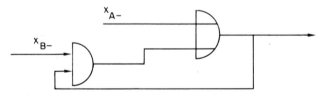

Fig. 9.8 Secondary circuit for sequence A+ B+ B− A− using comination logic (Based on an example given in Retallick, Ref. 3)

[9.3] SECONDARY AND OUTPUT EXCITATION

If the memory is to be implemented by the use of the Flip-Flop, the excitation maps are a little different where two maps are now necessary, one for the set signal (S_Y) of the Flip-Flop and the other for the re-set signal (R_Y) (Fig. 9.9).

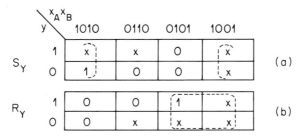

Fig. 9.9 Secondary excitation maps for cylinder sequence A+ B+ B− A− using flip-flops (Based on an example given in Retallick, *Ref. 3*)

The necessary switching logic is now:

$$S_Y = x_{A-}$$
$$R_Y = x_{B+}$$

and the logic circuit is very simply produced, as shown in Fig. 9.10.

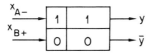

Fig. 9.10 Secondary circuit for sequence A+ B+ B− A− using flip-flops (Based on an example given in Retallick, *Ref. 3*)

Similarly, the output circuits are derived by drawing Karnaugh maps. Consider first the output Z_A to be implemented by combinational logic. The map for the output Z_A is drawn with reference to the primitive flow table of Fig. 9.4 and is shown in Fig. 9.11.

$x_A x_B$ y	1010	0110	0101	1001
1	1	1	1	x
0	0	0	1	x

Fig. 9.11 Output map for cylinder A using combinational logic (Based on an example given in Retallick, *Ref. 3*)

Hence at first sight:

$$Z_A = y + x_{B+}$$

But this allows for misinterpretation during the interval when cylinder B is travelling between sensors x_{B+} and x_{B-} since no signal is obtainable. The way of covering this error is by using the complement of the signal from x_{B-}, i.e.

$$Z_A = y + \bar{x}_{B-}$$

and the circuit becomes as shown in Fig. 9.12.

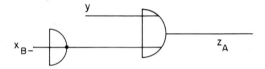

Fig. 9.12 Output circuit for cylinder A using combinational logic (Based on an example given in Retallick, Ref. 3)

Similarly the map for Z_B is shown in Fig. 9.13 and the circuit in Fig. 9.14.

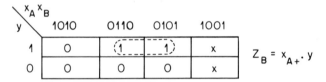

Fig. 9.13 Output map for cylinder B using combinational logic (Based on an example given in Retallick, Ref. 3)

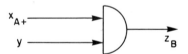

Fig. 9.14 Output circuit for cylinder B using combinational logic (Based on an example in Retallick, Ref. 3)

If, on the other hand, Z_A and Z_B are both to be implemented by Flip-Flops, the maps and switching functions are shown in Fig. 9.15.

Notice that in Fig. 9.15, alternative forms are given for S_A and R_B, and that these forms do not agree with the output excitation maps. This is

[9.3] SECONDARY AND OUTPUT EXCITATION

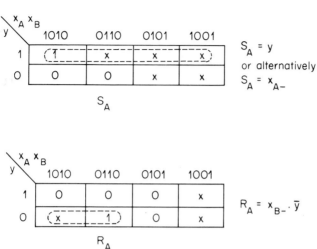

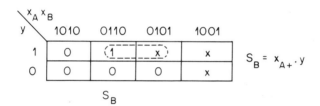

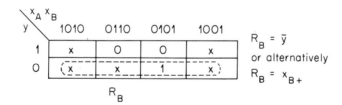

Fig. 9.15 Output excitation maps for cylinders A and B using flip-flops. Cylinder sequence A+ A+ B− A− (Based on an example given in Retallick, Ref. 3)

because the switching for Z_A can be initiated in either unstable state 1 before the secondary has changed or in stable state ① after the secondary has changed and the former gives $S_A = x_{A-}$. Strictly if this is done, there should be a zero in that same entry for the reset function for Z_A instead of a 'don't care', in order to avoid uncertainty. However, this is not necessary, because the signal $R_A = x_{B-} \cdot \bar{y}$ disappears before $S_A = x_{A-}$ appears, simply because of the physical nature of the circuit.

This may appear to be 'black magic', but the alternative approach was that arrived at intuitively before the application of the formal method, and it is this circuit which is given in Fig. 9.16. The circuit arrived at formally is given in Fig. 9.21.

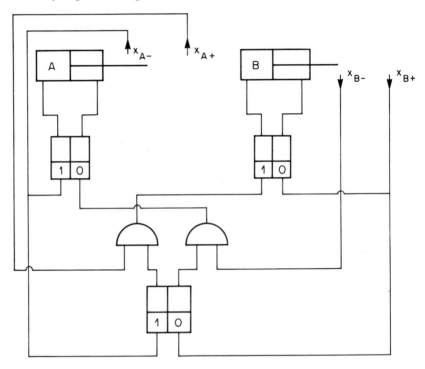

Fig. 9.16 Circuit for a sequence A+ B+ B− A− (Based on an example given in Retallick, *Ref. 3*)

9.4. Pneumatic Ladder—Type Sequence Circuits

9.4.1 General Ideas

The Cascade system for sequential circuits using pneumatic spool valves is, in concept, similar to the preceding method. The sequence is divided up into groups of operations in which no cylinder appears more than once. A three-port valve is provided to signal each cylinder movement, and one less five-port Flip-Flop valves than the number of groups is also provided.

The original example will be considered, i.e. that the sequence of the

[9.4] PNEUMATIC LADDER—TYPE SEQUENCE CIRCUITS

two cylinders is $A+ B+ B- A-$. The two groups are $A+ B+$ and $+B- A-$, and the layout of the circuit is shown in Fig. 9.17. The sensors are three-way valves and the first one, i.e. the one which would usually be

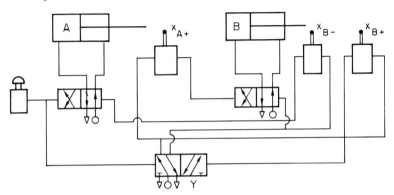

Fig. 9.17 Circuit for the sequence $A+ B+ B- A-$ according to the Cascade method using piston valves

x_{A-} is used instead as a start valve, so that the circuit does not continually cycle. In pressing the start button, cylinder A is caused to move and sensor x_{A-} is activated by receiving air from the five-port memory valve Y. At the end of the stroke, x_{A+} switches the three-port valve controlling cylinder B, which then extends. Sensor x_{B+} sends a signal to memory Y, which re-sets the latter, thus switching off the supplies to sensors x_{A+} and x_{B+} and switching it on to sensor x_{B-}. As cylinder B retracts, the sensor x_{B-} now allows a signal to pass to cylinder A to cause it to retract. Thus the memory valve Y serves to energize the sensors for the appropriate group of the sequence at the appropriate time whilst de-energizing those which are not required.

9.4.2 The Martonair Method of Design

In the Martonair cascade system of analysis for sequence circuits (given in a leaflet issued by the company), the method is formalized in the following way:

(a) Award a letter to each cylinder and write down the sequence in the usual way, i.e.

$$A+ B+ C+ B- D+ D- E+ E- C- A-$$

(b) Split the sequence into groups in such a way that:
  (i) any letter, regardless of sign, appears once only in any group, i.e. a cylinder must move in only one direction in any group;
  (ii) it breaks up into as few groups as possible. Try splitting it in the

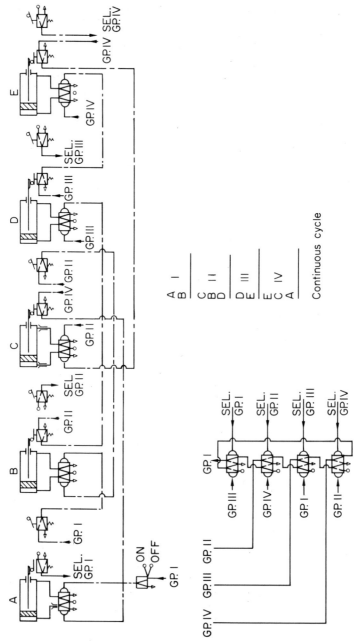

Fig. 9.18 Circuit for the sequence given above, designed by the 'Martonair' method (From leaflet issued by Martonair Ltd., Fig. 8)

[9.4] PNEUMATIC LADDER—TYPE SEQUENCE CIRCUITS 387

forward direction first, but remember that it can be split in the reverse order since it is a cyclic process;
(iii) the cylinder at the end of any group reverses in the next Group.

According to the way in which the sequences have been split in (b), if cylinder C is placed at the end of the first group, it does not reverse until the fourth group. However, splitting in reverse order gives the following groups.

$$\begin{array}{cccc} \text{I} & \text{II} & \text{III} & \text{IV} \\ A+\ B+\ / & C+\ B-\ D+\ / & D-\ E+\ / & E-\ C-\ A-\ / \end{array}$$

Now proceed as follows. Allocate a pilot operated five-port valve to each of the groups and one spring returned, manually operated (by a trip) three-port valve to each end position of each cylinder. Now the following connexions are made:

(a) The valve controlling any group is supplied with air by flow from the valve controlling the group immediately after it, when that next valve is OFF. The valve controlling the group is then switched ON by the sensor corresponding to the last position in the previous group. Hence the flow to the sensors will be ON if the last group of cylinders have completed their operations but the next group has not been initiated. When the next group is due to start, the valve controlling the supply of air to the sensors goes ON, thus switching the air OFF for the group which is currently operating.

(b) Within any group, the output from any sensor is connected appropriately to the control valve of the next cylinder to operate in the group, except for the output from the last sensor in the group which, as already explained, switches on the flow to the sensors in the following group.

Fig. 9.18 shows the circuit for the sequence under discussion and it may be seen that the layout is very orderly and easy to follow. The secondary circuit of memory valves is itself a cyclic sequence circuit and is thus simple to follow and connect. Some limitations of the method are:

(a) It does not always give the circuit of minimum complexity for a given application, particularly where cylinders have repeated operations during the cycle.

(b) It is not necessary for circuits that are already 'fully specified', because in that case no memories are required and combinational logic may be used.

(c) The circuit may go into continuous oscillations if any of the trip valves is operated by some external source. However, care taken in the interlocking circuit will allow external inputs into the basic cascade systems.

(d) For cylinders operating simultaneously extra logic has to be built in so that the sequence can only proceed when all the particular cylinders have completed their operations.

9.4.3 The 'Lucas' Method of Design

A more economical variant on the above methods proposed by Joseph Lucas Ltd.[1] uses limit switches as memories. Here, after listing all the desired motions alphabetically in the form of a continuing sequence, with the convention that a plus sign means that a cylinder has moved from its normal position, the sequence is divided into a minimum of groups so that:

(a) No letter must appear more than once in any group.
(b) The last letter in a group must appear again somewhere in the next group.
(c) The last letter in all groups must be plus; in the last group it can be minus if the sequence does not recycle.

Now the circuit is constructed by the following method:

1(a) Allocate a 2-position, 3-connexion, mechanically-operated, air-reset valve to each cylinder stroke extremity which is last in a group;
 (b) join inlets to mains air;
 (c) join outlets to the manifold of the next group in sequence;
 (d) join reset to manifold of the next but one group in sequence.

2. Join the end of the cylinder control valve applicable to the first letter in each group to its own group manifold.

3(a) To all other cylinder stroke extremities allocate a 2-position, 3-connexion, mechanically-operated, spring-return valve;
 (b) join inlets to their own group manifolds;
 (c) join outlets to the appropriate end of the cylinder control valve which applies to the next letter in sequence.

If the three instructions are followed, a working circuit will result in which all possibility of blocked signals has been removed. It only remains to insert a suitable start-stop valve in advance of the A+ end of cylinder control valve A and any necessary auxiliary functions. Rule 1 ensures that air is not applied to both ends of a cylinder at one time. Rules 2 and 3 are necessary because the last cylinder of a group operates a limit switch which feeds air to the next group of sensors. The mechanical signal on the switch must be removed during the working of the group so that at the start of the next group but one the limit switch can be reset.

In the circuit shown in Fig. 9.19 for the two cylinder example, the memory element of Fig. 9.17 is no longer required, since it is incorporated in limit switches A+ and B+. It is difficult to realize circuits in this way with pure fluid devices because of the energy loss when one element feeds air to the supply channels of other devices.

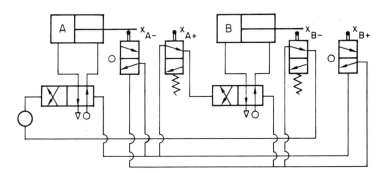

Fig. 9.19 Circuit for the sequence A+ B+ B− A− using the cascade system proposed in ref. 1, again using piston valves (*Ref. 1*, Fig. 7)

For overlapping functions, that is, when several cylinders operate at once, i.e.

$$
\begin{array}{ccc}
B+ & B- & D+ \\
& C+ & \\
& X+ &
\end{array}
$$

the completion of B+ will initiate B−, C+ and X+. The limit valves of these functions are connected in series so that D+ will only be initiated when all three are completed.

9.5 Sequential Logic and Some Ideas on a Universal Logic Block

9.5.1 Stancielescu's Sequential Logic Algebra

Stancielescu[2] proposes an alternative approach to the algebra of sequential systems. The difference in the approach from that of the normal combinational logic is that the output of a sub-circuit is defined in terms of both its present input and the input at a previous point in time; hence the term 'sequential logic'. Three basic functions are described, the 'sequential AND', the 'sequential OR' and the 'sequential NOT'. Consider two pulses x and y occurring at the two points in time $N - 1$ and N respectively, which produce an output at time $N + 1$, then:

x 'sequential AND' y

$$I_{N+1}(x, y) \equiv (x/y)_{N+1} = x_{n-1} y_N$$

i.e. $(x/y)_{N+1}$ = x at time N − 1 AND y at time N

x 'sequential OR' y

$$O_{N+1}(x, y) \equiv (x0y)_{N+1} = x_{N-1} + y_N$$

'sequential NOT' x

$$S_{N+1}(x) \equiv (x)_{N+1} = \bar{x}_N$$

Stancielescu infers that the functions I and 0 are nonidenpotent, noncommutative, nonassociative and nondistributive, but there are a number of identities useful in the simplification of sequential logic expressions.

By definition these functions I and 0 require a memory to record the occurrence of x until y appears, so that the sequential AND function would consist of an AND gate plus a memory element, the latter being a Flip-Flop in asynchronous circuitry or merely a delay in synchronous circuitry. The implementations shown in Fig. 9.20 are suitable for pulse

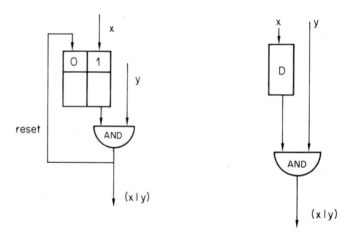

Fig. 9.20 Logic for Stancielescu's SEQUENTIAL AND function

inputs where x and y are not present at the same time. Stancielescu gives two examples using the sequential AND function, but does not show how the sequential OR could be used so that the practical significance of the latter remains obscure.

9.5.2 Modification to the Sequential Logic Method

Retallick[3] suggests that this general idea may be adapted for specific use in a pneumatic sequencing circuit.

In Fig. 9.20 the output is used to reset the memory Flip-Flop so that a short pulse is obtained at the output. A pneumatic sequencing system always returns to the original starting point so that once a memory element is set it will always be returned to its original state, sometimes later in the cycle. Already we have described this type of system as a complicated oscillator, and to relate it to the sequential logic, two 'sequential AND' units may be combined, as shown in Fig. 9.21, where the resemblance to an oscillator may be clearly seen.

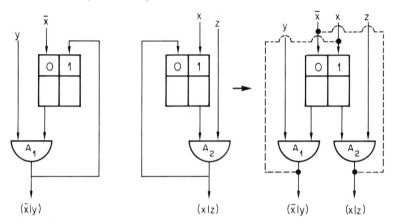

Fig. 9.21 The DOUBLE SEQUENTIAL AND unit suggested by Retallick (Retallick, *Ref. 3*)

The two feedback lines are not essential in addition to the set and reset signals x and $\bar{x}$, but one or both may be used instead of x and $\bar{x}$, as required. In complex circuit, x and $\bar{x}$, will be generally elsewhere in the circuit so that the set and reset takes place at the appropriate points in the cycle. Retallick omits the feedback lines and calls the resulting collection of elements a DOUBLE SEQUENTIAL AND unit (this should not be confused with one suggested by Stancielescu which is somewhat more limited in application).

The best way of proceeding in the design of a circuit is to arrive at the switching logic following the method outlined earlier and then group the switching functions in a sequential form. Thus, from Fig. 9.15, the switching logic is:

$$S_A = y \qquad S_B = x_{A+y}$$
$$R_A = x_{B-\bar{y}} \qquad R_B = \bar{y}$$

where $S_Y = x_{A-}$
$R_Y = x_{B+}$

To group these in sequential form, those following y are collected together in one group and those resulting from the signal $\bar{y}$ are collected in another group.

$$\begin{aligned} &\text{Time}\\ &1 = y\\ &2 = y|x_{A+}\\ &3 = \bar{y}\\ &4 = \bar{y}|x_{B-}\end{aligned}$$

where $S_Y = x_{A-}$

and $R_Y = x_{B+}$

The circuit is as shown in Fig. 9.22.

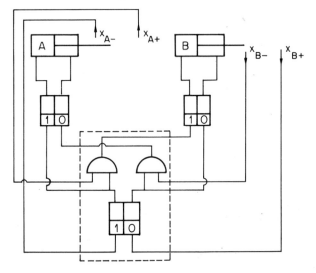

Fig. 9.22 Circuit for the sequence $A+ B+ B- A-$ with the memory arranged as a double sequential AND unit

The elements shown within the broken lines make up one DOUBLE SEQUENTIAL AND UNIT. Note the difference between this circuit and that of Fig. 9.16.

Ref. 2 extends the idea to more complicated problems involving more use of the two cylinders during a given sequence and also involving more

[9.5] SEQUENTIAL LOGIC 393

than two cylinders. In the problem already examined, not all the input combinations are used, so that the memory circuit is relatively simple. If the sequence is now B− A+ A− B+ A+ A−, the input combination map is as shown in Fig. 9.23, and the flow tables and logic are shown in Fig. 9.24. The circuit is shown in Fig. 9.25.

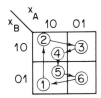

Fig. 9.23 State map, for sequence B− A+ A− B+ A+ A−

y	$x_A x_B$ 1001	1010	0110	0101	z_A z_B	
0	①	2	–	–	0 - 01	
0	–	②	3	–	10 0-	
1	–	4	③	–	01 0-	Primitive flow
1	5	④	–	–	0 - 10	table
1	⑤	–	–	6	10 - 0	
0	1	–	–	⑥	01 - 0	
0	①	②	3	⑥		Merged flow
1	⑤	④	③	6		table
0	0	0	1	0	$S_Y =$	Secondary flip-
1	–	–	–	0	$x_{A+} x_{B-}$	flop excitation
0	–	–	0	–	$R_Y =$	
1	0	0	0	1	$x_{A+} x_{B+}$	
	0-01	100-	010-	01-0		Output map
	10-0	0-10	010-	01-0		

Fig. 9.24 Flow table, secondary map, output maps and logic for sequence B− A+ A− B+ A+ A− (Retallick, Ref. 3)

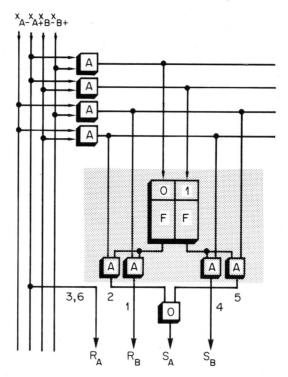

Fig. 9.25 Circuit diagram for sequence B− A+ A− B+ A+ A− (Retallick, *Ref. 3*)

In this circuit, the sub-circuit in the shaded area is again a Double Sequential AND. The Flip-Flop is the memory that divides the circuit into two separate sub-sequences, i.e.

$$A- B+ A+ \quad \text{and} \quad A- B- A+$$

so that the result is similar to that achieved by ladder-type (or cascade) methods of design. The Flip-Flop 'gates' the appropriate sensors for each half of the sequence through the AND gates connected on each output of the Flip-Flop. In the previous section this would be equivalent to switching the supply to the appropriate sensors, but in pure fluid logic terms, this must be achieved by the use of logic AND gates. On the input, the combinational logic is achieved by using two-input AND gates. Some sort of pattern may be seen to be emerging by comparing this circuit with the previous one. In the simpler circuit, only two movements occurred in each half of the sequence and one AND gate is necessary on the output of each Flip-Flop. In the present circuit, three movements

[9.5] SEQUENTIAL LOGIC 395

occur in each half and two AND gates are required in each half with two input AND gates at the inputs of the Flip-Flop. If n movements were to occur in each half of the cycle, (n − 1) AND gates would appear to be required on the memory outputs and AND gates with (n − 1) inputs on the inputs to the memory.

Thus it may be seen that the basic ideas for a 'Universal Logic Block' are emerging. Much more needs to be done in evaluating a variety of circuits before an ideal combination of elements appears; the arguments given here are rather superficial. In particular, the problem of 'fan-in' to large numbers of AND gates raises the problem of whether or not these should be active or passive devices; and so on. Any investigation of this kind must also take into account proposed manufacturing methods in order to decide on the optimum number of elements to include in a single block.

9.6 Number Codes

9.6.1 General

In any digital system the easiest representation of a number is in binary form; since any particular part of the system may be ON or OFF it is possible to represent the number 17 like this:

$$\underbrace{000000000}_{1}\underbrace{10001111111}_{7}$$

This, however, is insufficient in the use of storage space, and some form of code is much more economical. The number of different ways using K bits is 2^K. It is possible to work in various radixes, and the number of bits necessary to produce a 10 decimal digit number in terms of these is as shown in Fig. 9.26.

Hence a natural binary scale is the most economical, but little penalty results from using any other radix.

It is unusual for a natural binary code to be used for information transmission, because the order of 34 or more bits is too complex to remember. Since we normally think in decimal form, the binary bits are grouped in separate decimal groups, and a Binary Coded Decimal system is employed (B.C.D.).

The choice of binary code is now quite difficult since there are 16 ways of arranging the four bits necessary for a decimal digit and the 10 out of 16 ways can be used in any order to give (16!/6!) ways (=29,059,430,400). The number of alternatives is reduced by using a code in which the successive digits, usually from right to left, represent weights equal to successive powers of 2. Such a code is additive, for the sum of the representations of two digits is the representation of their sum. These are called Weighted Codes.

r	Bits
2	34
3	42
4	34
5	45
6	39
7	36
8	36
9	44
10	40
11	40
12	40
13	36
14	36
15	36

Fig. 9.26 Number of bits necessary for a 10 decimal digit number using various radisces

The weights may be chosen so that the sum of the weights is not greater than 15 or less than 9. Also one of them must be 1 and another either 1 or 2, as shown in Fig. 9.27.

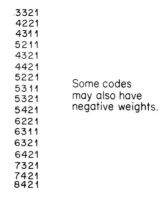

```
3321
4221
4311
5211
4321
4421
5221
5311       Some codes
5321       may also have
5421       negative weights.
6221
6311
6321
6421
7321
7421
8421
```

Fig. 9.27 A list of positive weights for a four-bit weighted binary coded decimal

A preferable code for computation is one which allows complementation, i.e. that the 9's complement of the number is given by changing all the 0's to 1's and all the 1's to 0's. One such code is given in Fig. 9.28.

[9.6] NUMBER CODES

	4	3	1	1
0	0	0	0	0
1	0	0	0	1
2	0	0	1	1
3	0	1	0	0
4	1	0	0	0
5	0	1	1	1
6	1	0	1	1
7	1	1	0	0
8	1	1	1	0
9	1	1	1	1

Fig. 9.28 A complementing weighted code

The only weighted systems which have this property are those in which the sum of the weights is 9. The positive weight ones for which the sum is 9 is as shown in Fig. 9.29.

```
3321
4221
4311    (the orders of the
5211     numbers can be
         changed).
```

Fig. 9.29 Other weighted codes that are complementing

9.6.2 Use of Redundancy to Avoid Errors

The use of 4 binary bits to represent 1 decimal digit leaves 6 combinations of 16 unused. If any one of these combinations appears in the computation, then there is an error. Recognition of these errors would provide a check on accuracy. However, by examining the possible 40 errors in the 8421 code, only 10 will provide the non-valid combinations, so that this method will detect only 25 per cent of the errors.

A more sophisticated way is to use a 2 out of 5 code, indicated in Fig. 9.30(a) where the check is merely on the presence of a number of 1's other than two. The 2 out of 7 code shown in Fig. 9.30(b) is less economical but is used in certain measuring systems because of the simple encoding patterns.

However, the most widely used check in machine tool controls is the

	0	0	1	1	0	0
1	1	0	0	0	1
2	1	0	0	1	0
3	0	0	0	1	1
4	1	0	1	0	0
5	0	0	1	0	1
6	0	0	1	1	0
7	1	1	0	0	0
8	0	1	0	0	1
9	0	1	0	1	0

(a)

Two-out-of-five code.

	50	4	3	2	1	0
0	01	0	0	0	0	1
1	01	0	0	0	1	0
2	01	0	0	1	0	0
3	01	0	1	0	0	0
4	01	1	0	0	0	0
5	10	0	0	0	0	1
6	10	0	0	0	1	0
7	10	0	0	1	0	0
8	10	0	1	0	0	0
9	10	1	0	0	0	0

(b)

Two-out-of-seven code.

Fig. 9.30 (a) Two-out-of-five code. (b) Two-out-of-seven code

parity check. A parity check is a redundancy code in which the number of holes punched in a line across the tape is always an even, or in some cases always an odd, number. This means that one column along the length of the tape must be reserved solely for the insertion of parity digits. The left hand column, i.e. that which has maximum significance, is generally but not always used for this purpose. Thus when parity codes are used only sixteen other code values are available with five-hole and one hundred and twenty-eight with a eight-hole tape. If the probability of reading an error without a parity code is 1 in 10^6, it is 1 in 10^{12} with it.

The use of odd parity codes for dimensional and even for management functions, or vice-versa, is usual so that differentiation between the two types of instruction may necessitate an extra code to indicate 'Dimensional Start'.

9.6.3 Block Codes

The information for contouring is generally read in blocks, with two sets of co-ordinates read in at a time. A code may signal the end of the block. From the two directions some form of interpolation signals the required feedrates along each axis.

9.6.4 Unit Distance Codes

For measurement purposes the natural binary code is not very attractive because two or more bits may change during the change from one number to the next. It is difficult to avoid one bit changing before the other so that a false number is momentarily generated. If the switching circuit will respond to this, an error is introduced. To overcome such a difficulty codes are used in which only one bit changes at any one time. A wide variety of these is possible, and one particular one, the Gray Code, is shown in Fig. 9.31(a). A disadvantage of this particular code is that on changing from 9 back to 0 three digits change, and an alternative Unit Distance Code, the Petherick Code, may be used, as shown in Fig. 9.31(b).

0	0	0	0	0
1	0	0	0	1
2	0	0	1	1
3	0	0	1	0
4	0	1	1	0
5	0	1	1	1
6	0	1	0	1
7	0	1	0	0
8	1	1	0	0
9	1	1	0	1

0	0	1	0	1
1	0	0	0	1
2	0	0	1	1
3	0	0	1	0
4	0	1	1	0
5	1	1	1	0
6	1	0	1	0
7	1	0	1	1
8	1	0	0	1
9	1	1	0	1

Fig. 9.31 (a) Gray code. (b) Petherick code

Any use of a Unit Distance Code produces difficulties in a feedback control circuit because computation is only possible with a code that is capable of complementation. Therefore, the Unit Distance Code must be converted into a natural binary code for use in the circuit, and the logic necessary for achieving this may be complex. An alternative arrangement is to convert to the so-called 'Excess 3' code, which is a weighted decade code with 0011 added to all the digits; this is a complementing code which can be used for arithmetic providing that the appropriate number of 0011's are deducted from the final result.

9.6.5 Codes for Punched Tape for Numerically Controlled Machine Tools

The future British Standards for codes for numerically controlled machine tools will be based on a series of recommendations from the International Organisation for Standardization (ISO). To begin with, the form of the coded characters punched on the tape is described[4] and this is compatible with an earlier recommendation for 6 and 7 bit coded characters sets for information processing interchange[5]. The code is shown in Fig. 9.32.

b_7							0	0	0	0	1	1	1	1
b_6							0	0	1	1	0	0	1	1
b_5						Column	0	1	0	1	0	1	0	1
Bits	b_7	b_6	b_5	b_4	b_3	b_2	b_1	Row	0	1	2	3	4	5	6	7
	0	0	0	0	0	NUL		SP	0		P		
	0	0	0	1	1				1	A	Q		
	0	0	1	0	2				2	B	R		
	0	0	1	1	3				3	C	S		
	0	1	0	0	4				4	D	T		
	0	1	0	1	5			%④	5	E	U		
	0	1	1	0	6				6	F	V		
	0	1	1	1	7				7	G	W		
	1	0	0	0	8	BS①		(③	8	H	X		
	1	0	0	1	9	HT		)③	9	I	Y		
	1	0	1	0	10	LF②		.④		J	Z		
	1	0	1	1	11			+		K			
	1	1	0	0	12					L			
	1	1	0	1	13	CR②		—		M			
	1	1	1	0	14					N			
	1	1	1	1	15			/		O			DEL

Fig. 9.32 Method of punching 7 bit characters on tape for numerical control of machines (This table from ISO R, 1968, 'Codes for numerical control of machines', is reproduced by permission of the British Standards Institution, 2 Park Street, London W1Y 4AA, from whom copies of the complete ISO Recommendations may be obtained)

The fifteen rows are represented by four bits, $b_1, \ldots, b_4$, and the five columns are distinguished by some of the combinations of 3 bits b_5, b_6 and b_7. The last hole on the normal eight hole tape is used as a parity check.

The way the characters are used must also be laid down, because it is essential that when the machine is operating, the information about the position along an axis, a new feed or spindle speed, a new tool, and so on,

[9.6] NUMBER CODES   401

is entered in the correct way. To this end, three other ISO recommendations have been drawn up[6, 7, 8]. The first two are concerned with variable block formats, whilst the latter describes a fixed block format. These in turn refer to two other recommendations concerned with axis and motion nomenclature[9], and also the format for preparatory functions and miscellaneous functions[10].

On any tape some preparatory information is included in order to describe the type of machine, i.e. design, capacity, power and load etc. This is followed by specification of the type and accuracy of the control system, a list of preparatory functions, specification of the feed, spindle and tool functions, finishing with a list of miscellaneous functions. For economy, feed and spindle speed functions are coded using either an arithmetic progression that specifies the decimal point plus two, three or four digits, or alternatively a scheme that breaks the spindle speeds into a hundred ranges, using a form of geometrical progression. The format for the blocks is specified in a shorthand way and each block must then correspond to this format, so that it consists of a series of numbers, each preceded by an address letter that specifies the function to which the number refers. A list of address letters is contained in the ISO recommendations and is therefore, common to all machines—the machines will differ only in whether or not they use particular functions. An example of a block is given in ref. 7 for a boring machine with a cross-slide, work-table and rotary table, a horizontal spindle head sliding vertically on upright slideways and a manually positional quill. Here the classification detailed short-hand for each block is:

$$N3 \cdot G2 \cdot X + 42 \cdot Y + 32 \cdot Z\,31 \cdot B\,33 \cdot F\,2 \cdot S\,2 \cdot T\,2 \cdot M \cdot 2^x$$

The meaning being as follows:

- N3 — Three-digit sequence number
- G2 — Two-digit preparatory function
- X + 42 — Linear dimension on X axis, with either + or − sign, four digits to the left of the implicit decimal point, two to the right
- Y + 32 — Linear dimension on Y axis, with either + or − sign, three digits to the left of the decimal point, two to the right
- Z31 — Linear dimension on Z axis, three digits to the left of the decimal point, one to the right
- B33 — Rotary dimension about the Y axis, three digits to the left of the decimal point, three to the right
- F2 — Two-digit (geometric progression) feed function code

  S2   — Two-digit (geometric progression) spindle speed function code
  T2   — Two-digit tool function
  M2   — Two-digit miscellaneous function
  *   — End of block symbol
       — Shows a tabulation and should appear, where employed, before every word.

Thus a typical single block would be as follows:

  TAB N001 G 41 X + 125050 Y − 15300
                Z + 5410 F 99 S 00 M 13

and the tape would consist of a series of such blocks, all punched according to the code specified in Fig. 9.31, and each letter or number including the parity check.

Thus it may be seen that considerable standardization has taken place in the coding of punched tape for numerically controlled machines. In order to be widely applicable, any new control system should be able to use this type of information. Unfortunately, the logic for decoding would be rather complex and so far any attempts at making fluidic numerical control systems have fallen far short of this sophistication.

9.7 Counting Techniques

9.7.1 Introduction

The problem of counter design is included here, because counters are a useful class of sequencing circuit and may be either asynchronous or synchronous in operation. Once the type of counter has been specified, it will have to meet certain well defined conditions, so that it is a good example on which to demonstrate ideas of sequential system design using the flow table concept with final minimisation of the combinational logic. It is later shown in Chapter 12 that the properties of fluidic elements can be utilised to produce circuits that are simpler than at first sight supposed, but it is nevertheless useful to try to catalogue counter designs here without reference to fluidic circuits in particular.

First of all, consider the design of a counter which will count from 0 to 3 and reset itself back to 0. The input may be considered as a single level signal having the values 0 or 1, so that the circuit has 8 internal states as shown in Fig. 9.33.

[9.7] COUNTING TECHNIQUES 403

	x			
(a)	0	1	z_1	z_2
	①	2	a	00
	3	②	b	01
	③	4	c	01
	5	④	d	11
	⑤	6	e	11
	7	⑥	f	10
	⑦	8	g	10
	1	⑧	h	00

	x			
(b)	0	1	z_1	z_2
	①	2	0	0
	3	②	0	1
	③	4	0	1
	5	④	1	0
	⑤	6	1	0
	7	⑥	1	1
	⑦	8	1	1
	1	⑧	0	0

Fig. 9.33 Flow tables for a counter to count from 0 to 3, (a) is for unit-distance code whilst (b) is for a natural binary output (Foster and Retallick, *Ref. 11*, Fig. 1)

It will be assumed that the outputs are required to appear when x is ON and to be in binary form. Two flow tables are shown, differing in that (a) has a unit distance output coding whilst (b) has a natural binary output. For case (a) it is possible to make an assignment such that $z_{1,2} = y_{1,2}$ by taking outputs directly from two of the three secondary bistables dictated by the 8-row flow table, as shown in Fig. 9.34. It may be seen that the result is a relatively complex circuit, and example (b) proves to be even more so because the z, or output equations are more complicated.

9.7.2 The T-Flip-Flop

In the previous example all the logic circuitry operates when an input pulse arrives at the circuit, and the resulting 'parallel mode' design is relatively complicated. Simpler circuits may be produced working in a 'serial mode' by use of the T-Flip-Flop, a component which changes state every time an input signal is received. By considering the switching equations for one binary bit, several arrangements are possible, depending upon whether the secondary memories are achieved using bistable elements or combinational logic elements, and also depending on whether the input is level or pulsed.

The flow table is shown for the T-Flip-Flop in Fig. 9.35, together with the secondary assignment: the switching equations for four alternative cases are given in the following text.

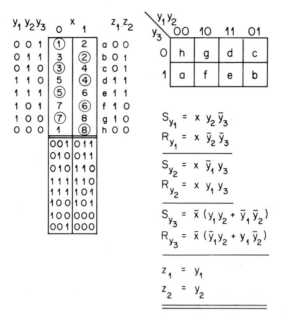

$$S_{y_1} = x\, y_2\, \bar{y}_3$$
$$R_{y_1} = x\, \bar{y}_2\, \bar{y}_3$$

$$S_{y_2} = x\, \bar{y}_1\, y_3$$
$$R_{y_2} = x\, y_1\, y_3$$

$$S_{y_3} = \bar{x}\,(y_1 y_2 + \bar{y}_1 \bar{y}_2)$$
$$R_{y_3} = \bar{x}\,(\bar{y}_1 y_2 + y_1 \bar{y}_2)$$

$$z_1 = y_1$$
$$z_2 = y_2$$

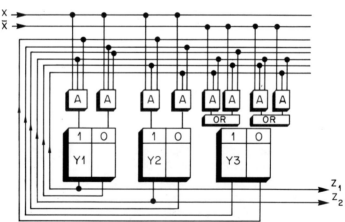

Fig. 9.34 Flow table, secondary map, output map, logic equations and circuit for the unit-distance code counter of Fig. 9.33 (a) (Foster and Retallick, *Ref. 11*, Fig. 2)

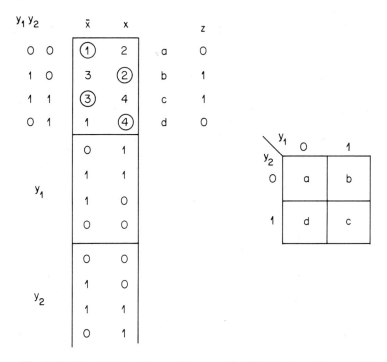

Fig. 9.35 Flow table and secondary map for 'T' flip-flop (Foster and Retallick, *Ref. 11*, Fig. 3)

Case I Y_1, Y_2 bistable memory, level input

$$S_{y_1} = x\bar{y}_2$$
$$R_{y_1} = xy_2 \quad z = y_1$$
$$S_{y_2} = \bar{x}y_2$$
$$R_{y_2} = \bar{x}\bar{y}_1$$

Case II Y_1 bistable, Y_2 logic memory, level input

$$S_{y_1} = x\bar{y}_2$$
$$R_{y_1} = xy_2 \quad z = y_1$$
$$y_2 = \bar{x}y_1 + xy_2 + (y_1 y_2)$$
$$\bar{y}_2 = \bar{x}\bar{y}_1 + x\bar{y}_2 + (\bar{y}_1 \bar{y}_2)$$

(The bracketed terms are necessary to overcome the transition hazard between $\bar{x}$ and x).

The last two equations may be written in an alternative form:
By De Morgan's theorem,

$$y_2 = y_1\overline{(x\bar{y}_2)} + xy_2$$

$$\bar{y}_2 = \bar{y}_1\overline{(xy_2)} + x\bar{y}_2$$

Case III Y_1, Y_2 *logical memory, level input*

$$y_1 = x\bar{y}_2 + \bar{x}y_1 + y_1\bar{y}_2$$
$$\bar{y}_1 = xy_2 + \bar{x}\bar{y}_1 + \bar{y}_1 y_2 \qquad z = y_1$$
$$y_2 = \bar{x}y_1 + xy_2 + y_1 y_2$$
$$\bar{y}_2 = \bar{x}\bar{y}_1 + x\bar{y}_2 + \bar{y}_1\bar{y}_2$$

or alternatively,

$$y_1 = \bar{y}_2\overline{(xy_1)} + \bar{x}y_1$$

$$\bar{y}_1 = y_2\overline{(xy_1)} + \bar{x}\bar{y}_1$$

$$y_2 = y_1\overline{(x\bar{y}_2)} + xy_2$$

$$\bar{y}_2 = \bar{y}_1\overline{(xy_2)} + x\bar{y}_2$$

Case IV *Pulse input, 1 secondary*

$\bar{x}$ x $\quad S_{y_1} = x\bar{y}_1$
$\qquad\qquad\quad R_{y_1} = xy_1$

	$\bar{x}$	x
	1	2z
	3z	4
0	—	1
1	—	0

$z = y_1$

It will be seen that of the four cases represented three are for level input signals and one for pulse inputs. Case 1 uses a bistable output device and a bistable intermediate storage device whilst Case II uses bistable output storage but logical-memory intermediate storage and Case III uses logical-memory only. Case IV uses pulse inputs and a single bistable but needs the addition of feedback delays. The circuits derived from the equations of the four cases are illustrated in Figs. 9.36, 9.37, 9.38 and 9.39.

The equations derived may be written in very many different forms and

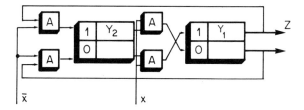

Fig. 9.36 'T' flip-flop using a bistable output device and a bistable intermediate storage element (Foster and Retallick, *Ref. 11*, Fig. 4)

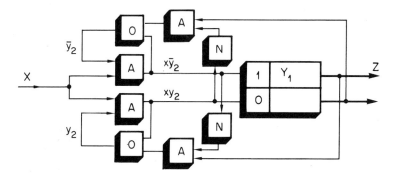

Fig. 9.37 'T' flip-flop using a bistable output device but a logical-memory intermediate storage (Foster and Retallick, *Ref. 11*, Fig. 4)

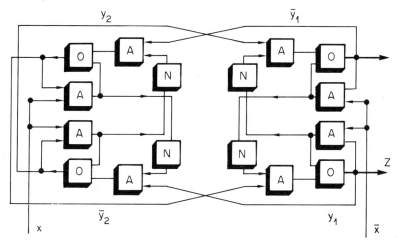

Fig. 9.38 'T' flip-flop using logical-memory output and logical-memory intermediate storage (Foster and Retallick, *Ref. 11*, Fig. 5)

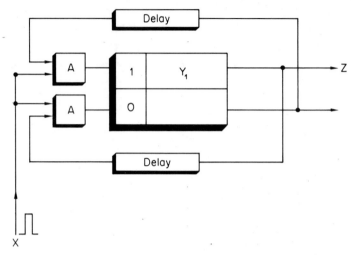

Fig. 9.39 'T' flip-flop using bistable memory output, pulsed input and delay line feed back (Foster and Retallick, *Ref. 11*, Fig. 5)

the examples illustrate only one way of expressing the equations. Case III is of particular note because no bistable memories are used, and if the equations are written in NAND/NOR logic they represent counters constructed of turbulence amplifiers and in fact all T.A. counter circuits must be designed on the equations of Case III. The reader may like to refer to a paper by J. Bulk[12] for a more comprehensive presentation of T.A. counter circuits.

A preliminary assessment of Case I indicates that it requires two bistable elements plus an active complement regenerator or monostable to provide $\bar{x}$, making 3 active elements in all. Case II looks complex but it can be shown that a fluid logic design is made feasible by using fluid flow phenomena to realise the logic functions (Warren type counter[13]). Case IV uses one active bistable but would probably need at least one more active element for the pulse shaping function.

Using a T-Flip-Flop it is easy to produce a multi-bit counter operating in a pure binary code if the T-Flip-Flops are connected in series, as shown in Fig. 9.40.

The T-Flip-Flop counter may be made to operate in other codes by using the S and R inputs as well as the T inputs and by providing extra feedback loops from more significant stages.

9.7.3 Decade Counters, Serial Operation

All serial counters use the T-Flip-Flop, as derived in the introductory part of this section, and to count in decades four stages per decade are

[9.7] COUNTING TECHNIQUES

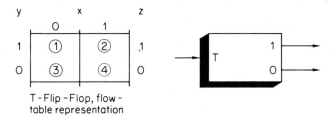

T-Flip-Flop, flow-table representation

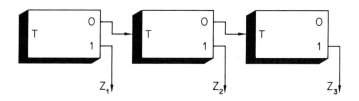

$$Y_1 = Tx$$
$$Y_2 = T\bar{y}_1$$
$$Y_3 = T\bar{y}_2$$

$$Z_1 = y_1$$
$$Z_2 = y_2$$
$$Z_3 = y_3$$

Fig. 9.40 A serial binary counter using 'T' flip-flops (Foster and Retallick, *Ref. 11*, Fig. 6)

needed. This arrangement would normally count to 15 before resetting to 0, and so in order to operate in a B.C.D. code reset logic has to be used to obtain the desired output combinations. This section gives a number of examples of B.C.D. codes and the associated counter circuitry. This is done for both directions of counting, up-counting to 9 or down-counting to 0 because in a B.C.D. code, particularly if the counter is pre-settable, a different six combinations of digits could be regarded as redundant. The examples given are counters used in a pre-set mode with the result that it is desirable to finish the count on 0000 or 1111 because these numbers are easily recognised with a four-input AND.

A. 8421 B.C.D. Up-Counting (Fig. 9.41)

This counter has to be reset to 0000 at the end of the decade, and the necessary but sufficient condition to recognise 10 or 1010 is that A.C. = 1 which is a simple AND function. The up-counting method cannot be used here, however, because the code is NON-COMPLEMENTING.

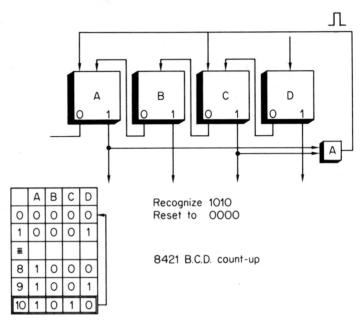

Fig. 9.41 8421 B.C.D. count up (Foster and Retallick, *Ref. 11*, Fig. 8)

B. 8421 B.C.D. Down-Counting (Fig. 9.42)

The next number after 0000 is 1111 and this has to be recognized and used to reset the counter to 9 or 1001. The 1 output is taken as a carry, the 0 output as a readout; so the reset is recognized as 0000 by a 4 input NAND, and the final position is read as 1111. This arrangement gives a manageable reset function and the desired end-state.

C. 2421 B.C.D. Up-Counting (Fig. 9.43)

2421 binary coded decimal gives the desired end-states for up- and down-counting. In the up-counting case 1000 is recognized simply by A = 1 and the counter is reset to 1110 = 8. Again the code is non-complementing so the down-counting method is more applicable.

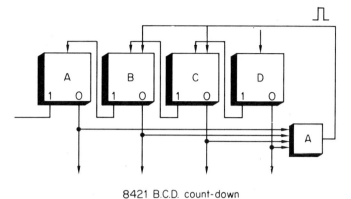

8421 B.C.D. count-down

Fig. 9.42 8421 B.C.D. count down (Foster and Retallick, *Ref. 11*, Fig. 9)

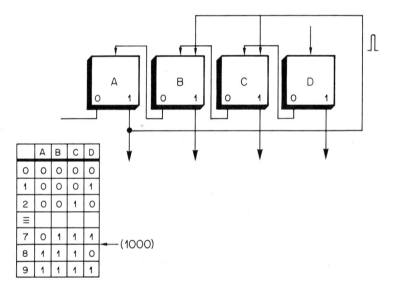

	A	B	C	D
0	0	0	0	0
1	0	0	0	1
2	0	0	1	0
≡				
7	0	1	1	1
8	1	1	1	0
9	1	1	1	1

← (1000)

Fig. 9.43 2421 B.C.D. count up (Foster and Retallick, *Ref. 11*, Fig. 10)

D. 2421 B.C.D. Down-Counting (Fig. 9.44)

The circuitry is the same as for the previous case but the code is different and more suitable for use in a data system. The complement of 0111 is recognized as $\overline{A} = 1$ and the counter is reset to 0001. This case gives a simple reset, a suitable end-state and a suitable code. It is the best of the arrangements described so far, but one more arrangement, a special case of a 2421 code, is interesting.

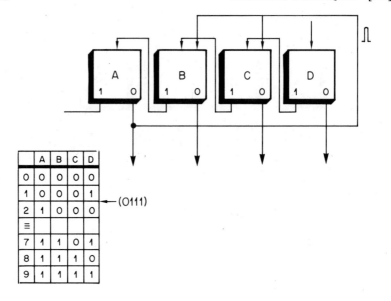

Fig. 9.44 2421 B.C.D. count down (Foster and Retallick, *Ref. 11*, Fig. 11)

E. 2421 Complementary B.C.D. (Fig. 9.45)

A complementing code is versatile in that it allows up- or down-counting; in fact the two cases are now indistinguishable.

The conclusions are that E is the more versatile system but the reset logic is liable to give problems. B and D are both possible and although the reset logic of D is superior B uses the standard 8421 code. If a 2421 code be tolerated then D is by far the best arrangement because the reset logic is simple to arrange.

A fluidic T-Flip-Flop is capable of operating at frequencies in excess of 1000 Hz, but when four are connected in series a definite limit to operating speed is imposed. If one stage switches in a time of 1 m s approximately 4 m s are needed for the worst case of the 4 stages changing state. If in a B.C.D. counter the resetting time is added to this, a decade counter could take 5 m s to cycle through unstable states before attaining stability. This limits the maximum usable speed to an input frequency of only 200 Hz and furthermore the counter could be operated safely in a synchronous mode with a clock pulse to gate the outputs. Theoretically the parallel mode counter offers a better solution because all stages receive the x input simultaneously.

[9.7] COUNTING TECHNIQUES

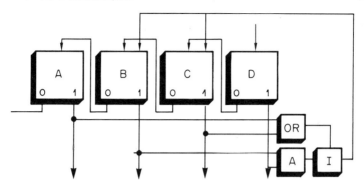

Fig. 9.45 2421 complementing B.C.D. (Foster and Retallick, *Ref. 11*, Fig. 12)

9.7.4 Decade Counter, Parallel Operation, B.C.D. Code

If a counter is run in a parallel mode, all stages receive the input pulse at the same time via gating, and this ensures that all stages switch simultaneously and present no output hazards. An example of this type of arrangement is shown in Fig. 9.46, and it will be seen that T-Flip-Flops are used as in the series counter, but that '1' outputs are used as feedforward loops. The difference in derivation may be seen by comparing the switching equations in Fig. 9.46 with those for the series counter of Fig. 9.40.

Because T-Flip-Flops are used in this counter it must operate in a weighted code. If converted to B.C.D. operation by means of reset logic, the maximum usable speed would be reduced by 50 per cent. This still represents an inherently faster design than the serial counters, but the

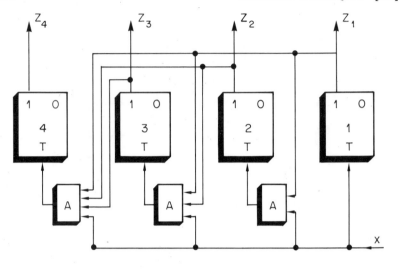

$$Y_1 = Tx$$
$$Y_2 = Tx\, y_1$$
$$Y_3 = Tx\, y_1 y_2$$
$$Y_4 = Tx\, y_1 y_2 y_3$$

$$Z_1 = y_1$$
$$Z_2 = y_2$$
$$Z_3 = y_3$$
$$Z_4 = y_4$$

Fig. 9.46 Parallel counter using 'T' flip-flops (Foster and Retallick, *Ref. 11*, Fig. 13)

reset logic reintroduces output hazard problems. A sequential circuit using a unit-distance code may offer a better solution because extra complexity may be more tolerable than synchronous operation.

9.7.5 Decade Counter, Parallel Operation, Unit Distance Code

If S/R Flip-Flops are used instead of T-Flip-Flops, counters can be designed to use any desired code. If the secondary assignment is made to conform to the output coding then the output equations reduce simply to $z_1 = y_1$, $z_2 = y_2$, $z_3 = y_3$ and $z_4 = y_4$. All the logic is contained in the secondary switching equations.

Example—Gray Code

This example uses pulsed inputs and the Gray code is modified in the binary 9 coding to convert it into a unit distance Lippel code. The input, x, must appear in every switching equation so the normal flow table is dispensed with and a Karnaugh map is used instead for ease of manipulation.

This circuit, shown in Fig. 9.47, has an input pulse supplying 10 AND gates, 2 cases of 4-input AND functions and several cases of 3-input AND functions. The circuit may be physically realizable electronically, but it is not feasible as far as fluidics is concerned.*

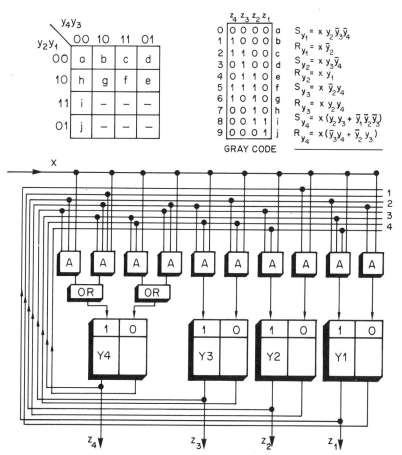

Fig. 9.47 GRAY code counter (Foster and Retallick, *Ref. 11*, Fig. 14)

*Slight improvements can be made to the circuitry by taking a different assignment to produce a slightly modified code. The best result achieved was a circuit using 10×3 input AND gates and 2×2 input OR gates.

9.7.6 Decade Ring Counters

There is a class of counters which use a larger number of secondaries, with very simple logic, giving a decimal output and known as ring counters. One type in particular uses 10 secondary memories with a level input and gives 10 separate outputs, and the derivation of the first four stages of such a ring counter is shown in Fig. 9.48.

The circuit consists of a chain of bistable devices and AND-gates, fed by x and $\bar{x}$ alternatively. It will be noticed that the S_{Y_1} term has not been

y_{1-10}	$\bar{x}$	x	y_{1-10}
1 0 0 0	①	2	1 0 0 0
0 1 0 0	3	②	0 1 0 0
0 0 1 0	③	4	0 0 1 0
0 0 0 1	5	④	0 0 0 1

$$S_{y_1} = ---$$
$$R_{y_1} = x\, y_1$$
$$S_{y_2} = x\, y_1$$
$$R_{y_2} = \bar{x}\, y_2$$
$$S_{y_3} = \bar{x}\, y_2$$
$$R_{y_3} = x\, y_3$$

$$\left.\begin{array}{l}S_{y_n} = x\, y_{n-1}\\R_{y_n} = \bar{x}\, y_n\end{array}\right\} \underline{n\ \text{even}}$$

$$\left.\begin{array}{l}S_{y_n} = \bar{x}\, y_{n-1}\\R_{y_n} = x\, y_n\end{array}\right\} \underline{n\ \text{odd}}$$

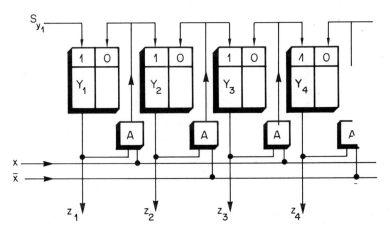

Fig. 9.48 Ring counter fed by signals x and $\bar{x}$ coming a binary divider (Foster and Retallick, *Ref. 11*, Fig. 15)

[9.7] COUNTING TECHNIQUES 417

entered in the switching equations and that if there is an even number, n, of stages, $S_{Y_{n+1}} = \bar{x}y_n$. Thus if n = 10, $S_{Y_{n+1}}$ becomes S_{Y_1} if the ends of the chain are connected together; hence the term ring counter. Instead of the AND-chain ring counter shown, a NOR-chain ring counter is obtained if the equations are transformed into NOR-logic.

There are two ways of using the chain-ring counter, and as shown above the '1' signal is passed through two stages from every time $x \to 1$ and then back to 0. This would need 20 stages for a decade ring. If, alternatively, the input is passed through a binary counter stage acting as a frequency divider, then only 10 stages are needed for a decade counter.

In decade form, the output coding is essentially a 10-bit non-unit-distance format which is excellent for batch counting or frequency dividing but wasteful of space in a large numerical system. The speed is theoretically high, over 500 Hz, and spurious signals due to the non-unit-distance nature of the code do not present a serious problem because two stages switch simultaneously.

9.7.7 Twisted Ring Decade Counters

The twisted ring counter uses a different secondary assignment from the chain counters described because only five stages are used, thus operating in a five-bit unit distance code. In this design, pulse or level input signals may be used, but for level signals the number of secondaries must be doubled. The pulsed input design is given in Fig. 9.49.

The secondary circuitry uses 5 bistables with 2-input AND gates on each input and the secondary equations represent a shift-register. The 5 shift-register stages are connected tail to head with a twist in the connecting leads, which causes the stages to fill up successively with 1 states and then successively empty again, producing the '5-bit creeping code' or 'quinary code'. A decimal readout may be accomplished by a decoding matrix consisting of 10 AND-gates.

Parallel counters differ from each other only in the number of secondaries chosen, and the secondary assignment made and the design of a counter must be a compromise between circuit complexity and the practicability of the associated code.

9.7.8 Pulse Shaping

For pulse input operation a pulse shaper must be an integral part of the design and should produce a pulse width of the same order as the switching time of an element, i.e. less than 1 m s. The only fluidic means known for timing in this range is the delay line, and so this will form part of the pulse shaper circuitry, so that two inputs are now available, x_1 and x_2, where x_2 represents x_1 delayed by time t. It will be assumed that the output z will

$y_1 y_2$ \ $y_3 y_4$	$\bar{y}_5$				y_5			
	00	10	11	01	01	11	10	00
00	a	b	c	—	—	—	—	j
10	—	—	d	—	—	—	—	—
11	—	—	e	—	g	f	—	h
01	—	—	—	—	—	—	—	i

x pulse present in all logic

$S_{y_1} = x \bar{y}_5$
$R_{y_1} = x y_5$
$S_{y_2} = x y_1$
$R_{y_2} = x \bar{y}_1$
$S_{y_3} = x y_2$
$R_{y_3} = x \bar{y}_2$
$S_{y_4} = x y_3$
$R_{y_4} = x \bar{y}_3$
$S_{y_5} = x y_4$
$R_{y_5} = x \bar{y}_4$

$z_0 = \bar{y}_1 \bar{y}_5$
$z_1 = y_1 \bar{y}_2$
$z_2 = y_2 \bar{y}_3$
$z_3 = y_3 \bar{y}_4$
$z_4 = y_4 \bar{y}_5$
$z_5 = y_1 y_5$
$z_6 = \bar{y}_1 y_2$
$z_7 = \bar{y}_2 y_3$
$z_8 = \bar{y}_3 y_4$
$z_9 = \bar{y}_4 y_5$

decimal output decoding

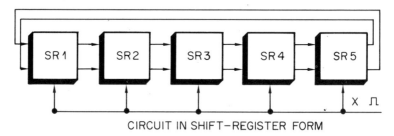

CIRCUIT IN SHIFT-REGISTER FORM

Fig. 9.49 Twisted ring counter using pulsed-input (Foster and Retallick, Ref. 11, Fig. 16).

appear at the same time as x_1. The time chart is drawn, the primitive flow table obtained from it and the resulting switching equations are illustrated in Fig. 9.50. Assuming that the complements $\bar{x}_1$ and $\bar{x}_2$ are available if necessary from an active monostable element, then the circuit $z = x_1 \bar{x}_2$ may be implemented in two ways:

1. AND logic is used with the equation $z = x_1 \bar{x}_2$ and a monostable, delay line and AND gate are needed.

2. By De Morgan's theorem, $z = \overline{(\bar{x}_1 \times x_2)}$ and this function is given by active NOR device plus once again the monostable and delay line.

[9.7] COUNTING TECHNIQUES 419

These circuits are illustrated in Fig. 9.50, parts (a) and (b).

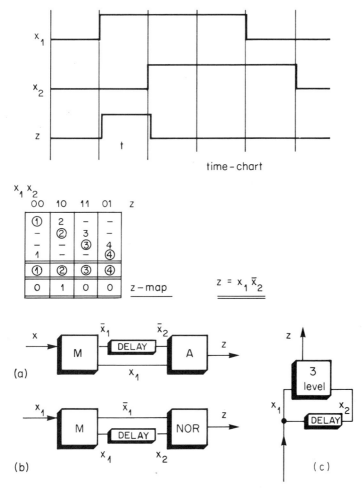

Fig. 9.50 Three alternative pulse shapers (Foster and Retallick, *Ref. 11*, Fig. 17)

A further method which has often been used is to run a wall attachment element in a three-level switching mode such that if x_1 is applied to the set input, a 1 is obtained, and if x_2 is applied to the reset input whilst x_1 is still present a 0 is obtained. This implies that the inputs have the truth values, $\bar{x}_1 = 0$, $x_1 = 1$, $x_2 = \infty$. The difficulty with this circuit, shown in Fig. 9.50(c), lies in obtaining the ∞ signal level, and even then it is difficult to operate a wall attachment element satisfactorily under these conditions.

References

1. Littlehales, J. S., The lucas system of pneumatic circuit control, *Machine. Production Eng.*, 82–92 (1969).
2. Stancielescu, F. S., Sequential logic and its application to the synthesis of finite automata, *I.E.E.E. Trans. Electronic Computers*, **FC-14**, No. 6, 786–791 (1965).
3. Retallick, D. A., *Switching Theory Applied to Fluidic Digital Systems*, Ph.D. Thesis (Birmingham), 1968.
4. ISO R 840 'Codes for numerical control of machines' 1968.
5. ISO R 646 '6 and 7 bit coded character sets for information processing interchange', 1967.
6. ISO R 1057 'Interchange punched tape variable block format for positioning and straight-cut numerically controlled machines', 1969.
7. ISO R 1058 'Punched tape variable block format for positioning and straight-cut numerically controlled machines', 1969.
8. ISO R 1059 'Punched tape fixed block format for positioning and straight-cut numerically controlled machines', 1969.
9. ISO R 841 'Axis and motion nomenclature for numerically controlled machines', 1968.
10. ISO R 1056 'Punch tape block formats for the numerical control of machines—coding of preparatory functions G and miscellaneous functions M', 1969.
11. Foster, K. and Retallick, D. A., 'Fluidic counting techniques', *B.H.R.A. Fluidics Feedback*, Aug, Dec 1969 and Feb 1970.
12. Bulk, J., Counting circuits with turbulence amplifiers, *C.F.C.* (3), Paper K8.
13. Warren, R. W., Fluid amplification—3. Fluid flip flops and a counter, *Diamond Ordnance Fuse Lab.*, Aug. 1962. U.S. Department of Commerce: Office of Technical Services AD285572.

   See also,
   H. C. Torng, *Introduction to the Logical Design of Switching Systems*, Addison-Wesley, 1964.
   M. P. Marcus, *Switching Circuits for Engineers*, Prentice-Hall (1962).

10
Fluidics Applied to Sequence Circuits and Simple N.C. Systems

10.1 Introduction

A discussion of the ideas behind the design of sequence circuits was given in Chapter 9, in which two classes of such circuits were considered; asynchronous and synchronous. The former type are circuits in which an operation takes place only after the completion of all the scheduled previous operations and a signal has been received indicating that the sequence can proceed, whilst the latter type change state each time an external gating pulse is received—for example, binary or decimal counters. In designing circuits for industrial machines, it is convenient to utilize the idea of asynchronous logic for the simpler systems, but for more complex ones it is normal to use a punched tape or card input, and within such circuits part of the operation may be considered to be synchronous. In this chapter the methods are separated into the following groups:

(*a*) Asynchronous systems where the signals from limit switches pass into a combinational logic circuit which then decides the next output state of the system. Some memory elements may be included if any combinations of the sensor signals are repeated, and external signals may be introduced to prevent the sequence continuing if a requirement is not satisfied; for example, if a guard is not down, or if a measurement is outside a gauging limit. The block diagram of such a circuit is given in Fig. 10.1.

The primary memories control the cylinders in a sequence depending on the limit switch combinations; the secondary memories are necessary only in order to distinguish between limit switch combinations that appear more than once.

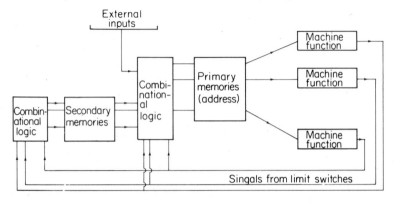

Fig. 10.1 Block diagram of asynchronous sequential circuit controlled by combinational logic and secondary memories

It is worth reminding the reader that the logic circuits may be constructed from either AND logic or OR/NOR logic and that an easy transposition from one to the other may be made by using De Morgan's theorem, i.e. that $\bar{x} + \bar{y} = \bar{x} \cdot \bar{y}$. For example, if an operation z can continue when a signal x appears from the combination circuit AND a signal y is produced by a gauge to say that the workpiece is within the required tolerance, i.e.

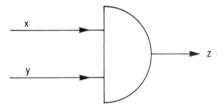

then this is equivalent to saying that the operation can only continue if NEITHER signal from the combinational logic *preventing* the sequence continuing is present OR a signal from the gauge to say the workpiece is *outside* the required tolerance is present, i.e.

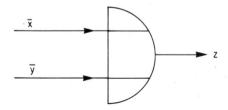

[10.1] INTRODUCTION 423

If Flip-Flops are used as memories, either monostable wall attachment devices may be used to provide the OR or NOR function for the logic or passive AND gates may be chosen as alternative logic which gives less air consumption but some loss in signal strength. If only NOR elements are available, as, for example, in turbulence amplifier systems, then there is no choice; both memories and interlocks must be performed by combinational logic of this type.

(b) Asynchronous systems where signals from limit switches pass to a multi-input gating circuit, the output from which is the input to a counter circuit (ring, binary or binary-coded-decimal). The appropriate operations are then carried out on the appearance of the correct count, which is decoded in the logic of an 'Address' circuit which distributes the information to the correct functions. The outputs from the ring counter are used where necessary to gate the signals from the limit switches in order to avoid ambiguity should the limit switch signal appear more than once. A block diagram of such a system is shown in Fig. 10.2.

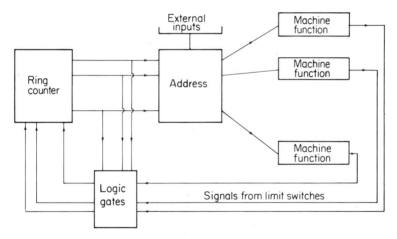

Fig. 10.2 Block diagram of asynchronous sequential circuit controlled by a counter together with combinational logic

The information labelled 'External Inputs' is additional information required from the machine before the sequence can continue (in the case to be quoted later this is from a pneumatic gauge). Having External Inputs implies having some interlocking functions in a circuit which is basically counter controlled, but some overlap is often unavoidable. Within this asynchronous circuit, the counter is almost certainly a synchronous sub-circuit operating only from the signal from the multi-input OR element which is external to it. The theoretical considerations of counter circuits was given in Part 9 and the fluidic implementation is covered later in Part 12.

(c) Asynchronous circuits which are controlled by plugboards or from punched cards or tape. In this method the plugboard, etc., is acting as a mechanical ring counter (or is a device operated by a ring counter) but has the advantage that information may be included on the plugboard, etc., in order to individually control any one operation of the machine. In other words, there are individual and variable instructions within the control loops, the block diagram of which is shown in Fig. 10.3.

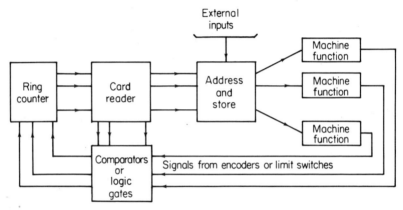

Fig. 10.3 Block diagram of asynchronous counter-controlled circuit with the addition of a card reader to provide additional information for each operation

For example, in addition to initiating the next turning operation in an automatic lathe, it may be necessary to select the appropriate tool and the correct spindle speed according to the information on the plugboard. In most cases this is done without feedback, but in certain applications, feedback from an encoder must be compared with the information on the card to be sure the correct operation is being carried out. For this a comparator is required and an example is given later in the chapter where a stop on a stop-bar is selected for a drilling operation. The choice of a punched tape in preference to a punched card for information storage is generally determined by the quantity of information to be handled, the capacity of a tape being significantly higher than that of a card. Plugboards, card readers and tape readers are discussed more fully in Part 11.

In this chapter, examples of each of these types of system will be given in order to develop the ideas more fully. In each case, fluidic

[10.1] INTRODUCTION 425

circuits rather than logic diagrams will be used to illustrate the principles so that the presentation will be somewhat different from that of Part 9.

10.2 Simple Asynchronous Circuits

A relatively simple circuit of this type[1], has been in operation in the laboratory since 1964 and demonstrates the possible simplicity, and low cost, of pure fluid logic control circuits. The circuit consists of four turbulent jet re-attachment devices used as two passive 'AND' gates (originally the 'Half-Adder' of Greenwood), four flapper valves performing the functions of micro-switches, and a resistance capacitance delay. Fig. 10.4 shows the complete circuit.

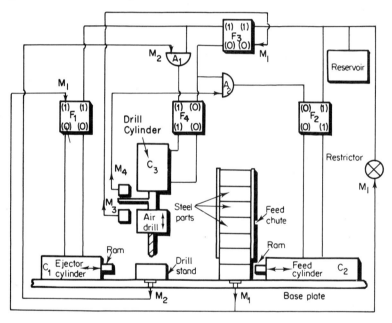

Fig. 10.4 A schematic diagram of an asynchronous sequential control circuit for a drilling operation (Foster and Parker, *Industrial Electronics*, Oct. 1966, Fig. 4)

The Flip-Flops are initially all in position 0. When a block passes from the feed chute to the base plate, flapper-valve (microswitch) M_1 operates, and sets Flip-Flops F_1, F_2 and F_3 to the 1 position, F_2 and F_3 being set after a time delay caused by a capacitance and resistance (an air volume and needle valve). A monostable element is used before the capacitance in order to amplify the signal from the microswitch. F_1 causes the cylinder

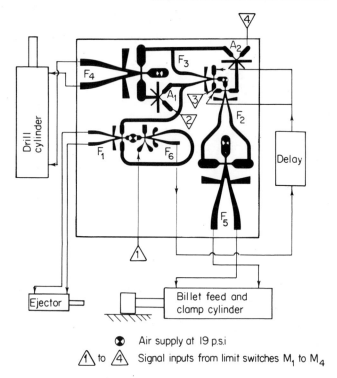

Fig. 10.5 Labelled silhouette of asynchronous drilling circuit of Fig. 10.4 (actual size 4 × 4 in.) (Foster, Jones and Mitchell, *Ref. 2*, Fig. 2)

C_1 to operate and eject any block on the drill stand. After F_3 has been set to position 1 a signal goes to Flip-Flop F_1 so that this returns to position 0 and the ejector is returned. At the same time, the feed cylinder C_2 pushes the next workpiece into position. When the workpiece has been clamped by C_2, microswitch M_2 operates, sending a signal to 'AND' gate A_1 which is already receiving a signal from F_3. A_1 can therefore operate to set F_4 to the 1 position and C_3 then moves the drill to the workpiece.

When the drill has travelled the correct distance, microswitch M_3 operates and resets F_3 to position 0. F_3 sends a signal to 'AND' gate A_2 and resets F_4 to return the drill. When the drill has returned, microswitch M_4 completes the operation of A_2 which resets F_2. Thus ram C_2 returns to accept the next workpiece.

The original circuit was 'breadboarded' and then integrated into one layer of 'Dycril', 4 × 4 in.². A silhouette of this circuit is shown in Fig. 10.5, labelled so that, in conjunction with the diagram of Fig. 10.4, the

[10.2] SIMPLE ASYNCHRONOUS CIRCUITS 427

operation should be clear. In the silhouette ⊗ denotes the constant supply pressure to the Flip-Flops F_1 to F_5 and an amplifier F_6, the other two elements being passive AND gates. The system is switched by signals labelled ⚠, etc. from flapper valves (microswitches) M_1, etc.

A point of interest in this circuit is that the memory Flip-Flops F_3 and AND gates A_1 and A_2 constitute a DOUBLE SEQUENTIAL AND gate as suggested in Section 9.5. The actual circuit connexions are very slightly different from those of Fig. 9.21, but there is no difference in principle. This particular circuit is made more complicated by the addition of the ejector cylinder, but since this is operated after a time delay from the main circuit, the basic sequence is still only a two cylinder one.

A slightly more complex circuit is that required for the example given in Section 9.5.2 for the two cylinder sequence $B-\ A+\ A-\ B+\ A+\ A-$. Such a circuit is perhaps more representative of a typical industrial system in the amount of logic required relative to the output functions. For this reason, it is instructive to comment on the relative costs of circuits using (a) turbulence amplifiers, (b) wall attachment logic and (c) diaphragm elements. In the first case, the logic must be run at very low pressures so that two step-up relays are required to operate each cylinder or one step-up relay is required before the output piston valve; in case (b) the logic could be run at a high enough pressure to operate piston valves directly, but because the air consumption would then be high such a situation is not so desirable and step-up relays must be used as in the previous case; and in case (c) the logic easily operates at a high enough pressure to switch conventional piston valves. In comparing costs, the total cost of the logic and power valves must be taken into account, but the rest of the system, i.e. power cylinders, pressure reducing valves and so on, may be neglected because these will be common to each type. The approximate cost is as follows:

Type	No. of logic elements	No. of step-up relays	No. of piston valves	Cost £
		4	0	34
(a)	10	2	2	26
		4	0	60
(b)	9	2	2	52
(c)	10	0	2	24

N.B. Based on:

Turbulence amplifier	£1	0s.
2-input wall attachment gate	£4	0s.
2-input diaphragm gate	£2	0s.
Step-up relay	£6	0s.
Power valve	£2	0s.

Clearly the high cost of step-up relays is an important factor so that these should be kept to a minimum. The high cost of wall attachment elements puts them at a real disadvantage compared with either turbulence amplifiers or diaphragm elements.

A rather more complex sequence is that described by Levesque and Loup[3] for the control of a twist drill flute milling machine. The basic circuit contains only three limit switches plus 'start' and 'stop' buttons, so that the information input is only slightly greater than in the previous example, but two extra functions are added which may be of interest to the reader:

(a) When the air is switched on, the system must start up in the correct mode. This is achieved by using a preferentially based Flip-Flop which always starts up so as to hold the air supply OFF until the 'return to start' button has been depressed.

(b) The basic operation has to be completed twice during one cycle because there are two flutes per drill, and the logic for this is provided by one stage of a binary counter.

10.2.1 Universal Logic Blocks

In Chapter 9, some discussion of a Universal Logic Block is given, but from a logical design point of view. It is worth while, at this point to consider it from a practical point of view. The costing in the case of the two cylinder sequence shows that wall reattachment elements are expensive relative to the other types and therefore the potential market may not be so great. However, one advantage of this type of device is that it lends itself to integrated circuit manufacture, and it is apparent that the cost of manufacturing a block of elements is not so very much greater than the cost for one element so that the cost per gate would come down.

The problem is to decide on the most appropriate collection of elements to include in one integrated circuit, bearing in mind that it will be necessary to have some redundancy in order to cover a wide variety of cases. The ideas of Chapter 9, and indeed of the example given in this chapter, suggest that the memory function and interlocks form the basis of the logic used in sequencing circuitry. The Double Sequential AND combines the memory and some logic and may well be a suitable basis for the design. The number of AND gates that may be supplied by a Flip-Flop depends upon whether they are active or passive. If passive, the number will be limited to 3 or 4, and therefore on the input to the Flip-Flop either 3 or 4 input AND gates will be required. One or two other elements may be included for amplification of signals where necessary.

Whether such an approach would greatly affect the utilization of fluidic systems depends on how much the price of individual elements comes

[10.2] SIMPLE ASYNCHRONOUS CIRCUITS

down in the future and also on how generally applicable such a collection of elements would be in the more complex systems required by industry.

10.3 Counter Controlled Circuits

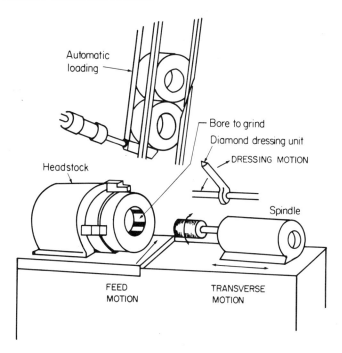

Fig. 10.6 General arrangement of a grinding operation (Monge, *Ref. 4*, Fig. 12)

A counter controlled circuit, shown in Fig. 10.6, for the control of the grinding operation and which is described by Monge[4], consists of the following steps:

1. Loading.
2. Wheel traverse to workpiece.
3. Feed of wheel into work.
4. Wheel dressing.
5. Repeat operations 2 and 3 for final grind.

These operations are shown diagrammatically in Fig. 10.7 and the circuit diagram of the system is given in Fig. 10.8. Pressing the 'start' button actuates the loader, which feeds a workpiece into the machine.

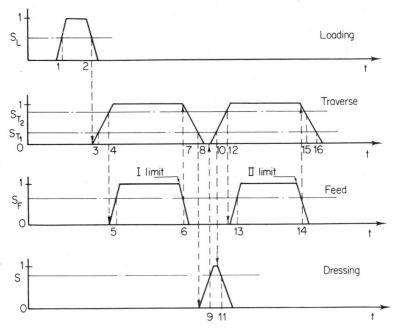

Fig. 10.7 Timing diagram for the grinding process of Fig. 10.6. The numbers 1 to 16 indicate the state of the counter shown in the circuit diagram of Fig. 10.8 and listed in the truth table of Fig. 10.9. Letter S denotes sensor
(Monge, *Ref. 4*, Fig. 13)

After a time lag, the loader returns to its initial position, and simultaneously the wheel starts traversing towards the workpiece. After completion of this operation the wheel moves into the workpiece at a rough feeding rate, and when the rough grinding is finished the wheel retracts from the workpiece and the traverse return stroke starts. During this return stroke the diamond dressing unit dresses the wheel. The wheel is then fed back into the workpiece to finish-grind it and, on the return of the wheel to its initial position, the cycle stops.

In the circuit diagram of Fig. 10.7 the counter is clearly defined, and below the counter are three separate circuits which form the 'address' logic. The signals into the counter are from pneumatic sensors on each of the feed motions passing through the OR gates at the top left of the circuit diagram. Two additional signals, I limit and II limit, come from the automatic gauging system. For clarity, an indication of the points at which the sensors give a signal are indicated on the timing diagram by levels S_L, S_{T_1}, S_{T_2}, S_F and S_D—where suffices L, T, F and D denote loading, traverse, feed and dressing respectively. On the time axes, numbers are

[10.3] COUNTER CONTROLLED CIRCUITS

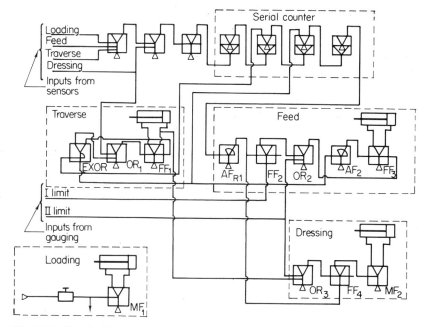

Fig. 10.8 Circuit diagram of counter-controlled sequence for machine of Fig. 10.6 (Monge, *Ref. 4*, Fig. 15)

given each time a sensor is activated and thus correspond to the numbers recorded by the counter. These numbers cross refer to the number of the truth table of Fig. 10.9 so that a check on the operation of the circuit can easily be made. Notice that two sensors are allocated to the traverse motion because the feed must be set in motion at a different point from the dressing motion. Also note that at certain points—1, 3, 5, 7, 13, 15 and 16—no action is required of the circuit, even though the sensor signals cause the counter to change state. The circuit is designed to ignore these counter states. It may be seen that the decoding from the counter has been done with some skill so that a minimum of logic is required.

As an illustration of the operation, let us consider the feed motion governed by Flip-Flop FF_3. It is required to go ON at counts 4 and 12 and go OFF on receipt of signals from the gauges I and II, as may be seen from Fig. 10.7. At count 4, AF_{R2}* switches, but when limit signal I comes on, AF_{R2} is overridden. The signal to AF_{R2} goes off at a count of 8 and at the same time the effect of the gauge signal I is removed because the signal

---

* The notation AF is for a particular type of monostable element that switches quickly at a set pressure level.

THE FLUID LOGIC 'TRUTH TABLE'

Input pulses	Binary counter				External inputs		Address														
							Loading	Traverse Table					Feed-motion					Dressing motion			
	1' stage	2' stage	3' stage	4' stage	1' lim.	2' lim.	MF$_1$	EX-OR		OR$_1$	FF$_1$		AF$_{R1}$	FF$_2$	OR$_2$	AF$_{R2}$	FF$_3$		OR$_3$	FF$_4$	MF$_2$
							1 / 0	A$_1$	A$_2$		1 / 0						1 / 0				1 / 0
1	1	0	0	0	—	—	1 / 0	0	0	0	0 / 1		0	0	0	0	0 / 1		0	0	0 / 1
2	0	1	0	0	—	—	0 / 1	1	0	1	1 / 0		0	0	0	0	0 / 1		1	0	0 / 1
3	1	1	0	0	—	—	0 / 1	1	0	1	1 / 0		0	0	0	0	0 / 1		1	0	0 / 1
4	0	0	1	0	—	—	0 / 1	1	0	1	1 / 0		0	0	0	1	1 / 0		1	0	0 / 1
5	1	0	1	0	—	—	0 / 1	1	0	1	1 / 0		0	0	0	1	1 / 0		1	0	0 / 1
6	0	1	1	0	1	—	0 / 1	1	1	1	0 / 1		0	1	1	1	0 / 1		0	0	0 / 1
7	1	1	1	0	1	—	0 / 1	1	1	0	0 / 1		0	1	1	1	0 / 1		0	0	0 / 1
8	0	0	0	1	1	—	0 / 1	1	0	0	0 / 1		1	1	1	0	0 / 1		0	1	1 / 0
9	1	0	0	1	1	—	0 / 1	0	0	1	1 / 0		1	0	0	0	0 / 1		1	1	0 / 0
10	0	1	0	1	1	—	0 / 1	1	0	1	1 / 0		1	0	0	0	0 / 1		1	0	0 / 1
11	1	1	0	1	1	—	0 / 1	1	0	1	1 / 0		1	0	0	0	0 / 1		1	0	0 / 1
12	0	0	1	1	1	—	0 / 1	1	0	1	1 / 0		1	0	0	1	1 / 0		1	0	0 / 1
13	1	0	1	1	1	—	0 / 1	1	0	1	1 / 0		1	0	0	1	1 / 0		1	0	0 / 1
14	0	1	1	1	1	1	0 / 1	0	1	0	0 / 1		1	0	0	1	0 / 1		—	0	0 / 1
15	1	1	1	1	1	1	0 / 1	0	1	0	0 / 1		1	0	0	1	0 / 1		—	0	0 / 1
16	0	0	0	0	1	1	0 / 1	1	0	1	0 / 1		0	0	1	1	0 / 1		—	0	0 / 1

Fig. 10.9 Truth table for the counter and logic states of the circuit of Fig. 10.8. The counter states correspond to numbers 1 to 16 on the timing diagram of Fig. 10.7 (Monge, *Ref. 4*, Fig. 16)

[10.3] COUNTER CONTROLLED CIRCUITS 433

from the last stage of the counter switches FF_2. At a count of 12 AF_{R2} again causes the feed motion to start until overridden by the gauge signal II operating through OR_2. A truth table is essential for the checking of a system as complex as this one.

Note that the circuit does not contain any gating of the feedback signals by the outputs from the counter, so that it is possible that the counter could get out of step with the sequence.

In this case, a natural binary counter was used to provide the operating signals, but in a sequence with a greater number of events a decade counter is the more convenient to use, since decades may be cascaded to give as big a count as required. As discussed in Chapters 9 and 12, a decade counter stage may itself be built up from a four stage binary counter with appropriate gating logic. To consider the counter itself, however, provides only part of the answer to a problem and the decoding and readout present additional parts of the circuit that may require a number of extra elements. On the whole, there is some justification in going to ring or twisted-ring counters where the decoding logic is very straightforward. At the University of Birmingham some success has been achieved with twisted-ring decade counters in a variety of applications, of which one will be described in the next section. It is, however, relevant to mention one of them here, which is the timing control for a sequence of cylinders operating the valves which govern the operation of a high energy rate forming machine (Petro-Forge).

A block diagram of this control is shown in Fig. 10.10. A fluidic timer is set up by using an oscillator to drive a two stage counter which will count up to 40; the second stage provides one decade whilst the first one counts only up to four. Thus the time for any event may be selected to be any of the forty intervals, and the sequence will stop or start according to the button pressed, the multi-stop signal 'freezing' the sequence at the position it happens to be in. Single cycle or multi-cycle operation may be chosen at will. The choice of a pure fluidic timer was because the system must work in an environment of shock and vibration, and there is a possibility that a pure fluid circuit will provide a better life and reliability than any other alternative.

10.4 A Punched Card Controlled Sequence

The fluid logic circuits for a card controlled sequence circuit are given by Foster, Mitchell and Retallick[5], and a diagram of the control circuit for one axis is given in Fig. 10.11.

The position of the drill along one axis is fixed by pre-set stops, and a choice of six are available on the six sides of a hexagonal bar. The infor-

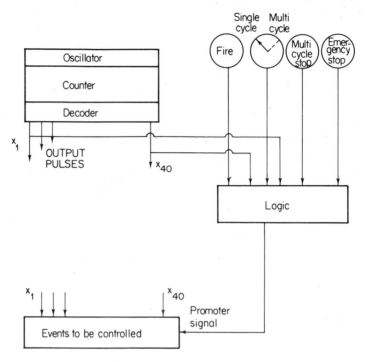

Fig. 10.10 Block diagram of pure fluidic, timed-sequence control for 'Petro-Forge' high energy rate forging machine

mation on the card indicates in binary code the required side of the bar and therefore the stop. The bar indexes until the encoder disc gives the same binary number, at which point a signal appears out of the comparator. For more information on encoder discs see Chapter 13. The bar then stops indexing, the drill moves up to position, slowing down to half speed before finally stopping. The machine is clamped, a drilling operation is performed and the ring counter indexed round again.

The point about this type of sequencing is that information about the choice of stop is required before each operation can be started, and this information is best stored on a plug board, a punched card or a paper tape.

The original circuit had a card reader using turbulence amplifiers gated in rows by a pure fluid ring counter. Because of the excessive number of turbulence amplifiers required, this has now been superseded by a 'Techne' card reader which is a rotating drum type pneumatic mechanical device, as described in Chapter 11. Brewin[6] describes two sequence circuits built up from the Techne spring NOR unit, one of which is punched card controlled.

[10.4] A PUNCHED CARD CONTROLLED SEQUENCE

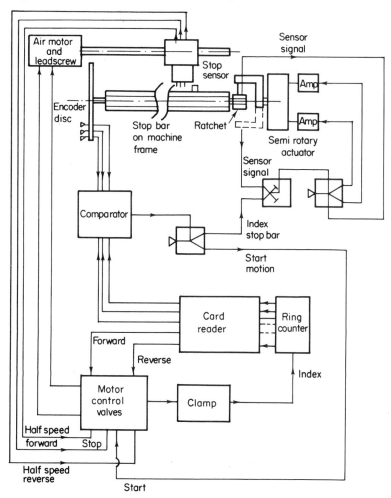

Fig. 10.11 Schematic diagram of card controlled radial drill at University of Birmingham

The comparator is a simple coincidence gate which is very easily built into an integrated pure fluidic circuit, and is one of a class of sub-circuits which would use pure fluid devices in the best possible way. The one chosen, described in Chapter 14, had only one active element per stage, and stages could easily be stacked together in a relatively small volume.

10.5 Air Consumption of Pure Fluid Devices and Moving Part Circuits

The use of pure fluid elements for machine control presents the problem that because the devices are open-centre the air consumption of the control circuit of a machine may be a significantly large proportion of the total consumption for the control and power circuits. Thus, whilst the consumption figures for one machine may be dismissed as unimportant compared with the total requirements of a factory, if all the machines in the factory were equipped with 'fluidic' control, an additional source of compressed air would have to be provided. Parker[7] discusses this problem, and Fig. 10.12 shows the capital cost of different types of air supply per s.c.f.m. of air supplied at different maximum pressures.

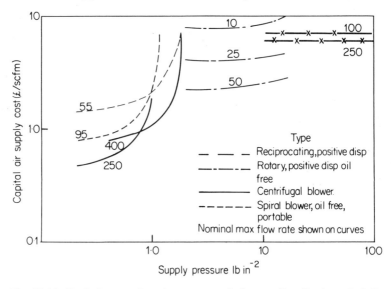

Fig. 10.12 Capital cost of various types of air supplies (Parker, *Ref.* 7, Fig. 4)

The conclusion is that it may be worth while buying separate small blowers to operate the logic circuit of each machine (or group of machines), and the logic functions would have to operate at pressures lower than 1 lb. in^{-2}. Fig. 10.13 is also reproduced from this paper and gives the air consumption for turbulent re-attachment and turbulence amplifiers. It is easy to see that for a circuit using more than, say, 50 turbulent re-attachment elements, the air consumption becomes significant.

From Fig. 10.13 it would appear that the use of turbulence amplifiers is attractive, but this is not necessarily so because two T.A's are required to

[10.5] AIR CONSUMPTION 437

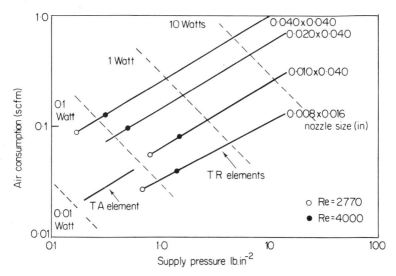

Fig. 10.13 Air consumption of turbulence amplifier and various sizes of turbulent reattachment elements over a range of supply pressures (Parker, *Ref. 7*, Fig. 3)

make one Flip-Flop (and on occasions an additional two are required as buffer amplifiers). In consequence, a greater number of T.A's are required for any given circuit and the total air consumption may be little different. An example of this is given in the later discussion on comparators.

The use of passive AND gates in combination with wall attachment elements provides a further way of reducing air consumption in sequence circuitry, since the number of active NOR elements is then reduced. It is not always feasible to do this, however, since passive AND gates only have two inputs, and gates with more inputs can only be provided by cascading elements, with the attendant problem of loss of pressure. Apart from this pressure loss due to the use of passive devices, there is a problem that because a passive AND device requires a different operating pressure level from that needed to give switching of a turbulent reattachment element, the fan-out of a preceding wall attachment device may be limited unless care is taken in impedance matching. This is pointed out in Ref. 2, the example being that shown on the silhouette of Fig. 10.5, where F_3 fans out to F_4 and A_2, i.e. to a Flip-Flop and an AND gate. A restrictor has been introduced into the channel with the least impedance in order to reduce the flow through the channel at any given pressure. In this way an element switching at a lower pressure than that required to switch a second will be drawing less flow once the highest pressure has been reached. Fig. 10.14 shows this situation diagrammatically.

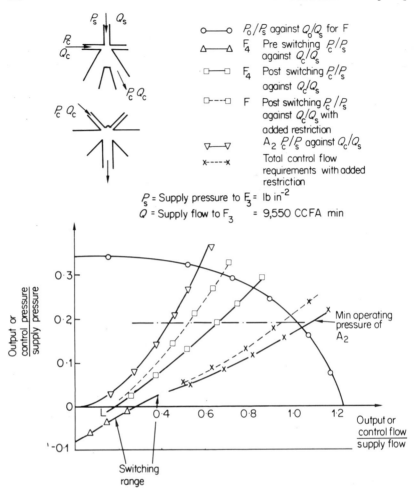

Fig. 10.14 Curves showing the effect of restrictors in improving the fan-out from F_3 to F_4 and A_2 of the silhouette of Fig. 10.5 (Foster, Jones and Mitchell, *Ref. 2*, Fig. 4)

The output pressure-flow characteristic of F_3 is plotted, and superimposed on this are the requirements of the elements driven by F_3, i.e. F_4 and A_2, plotted non-dimensionally with respect to the supply pressure and flow of F_3. The control port characteristic of F_4 is typical of a turbulent re-attachment device. Firstly, the curve is displaced upwards discontinuously after switching, because the vacuum existing in the re-attachment bubble is no longer assisting the introduction of control flow into that port. Secondly, the range of flow over which the element may switch is rather broad, and the switching point depends upon the output load on the element. For a blocked output, the switching flow requirement is at the low end of the range, and for an open output channel it is at the high end of the range. The characteristic of one input of the 'AND' gate A_2 is simply that of orifice flow.

Without a restrictor, Flip-Flop F_4 will switch at well below the minimum operating-pressure of the 'AND' element, so that, following the post-switching control port characteristic upwards, excessive flow is going into F_4 when the required operating pressure level is reached. When the flow requirements of A_2 and F_4 are added together, it may be seen that the operating point is outside the available output from F_3, i.e. the system will not work. Putting a restrictor in the control line to F_4 ensures that the minimum flow is drawn by that element at the correct operating pressure. The system now works so that the 'fan-out' has been improved. It may therefore be concluded that although problems in matching occur if passive devices are combined with active ones, the problems may be overcome with care and a considerable saving in air consumption may be achieved.

In circuits using moving part logic, air consumption is much less likely to cause concern, and although reliability may potentially be better with pure fluid devices, the superiority of either one is not yet proven. It may be, for example, that whilst a moving part device is prone to fatigue failure, a pure fluid device is more susceptible to failure due to contamination of the air. A good example of moving part devices are the miniature high speed devices from East Germany which are attractive on account of the well-developed manifolding arrangements which minimize external piping. Bouteille[8] suggests that piston valves have some advantages because although slower in operation they have the advantage of higher logic power and will retain their state even when the air is switched off, which is not a property of either pure fluidic devices or the miniature diaphragm elements. A great advantage of using either piston devices or certain of the diaphragm valves is that little loss occurs through an element so that one valve may pass the supply flow to one or more succeeding elements. If these succeeding elements are limit switches, then ladder type sequence circuits, as described in Section 9.4, may be easily built up. The circuits described by Bouteille are not ladder circuits, but they do illustrate the logic power of piston valves.

10.6 Point-to-point Position Control

Among the most complex digital circuits either proposed or actually implemented to date are point-to-point positioning systems, primarily aimed at the machine tool market. Continuous path control has been contemplated but the relatively long processing time of the fluid logic results in a poor frequency response of a complete control loop. If any instruction is processed, this delay results in the action being taken some time later, but if the speed of the motion is reduced sufficiently, the error in terms of position will be acceptable. Thus a point-to-point control with a large speed reduction for the fine positioning may have an acceptable accuracy but may yet have a high enough fast speed to be acceptable for machine tool use.

In view of the previous section, it is clear that it is necessary to keep the air consumption of such a circuit down to a minimum, and a paper by Foster, Parker, Jones and Retallick discusses this problem[9]. Unless the system is completely open loop, some feedback is necessary, a summing junction is required and a code must be decided on. These items are discussed more fully in Chapters 13, 14 and 9 respectively, but the main points from Ref. 9 are summarized here.

10.6.1 Types of Feedback

Digital feeback of position may be provided in two basic ways. In one method, an incremental pattern, sensed pneumatically, may be used to give a pulse for each small increment corresponding to the required resolution and the pulses counted in an up-and-down counter. Taft and Wilson[10] have used this method to give a measure of the total distance travelled and also the direction of motion. In the second method, a binary coded pattern, again sensed by air jets, gives an absolute indication of position. In this case, no counter is required, but the number of sensing elements will be a function of the total number of increments required. Systems of the latter type have been reported by Parker[11], by Ramanathan and Bidgood[12], by Stal and Bulk[13], and by Sharp, MacLean and McClintock[14]. The patterns may be either linear for the direct measurement of table position or rotary for the measurement of the rotation of a leadscrew.

If several decades of coded patterns are used to represent position, care must be taken to avoid errors, for example, due to backlash in a gear train, in changing from one decade to another. Either a form of 'U' or 'V' scanning must then be employed, or, alternatively, an overlap of code[13].

10.6.2 The Summing Junction

The binary number produced by the counter or coded pattern must be subtracted from that which represents the desired position, and the result passed through a Digital-to-Analogue converter to provide a signal proportional to the error in position. This signal will then be amplified

[10.6] POINT-TO-POINT POSITION CONTROL 441

and used to position the table. Alternatively, the subtractor may be replaced by a comparator which gives out a signal for the table to stop when the two numbers are equal. The comparator will also generally give output signals indicating the sign of any error, so that the table is automatically caused to move in the correct direction to reduce the error. This will be called an 'inequality' comparator.

If the direction signal is put into the machine from a tape, the comparator need only provide a signal when the desired and actual positions are identical, and a significant simplification in the logic results. Some allowance has, however, to be made for the distance moved by the machine before coming to rest, and the direction of motion has also to be taken into account.

A further simplification may be made by using a counter which is preset with the number required and is then arranged to count down to zero. Thus the counter becomes the summing junction, and the comparator, which need now only recognize the number zero, is very simple. A very useful additional advantage is that signals may be taken at various stages in the count down process, by again using simple comparators, so that D/A conversion presents little problem.

10.6.3 Types of Code

The binary code is most easily used in computation, but it is accepted that the information at all human access points should, for convenience, be in a binary coded decimal form (B.C.D.). For the computational part of the circuit, this may be acceptable, although for high speed of computation in a subtractor circuit there is some point in converting to a straight binary code.

In both the straight binary and B.C.D. codes, several digits may change at any one time and spurious signals may be generated both in the computation (see Parker and Jones[15]) and in the feedback, if a coded pattern is used, and in both cases these must be suppressed or eliminated. In the feedback path the spurious signals may be suppressed by using 'U' or 'V' scanning techniques in conjunction with a binary position encoder. Alternatively, special codes, called Unit Distance codes which have the property of only one digit position change at each increment change, may be used on the encoder. The same technique may also be used for decade codes which are generally called Unit Distance Lippel codes (U.D.L.). Many compromises between decimal display and binary computational requirements are possible, but, in general, any particular code selected must be converted to a weighted code form for use in a subtractor or inequality comparator. The code reported in Ref. 12 is the Petherick code which was converted to a weighted binary code for comparison purposes, and the circuit for the conversion used 12 active OR/NOR elements. Details of the comparator are given in Chapter 14. In order to economize on air consumption and number of elements required, a binary counter was used to gate the signals from the sixteen reading nozzles so that only the four of the appropriate decade are allowed through at any one time. An extra 19 active elements were required for the gating and counting, but this is less than would be required for the processing of all the decades together. In Ref. 13, an unconventional code was used that is progressive within one decade and also is made progressive to the next decade by inverting part of the most significant bit pattern. No details are given of the decoding logic except that the conversion is not a normal weighted code because that would have resulted in too long a time delay in the feedback path. Only the complete decades are weighted, the bits within the decade being pseudo-weighted.

10.6.4 General Comparison

In order to make an objective assessment of the various possible systems, a specification was laid down[9], for the resolution and length of travel of a simple point-to-point control system as follows:
(a) The resolution should be ± 0.001 in., since, for a large proportion of machine shop applications, this would be more than adequate.
(b) A travel of 10 in. was stipulated with the reservation that it may be necessary to increase this figure in practical systems.
(c) A decade code was to be used, thus needing four binary decades with a minimum of four bits each.

The possible combinations were then listed as shown in Fig. 10.15, and the maximum possible positioning speed and the approximate air consumption were calculated for each arrangement. All the circuits were assumed to use wall attachment Flip-Flops or NOR gates in conjunction with passive AND elements and foil diodes. The air consumption of each active element was taken to be 0·2 s.c.f.m. It was assumed that the sensors each used a similar quantity of air and that each was connected to an active monostable element in order to provide the complement of the encoder signal for use in the summing junction. The circuits considered were similar to ones already published by various authors[10, 11, 12], but it must be accepted that the assessment of number of elements and air consumption can only be approximate.

It may be seen that binary pattern encoder systems are faster than those which use increment pattern encoders. This is because the speed of a counter is limited to the rate at which pulses can be accepted by the least significant stage, whereas with the binary pattern encoder the most significant digits may be correct even if the operation is too fast for the less

[10.6] POINT-TO-POINT POSITION CONTROL

SYSTEM	BLOCK DIAGRAM	COMPONENT	NO. OF ELEMENTS		AIR CONSUM. c.f.m		MAX. TABLE SPEED ins./min.	REMARKS
			B.C.D.	U.D.L.	B.C.D.	U.D.L.		
COUNTER & COMPARATOR	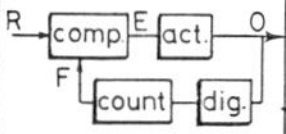 R–[comp]–E–[act]–O / F–[count]–[dig]	simple comparator	20	—	15		60	dig = digitiser counter limits speed
		inequality comparator	24	—				
		up & down counter	52	—				
			72–76	—				
COUNTER & SUBTRACTOR	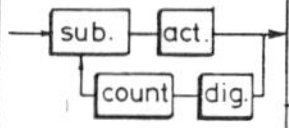 –[sub]–[act] / [count]–[dig]	subtractor	48	—	20		30–60	counter limits speed
		counter	52	—				
			100	—				
ENCODER & COMPARATOR	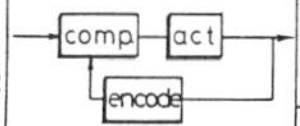 –[comp]–[act] / [encode]	simple comp.	20	20	18	7.2	nom. 60 up to 500 with D/A conversion	B.C.D. encoder figures include U or V-scanning circuitry
		inequ. comp.	24	16				
		encoder	68					
			88–92	36				
ENCODER & SUBTRACTOR	–[sub]–[act] / [conv]–[enc]	subtractor encoder + code converter	48	48	23	16	nom. 30–60 D/A – 500	U.D.L. code must be converted to B.C.D. for subtractor.
			68	—				
			—	32				
			116	80				
PRE-SET COUNTER	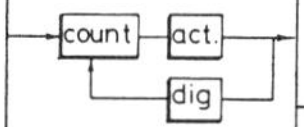 –[count]–[act] / [dig]	pre-set counter	36	—	7.2		60	counter limits speed
			36	—				

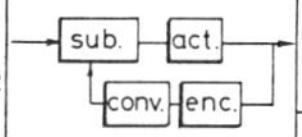

D/A = digital to analogue

Fig. 10.15 Comparison chart for air consumption and speed of response for various types of feedback circuit for point-to-point position control (Foster, Parker, Jones and Retallick, *Ref. 9*, Fig. 1)

significant ones to register. By using a D/A converter or a sequence of steps for a change of speed, the binary pattern system may be allowed to run at the speed appropriate to the resolution of each decade.

The advantage and disadvantages of the arrangements considered are summarized in Table 10.1. The simplest circuit is the encoder and comparator, but the difficulty of changing from one decade to another where there are gear reductions is a disadvantage. The pre-set counter is also relatively simple, but the speed of operation is slow if four decades of counting are being used.

If, however, the counter is used only to count the numbers of revolutions of a leadscrew on a table, then it would need to operate on only the two most significant decades and, therefore, the speed of response could be increased by a factor of 100. The fine positioning over one turn of the lead screw could then be done using two decades of counting, or by a comparator, operating at a low final approach speed. The outputs of these two least significant decades would be inhibited during the fast traverse.

A pre-set counter requires direction information to be put into the machine, so that the comparator need only provide an equality signal. Therefore, a U.D.L. code may be used without code conversion for use in the comparator. Since spurious signals may arise in the counter in the same way as from the encoder, it is convenient to design a counter working on the same U.D.L. code.

TABLE 10.1

System	Disadvantages	Advantages
Counter comparator	1. Slow 2. Needs up and down counter	1. Needs no encoder
Counter subtractor	1. Slow 2. Needs up and down counter 3. May miss counts	1. Needs no encoder 2. Can use D/A converter but speed still limited by counter
Encoder comparator	1. Encoder needs high quality gearing	1. Any code can be used 2. Simplicity
Encoder subtractor	1. Encoder needs high quality gearing 2. Code converter necessary	1. D/A conversion possible resulting in high average speed
Pre-set counter	1. Slow 2. May miss counts	1. Does not need large capacity 2. Simple 3. D/A conversion possible

The final system chosen used a pre-set twisted ring counter in decade form, as described in Chapter 12, to count the number of revolutions of a lead screw, and thus give a coarse position control. At the end of the count the motor speed is reduced through a gearbox operated by a pneumatic clutch and a simple comparator compares the pattern from an encoder with the required signal. Both the counter and the encoder use the quinary code, but the former is a relative measurement, whilst the latter is an absolute one. An exploded view of the mechanical arrangement is given in Fig. 10.16 and a block diagram of the control system in Fig. 10.17. The problem with such a system is that it uses a non-standard code and therefore is not compatible with conventional tape preparation.

Basically the same considerations of counting apply to stepping motor drive systems, except that since the whole of the information is to be counted, the stepping rate cannot be too high. However, the limitations of stepping motor design is such that high stepping rates cannot be achieved[16, 17, 18]. Because of the low stepping rate, it is unlikely that they

[10.6] POINT-TO-POINT POSITION CONTROL

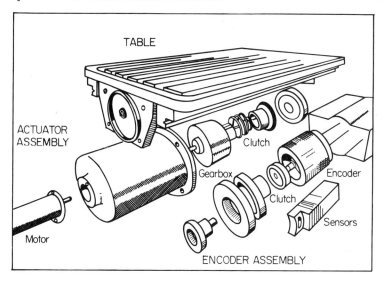

Fig. 10.16 Mechanical components of point-to-point position control (University of Birmingham)

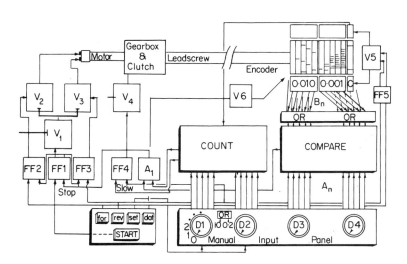

Fig. 10.17 Block diagram of fluidic circuit for point-to-point position control (University of Birmingham)

will find a use in machine tool systems, but it is still possible that there will be some advantage in their use for actuation in a difficult environment.

No mention has been made of completely open loop systems, although a stepping motor drive is semi-open loop. A digital positioning system may be made up from a number of length-weighted cylinders acting in series. The logic circuit is necessary only to obtain the correct signals to the cylinders, either direct from a tape or decoded from a decimal dialed input. A problem lies in the fact that the output may overshoot considerably depending on the order in which signals appear or disappear, but in many cases this is not a serious drawback.

References

1. Foster, K. and Jones, N. S., A simple logic circuit using fluid jet devices, *Int. J. Machine Tool Design Res.*, **5**, 35–42 (1965).
2. Foster, K., Jones, N. S. and Mitchell, D. G., Improvements to a pure fluid sequence control circuit for machine tool application, *Proc. 6th Int. M.T.D.R. Conf.*, 393–403 (1965).
3. Levesque, G. N. and Loup, R. L., Brown and Sharpe applied 'fluidics' to a battery of twist drill flute milling machines, *Seminar Industrial Appl. Fluidics* (Milwaukee School of Eng.) (1966).
4. Monge, M., Handling 'fluidics' for practical applications, *C.F.C.* (1), Paper E5.
5. Foster, K., Mitchell, D. G., and Retallick, D. A., Fluidic circuits used in a drilling sequence control, *C.F.C.* (2), Paper H4.
6. Brewin, G. M., The application of spring controlled NOR units to machine switching operations, *C.F.C.* (2), Paper J2.
7. Parker, G. A., Some aspects of fluidic sequencing and digital control systems for machine tools, *Proc. 7th Int. M.T.D.R. Conf.*, 587–599 (1966).
8. Bouteille, D., Recent development in piston fluid logic and applications in general automation, *C.F.C.* (2), Paper D3.
9. Foster, K., Parker, G. A., Jones, B., and Retallick, D. A., A digital fluidic point-to-point positioning system, *C.F.C.* (3), Paper A1.
10. Taft, C. K., and Wilson, J. N., A fluid encoding system, *H.D.L.*, **4**, 73–100 (1964).
11. Parker, G. A., Digital fluid position encoders, *C.F.C.* (1), Paper E2.
12. Ramanathan, S., Bidgood, R. E., and Aviv, I., The design of a pure fluid shaft encoder, *C.F.C.* (2), Paper E4.
13. Stal, H. P., and Bulk, J., The application of fluidics in numerical control of machine tools, *C.F.C.* (2), Paper J5.
14. Sharp, R., Maclean, R., and McClintock, A., A fluidic absolute measuring system, *C.F.C.* (2), Paper E3.
15. Parker, G. A., and Jones, B., A fluidic subtractor for digital closed loop control systems, *C.F.C.* (2), Paper H1.
16. Griffin, W. S., Development of high speed fluidic logic circuitry for a novel pneumatic stepping motor, *A.F.*, **1967**, 412–414 (1967).
17. Nomoto, A., and Shimada, J., Fluidic step motor, *C.F.C.* (3), Paper E3.
18. Blaiklock, P. M., Development of a pneumatic stepping motor, *IFAC Symp. Fluidics*, London, Nov. 1968.

11
Card Readers, Tape Readers, Decoding and Visual Displays

11.1 Plug Boards and Card Inputs

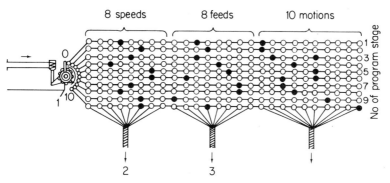

Fig. 11.1 A possible arrangement of a plug board for a hypothetical machine tool

The diagram of Fig. 11.1 shows a hypothetical arrangement of an electrical plug board. The possible connexions to a machine are arranged in 10 rows. The first eight holes of each row select spindle speeds, the second eight select tool feed rates (forward and reverse) and the last ten holes select the particular tool required. The indexing mechanism rotates to energize each row in turn and moves from one row to the next on receipt of a signal to say that the last operation has been completed. Each of the black dots connects the particular speed or feed which is required into the energized line. If the arrangement is a plug board, then the plugs will activate contacts, or if the arrangement is a card reader, sensing devices will decide whether or not there is a hole in the appropriate place in the card and will cause contacts to be made or broken accordingly.

In the diagram, a pawl and ratchet arrangement is shown as a means of indexing the energy source to each line. In a fluidic system this could be a rotary valve which carries an air supply to each of the air lines in turn. The problem in designing a pneumatic plug board is that it is not easy to design a plug that will leave the pneumatic channel sealed after it has been withdrawn; electrically there is no problem. It is much simpler to design a card reader in which a card could be placed over the 10 × 26 matrix of holes and the air would blow through the holes in the appropriate row. Each column of holes would have to be connected to the signal line appropriate to the required operation. Because any one operation may be called for several times during a cycle, it will probably be necessary to make the connexion to the line through diodes, or through a multi-input OR/NOR device, as shown in Fig. 11.2.

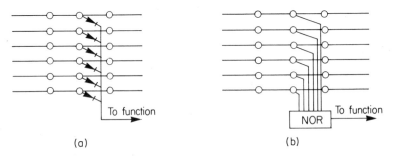

Fig. 11.2 Methods of taking outputs from card reader

Foil diodes are simple and convenient to use for this type of requirement. If, however, the multi-input NOR design is used, turbulence amplifiers are particularly attractive because they can be constructed with a large number of non-interacting inputs. In addition, the control flow required per element is very small so that the total flow requirements from the rotary valve would be small. One difficulty is that any particles blown out of the card would block the control channels of the turbulence amplifiers, but back pressure sensing could be arranged.

Instead of using mechanical indexing, it is possible to energize the rows of the card from the outputs of a pure fluid ring counter. The disadvantage of this arrangement is that a pure fluid device is continually using air, whether required or not, so that although air for only one row is required at any one time, the air consumption is effectively that for all the rows at once. It is possible to minimize the problem by using low flow devices for the ring counter and amplifying the outputs by moving part elements which shut off completely when no output is required. However, this solution reverts back to a moving part system which may be no more

[11.1] PLUG BOARDS AND CARD INPUTS 449

reliable than the simpler mechanical indexing. Of course, the flow requirement to switch to turbulence amplifier is small, so that if these are incorporated in the method of reading of Fig. 11.2(*b*) the use of a pure fluid ring counter for sequencing may be acceptable.

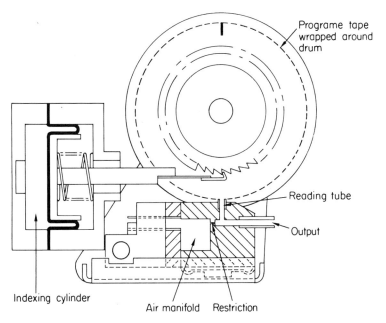

Fig. 11.3 The Techne card reader. The card is wound round the drum which rotates, carrying the information (Sales Literature—Techne Ltd.)

Another practical arrangement is that of the Techne Card Reader, shown in Fig. 11.3. Here, mechanical indexing is used to cause a drum to rotate. A card, wrapped round the drum, has holes punched in it and the presence of holes in a row is sensed when that row is opposite a stationary row of back pressure sensing devices; when a nozzle is covered, a signal appears at the output, otherwise there is no signal. This type of card reader has the advantage that the logic circuitry associated with it is kept to a minimum, since no diodes or multi-input NOR devices are required to process the signals; there is only one signal source per function, and that is connected directly to the valve controlling the function. Air consumption is also reduced to a minimum for the same reason. A point worth noting is that for good discrimination between the ON and OFF signals of a back pressure sensor the hole representing the restriction must be small compared with the hole in the card. Thus the amount of flow available may be considerably smaller than that from an interrupted jet,

11.2 Codes, Tapes and Tape Readers

The use of one hole per instruction is wasteful of space on a card. If combinations of holes may be allowed to have a meaning, then eight holes would have 2^8 ($= 256$) combinations, and therefore, meanings. In a binary-coded decimal, for example, four bits (holes) are used to represent figures up to 10 and another four bits represent 100, and so on. Of course, the logic required to interpret the meaning of a combination of holes is in itself complicated, but the interpreting logic is provided only once per block of information, and the saving in card space is important if a large number of instructions are required. A discussion of the methods of decoding is given in the next section.

If a great deal of information has to be fed into a machine, for example into a machine drilling a large number of holes, it may be more convenient to put this information on to a continuous tape. Standard tapes have either five or eight holes across the width of the tape, and with the large number of combinations possible, numerical information, address instructions and management functions may easily be coded on to the tape. Chapter 9 includes a section on the codes used for machine tools and the standard five and eight hole systems are given.

A pneumatic tape reader made by the Marconi Co. is described by Rosenbaum and Cant[1], and a photograph and diagram of it are reproduced in Fig. 11.4 and 11.5 respectively. The mechanism consists of a diaphragm actuator which operates a lever connected to a clutch which causes the tape sprocket to rotate a fixed distance on each input. A spring loaded roller and toothed wheel locates the tape accurately after each movement. In essence, the mechanism is similar to that of the Techne card reader, and like the latter it reads only one line at a time.

If the tape is read serially (i.e. line by line), a shift register must be used to collect all the information for one operation before that operation starts. A drilling machine, for example, requires the X and Y coordinates and spindle speed to be stipulated before the hole can be drilled. If the same (or nearly the same) amount of information is required for each operation, a 'block' tape reader is preferred. This reads several rows of the tape at each go, and the tape is moved on by that number of rows after each operation. In such a case, the tape acts as its own binary store, and a shift register is not required. A mechanism similar to that of Fig. 11.5 could be adapted for block reading simply by extending the stroke of the actuator and by changing the pitch of the toothed wheel. The speed of

[11.2] CODES, TAPES AND TAPE READERS

Fig. 11.4 Photograph of Marconi tape reader (Rosenbaum and Cant, *Ref. 1*, Fig. 1)

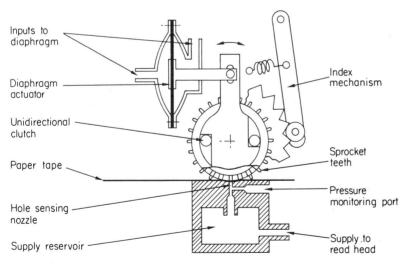

Fig. 11.5 A diagrammatic view of the Marconi tape reader (Rosenbaum and Cant, *Ref. 1*, Fig. 2)

indexing will be reduced but will probably still be satisfactory for most machine applications. If, however, speed is important, an alternative method of actuation is to have a motor continually driving a clutch which is air operated. A signal to index operates the clutch to connect the tape drive and this is automatically disengaged after the movement appropriate to the correct indexing of the block. In a tape reader supplied by the Plessey Co., an electric motor provides the drive, but this is easily replaced by an air motor to give completely pneumatic operation.

In both the Marconi and the Plessey tape readers, back pressure sensing is again used on the grounds that any particles of paper, or any dust, will be blown away from the inputs to any of the logic elements in the circuit. Information is also given[1] about the static and dynamic response of back pressure sensing nozzles of varying configurations, and it is shown that these sensors are good up to reading speeds of better than 300 bits/second, although the tape reader feeding arrangement works only at up to 25 bits second. Some general conclusions of interest were that rectangular nozzles had superior cut-off characteristics to circular ones and that the size of the reservoir behind the nozzles was important. Too small a reservoir created interaction between the signal lines; by increasing the size and also by putting baffles inside the reservoir between the outlets to the nozzles, this problem was reduced to negligible proportions. The effect of a transmission line from the sensing head on the signal waveform is also mentioned, and it is shown that the shape of a square wave becomes more distorted as the line length is increased.

Charnley, Bidgood and Ramanathan[2] report on a tape reader using

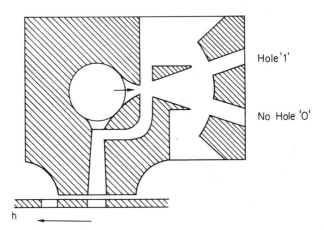

Fig. 11.6 Diagram of a back pressure sensing element used by I.B.M. [*Ref. 4*, Fig. 3 (I.B.M. Co.)]

[11.2] CODES, TAPES AND TAPE READERS 453

interrupted jet sensing and discuss the minimizing of 'cross-talk' between the bits. It is pointed out that the distance between the holes on a standard tape of 0·28 in. fixes the size of the receiver and that the size of the jet nozzle should be as large as possible to minimize pressure drop in the nozzle assembly; therefore, a diameter equal to that of the holes in the tape was chosen, i.e. 0·072 in. It was also pointed out that as the gap between the two faces of the sensing head is reduced, the interference between the holes increases; on the other hand, if the gap between the two faces is increased too much, the pressure recovered in the receivers falls to too low a level. It is recommended that the nozzles should be raised above the tape by having a cavity in that particular face the depth of which is between 0·063 in. and 0·093 in. Eisenburg[3] also uses interrupted jet sensing.

An interesting back pressure sensor which is combined with a wall attachment amplifier and used for tape reading is described in a patent filed by the I.B.M. Company[4], and a diagram of it is given in Fig. 11.6. The whole device can be made in a planar sheet and the close proximity of the nozzle and the tape means a minimum delay in the signal transmission.

11.3 Decoding

11.3.1 Matrix Decoding

To describe the process of decoding, the common example of a conversion from binary code to a decimal output is given. A straight binary code for numbers up to nine is given in Fig. 11.7, together with a Karnaugh map of the numbers.

The Karnaugh map is merely a convenient way of displaying the combinations of four variables which are used for the code and has the advantage that any unused combinations can be quickly picked out. In the figure, the unused combinations W, X, Y and Z are listed below the map in Boolean notation and the unwanted squares on the map have the decimal figures bracketed.

The total number of combinations of four binary bits is 16. One way of decoding them is to finish with a four × four matrix of two-input AND elements, where each horizontal line supplies four elements and each vertical line also feeds four. To do this, the binary numbers must be split into groups of two, and the four combinations of each of the groups provided by simpler logic. Such an arrangement is shown diagrammatically in Fig. 11.8. The internal AND gates of the matrix correspond with the squares of the Karnaugh map Fig. 11.8 and the unwanted combinations may be omitted by either omitting the appropriate AND gates or simply

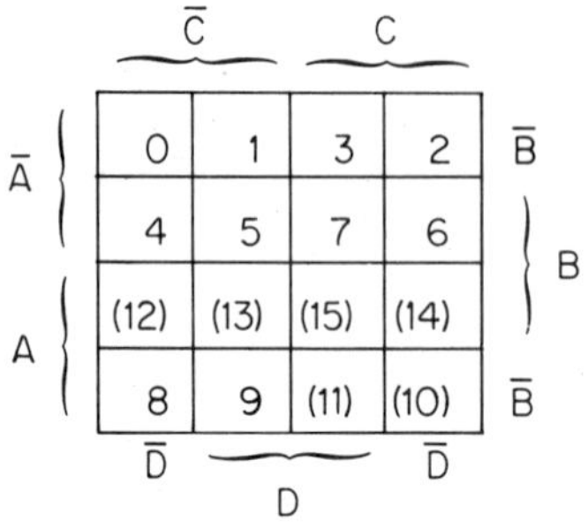

	$\bar{C}$		C	
$\bar{A}$ {	0	1	3	2	$\bar{B}$
	4	5	7	6
A {	(12)	(13)	(15)	(14)		B
	8	9	(11)	(10)	$\bar{B}$
	$\bar{D}$	D		$\bar{D}$

	A	B	C	D
0	0	0	0	0
1	0	0	0	1
2	0	0	1	0
3	0	0	1	1
4	0	1	0	0
5	0	1	0	1
6	0	1	1	0
7	0	1	1	1
8	1	0	0	0
9	1	0	0	1

U = $\bar{A}\bar{B}C\bar{D}$

V = $\bar{A}\bar{B}CD$

W = $\bar{A}BC\bar{D}$

X = $\bar{A}B\bar{C}D$

Y = $\bar{A}BC\bar{D}$

Z = $\bar{A}BCD$

Fig. 11.7 Binary-coded-decimal truth table and Karnaugh map

by not connecting up to them; the AND gates that may be omitted are indicated by the shaded area.

Alternatively, NOR logic may be used, since A.B is equivalent to $\overline{\bar{A} + \bar{B}}$. An example of this type of circuit using pure fluid wall attachment NOR elements but with the spare combinations omitted is given in Ref. 5 and reproduced in Fig. 11.9. It is not quite so easy to follow the layout because it is not arranged in a matrix form, but the method is the same. The outputs from the original four elements are arranged in two groups as indicated, one group consisting of combinations of C, $\bar{C}$, D and $\bar{D}$ and the other of A, $\bar{A}$, B and $\bar{B}$. The signals from the elements providing the combinations of these pairs are then grouped as though a matrix existed so that the resulting outputs are homogeneous groups of all the four literals. The right hand output from the bottom row of elements is given as an example, i.e. is $\overline{A + B + C + D} = \bar{A}.\bar{B}.\bar{C}.\bar{D}$. A disadvantage of this circuit is that only one output of each of the final stage Flip-Flops is used, so that some logic power is wasted and rather more elements than necessary are used.

[11.3] DECODING 455

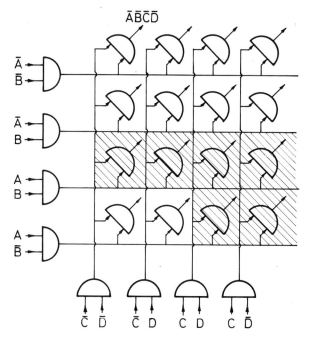

Fig. 11.8 Decoding a four-bit binary number using a matrix of two-input gates

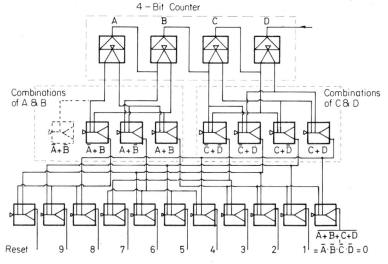

Fig. 11.9 The schematic of a binary to decimal decoder using wall re-attachment NOR gate logic (Bemal and Brown, *Ref. 5*, Fig. 15)

11.3.2 Decoding Trees

An alternative arrangement which would utilize the two outputs of each of the final stage elements is shown diagrammatically in the 'tree' configuration of Fig. 11.10.

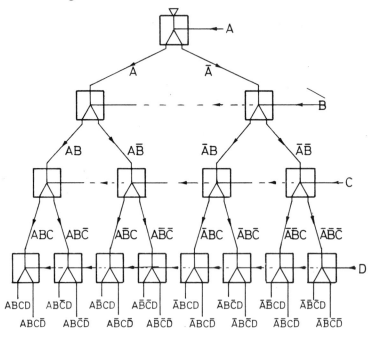

Fig. 11.10 A binary to decimal decoder using a tree of passive wall attachment elements

This uses both outputs of the elements in one stage as the possible supplies to two elements of the next stage, and thus follows the pattern of electrical relay decoding circuits. The outputs ABCD from a four bit binary number, i.e. a counter, are applied as control signals to the four stages of the tree. Only 18 elements are required for full decoding from four bits compared with 28 in the previous method. There are two disadvantages, however, in this case: one is that there is a pressure drop in each stage, which means that some intermediate amplification must be provided, so that the number of elements increases again; the second is that the fan-out from the element supplying the signal to a stage must increase the further down the 'tree' it is (e.g. the signal from D must feed eight elements). A 'folded tree' is preferable on this score, and an example of such an arrangement could be obtained simply by dividing the circuit of Fig. 11.10 into two halves; one half would be concerned with signals

[11.3] DECODING

containing A and the other with signals containing $\bar{A}$. The $\bar{A}$ group would remain as before, but the A signal would first be acted on by D rather than B; the outputs AD and $A\bar{D}$ resulting from the action would now be switched by the C signals. The final group of four elements in the A tree would be switched by the B signals. The outputs from the circuit would be the same, but B would now have to fan-out to five elements instead of one, whereas D would only have to fan out to five elements instead of to eight. A reduction in the fan-out requirement for any particular element is a useful advantage in a fluidic circuit. The circuit is called a 'folded' tree because the left-hand side of the figure could be turned upside down in order to keep the elements opposite the correct switching signal. An example of an arrangement similar to that just described is given by Bryan[6] and is reproduced in Fig. 11.11. Because of the

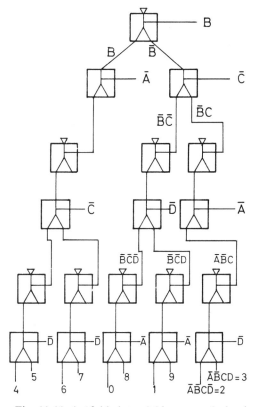

Fig. 11.11 A 'folded tree' binary to decimal decoder using active and passive wall attachment devices (Bryan, *Ref. 6*, Fig. 3)

signal loss in passive elements, active elements are used between rows to amplify the signals. In addition, unwanted combinations are omitted so that the circuit is not too easy to follow. However, one or two of the signal paths are indicated, and although the tree is not strictly 'folded', the inputs from the binary counter are distributed in such a way that the 'fan-out' for both $\bar{A}$ and $\bar{D}$ is only four.

11.3.3 Decoding with Moving Part Devices

A direct decoding arrangement which does not suffer from fan-out problems in the driver stages is one[7] built up from the Techne spring NOR element described in Chapter 8. The fact that when the spring of the element is bent it allows air to escape is used in the arrangement which is shown in Fig. 11.12.

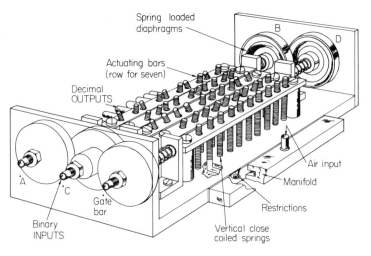

Fig. 11.12 A binary to decimal decoder using spring NOR elements (De Bruyne, *Ref. 7*, Fig. 12)

The operation is best understood by considering, for example, the row connected to the decimal number 'seven' output (binary 0111). This particular arrangement appears to use the complements of the binary digits, so that a signal must come from the spring element connected to the A line when the actuating bar for A is operated; thus this spring is initially bent. The springs connected to the B, C and D bars must be initially straight so that they are bent when the bars move on receipt of the binary signals. An additional bar appears to be used as a 'gate' for the operation, so that each row of springs acts as a five-input NOR gate.

This method of decoding is wasteful of elements, as discussed earlier in

[11.3] DECODING 459

the treatment of pure fluid decoders, and for combinations of more binary bits some other method is desirable. A compact arrangement of spool valves is described in the patent from the I.B.M. Company[4]. A six-bit number is split up into two blocks of three bits. Each of the three bits is decoded by use of a 'folded tree' of spool valves. The decoding for the group A, B, C is shown in Fig. 11.13.

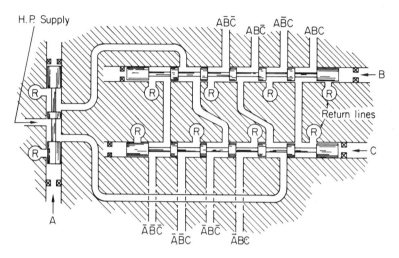

Fig. 11.13 A 'folded tree' decoder for three bits of a binary number using spool valves [*Ref. 4*, Fig. 8 (I.B.M. Co.)]

If A is present, the single output goes to the lower of the two long valves and gives one of two outputs from this, depending on whether or not C is present. The two lines go to the top long valve and become four possible outputs, depending on whether or not B is present. If A is not on (i.e. $\bar{A}$ is present), the output from A goes to the top of the long valves and finally appears as one output from the four possible ones from the lower long valve.

There are now eight outputs from the folded valve tree for A, B, C and eight from the one for D, E, F. These are passed to a matrix of AND gates (the use of piston valves as AND gates is described in Chapter 8) formed by more long spool valves connected to the A, B, C outputs and by the common supply channels which are connected to the D, E, F outputs.

The matrix is shown in Fig. 11.14 and was placed over the keyboard of an I.B.M. typewriter so that on decoding a six-bit binary input, the appropriate key of the typewriter would be depressed.

Another method of decoding from a six-bit number using foil two-

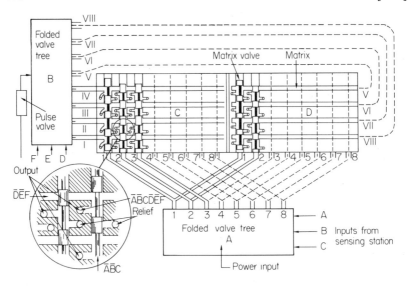

Fig. 11.14 An arrangement of decoding for a six-bit binary number using 'folded trees' and an 8 × 8 AND gate matrix of spool valves [*Ref. 4*, Fig. 9 (I.B.M. Co.)]

input OR gates and foil inverters[8] is shown in Fig. 11.15 (the individual devices are again described in Chapter 8). Two short trees are formed from small matrixes of the OR gates to give the combinations of the literals A, B, C and their complements plus those of D, E, F and their complements. A final 8 × 8 matrix of two-input OR gates gives all the combinations, and the inverters are only used for the amplification of the signal flow. Although this seems a complex arrangement, foil devices are compact, and the whole circuit may be integrated into a small unit.

It is also possible to make diode matrix decoders using foil diodes, and in schematic form these are identical to electronic diode decoders. A diagram of this is given in Fig. 11.16, where the first few rows of the decoding from a four-bit binary number is shown. Further details of the foil diode are given by Foster and Retallick[9]. To decode from a six-bit binary number, 12 columns and 64 rows would be required, which is a considerably more complicated arrangement than that just described. It still may be attractive because the basic devices are simple and not every junction requires a diode. The success of such a scheme would depend upon a very carefully designed mechanical arrangement of the interconnecting manifolds.

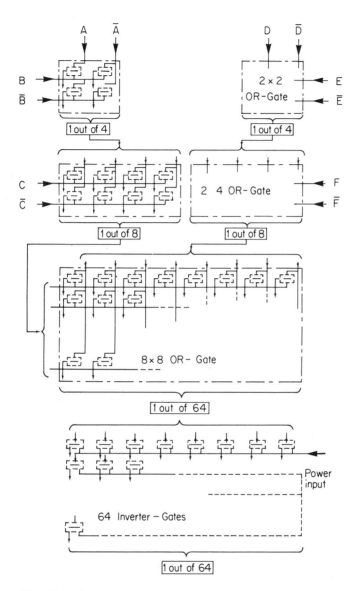

Fig. 11.15 An arrangement of active and passive elements for decoding a six-bit binary number (Bahr, *Ref. 8*, Fig. 23)

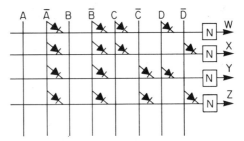

Fig. 11.16 Matrix decoder using foil diodes would have the same circuit as above with the diodes indicated by the electronic notation

11.4 Visual Displays

Once a number has been decoded, for example from the output of a binary counter, the read-out can be fairly simple. For example, numbers could be attached to ten miniature cylinders and the appropriate cylinder would be caused to push the number into view at the correct time. Alternatively, ten miniature cylinders could be arranged in a circle as in the older electronic counters, thus displaying a coloured dot in the correct space on receipt of the relevant signal. A more original indicator is that produced by Pitney Bowes Fluidics, in which the signal operates a selected number of miniature pistons from a matrix of 5 × 7 of them. These pistons form a pattern which represents the number. Fig. 11.17, from Ref. 10, shows the read-out and one of the methods of grouping the cylinders.

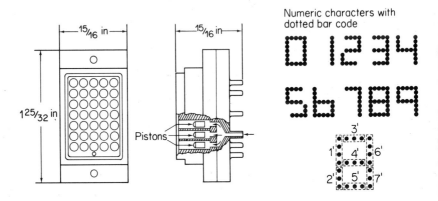

Fig. 11.17 Pitney-Bowes fluidic read-out using a 5 × 7 matrix of miniature cylinders with coloured ends (O'Keefe, *Ref. 10*, Figs. 5 and 6)

[11.4] VISUAL DISPLAYS 463

Because of the necessity to decode before displaying Decimal information, considerable interest has been shown in read-out arrangements in which the information is also decoded. Deason[11] describes a unit using

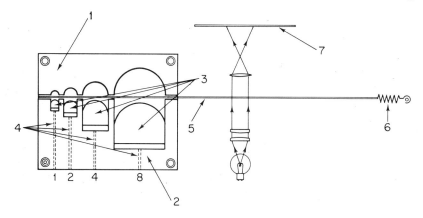

Fig. 11.18 Principle of read-out unit for a 4-bit natural binary number. (1) Block into which tape is deflected. (2) Block housing pistons. (3) Pistons. (4) Air passages to pistons. (5) Tape with numbers 0 to 9 cut in it. (6) Spring return. (7) Viewing screen (Deason, *Ref. 11*, Fig. 5)

weighted cylinders which move a spring-loaded tape into which the shapes of the numbers are cut. The arrangement is shown diagrammatically in Fig. 11.18, and the principle is fairly clear. The cylinders are arranged to move the tape 1, 2, 4 or 8 units of distance respectively, and combinations of the cylinders operating in a natural binary code give all numbers up to 9, with some unused combinations. In order to keep the size of the unit down, an optical arrangement is used to magnify the small letters on the tape. The precise arrangement of cylinders used in the actual device is not quite as shown in the figure, since a slight simplification is obtained from the fact that the binary digits 4 and 8 never occur together. The '8' cylinder is made the same size as the '4' cylinder and both are allowed to move together when the '8' bit gives out a signal.

Some work has been carried out at the University of Birmingham[12] on read-outs incorporating decoding. One possible arrangement is shown in Fig. 11.19, where, for simplicity, decoding is shown from a 3-bit number only, giving output numbers from 0 to 7. The viewing window has eight holes corresponding to these output numbers. Under the window are three masks containing eight holes, but these may be arranged to be either underneath the appropriate hole of the viewing window when the signal to the mask is ON or under the hole of the window only when the signal is

OFF, i.e. they will allow light through to the hole either when A is present or alternatively when A is not present, and so on. In the situation shown, the masks are set to allow light through to the hole representing 4, but light cannot pass to any of the other holes of the viewing window.

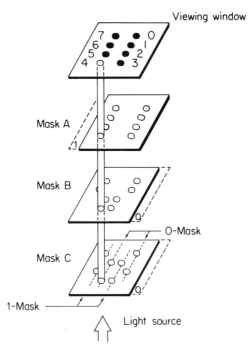

Fig. 11.19 Design of optical masks for a 3-bit binary counter. The figure shows how 100 (binary) is decoded to read 4 (decimal) (Cheng and Foster, *Ref. 12*, Fig. 4)

Because of the interest in twisted ring counters, work has also been put into designing visual decoders for the quinary code. Chapters 9 and 12 give information on the code and on the counters, but, to recap, the code requires five bits which may be designated, say, A_1, A_2, A_3, A_4, A_5. Table 11.1 shows the combination of signals of the five stages which represent the ten decimal numbers. The code is a progressive one and the list of ten codes may be split into two parts, namely, one containing $A_1 = 1$ and the other $A_1 = 0$. To obtain an economical read-out, cards are made as shown in Fig. 11.20, with some actuation means to move them to the left or to the right according to the signals received. Numbers are placed in two quadrants of a square or rectangle, as shown, and the other two

[11.4] VISUAL DISPLAYS

TABLE 11.1

Binary Decimal	A_1	A_2	A_3	A_4	A_5	
1	1	0	0	0	0	
2	1	1	0	0	0	
3	1	1	1	0	0	$A_1 = 1$
4	1	1	1	1	0	
5	1	1	1	1	1	
6	0	1	1	1	1	
7	0	0	1	1	1	
8	0	0	0	1	1	$A_1 = 0$
9	0	0	0	0	1	
10	0	0	0	0	0	

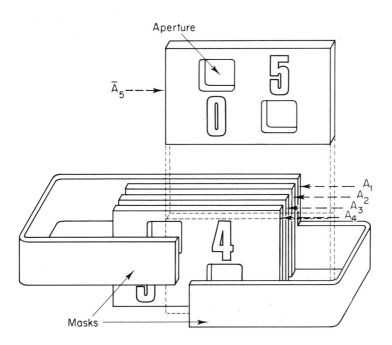

Fig. 11.20 Read-out for a quinary-code fluidic counter. Figure shows how 11110 (Quinary) is decoded to read 4 (decimal) (Cheng and Foster, *Ref. 12*, Fig. 12)

quadrants are apertures. The numbers for the A_1 card are 1 and 6; for the A_2 card 2 and 7, and so on until the A_5 card is 5 and 0.

The cards are stacked with the A_1 card at the bottom and the A_5 card at the top. When A_1 is ON and A_2–A_5 is OFF, the figure 1 will be seen through the top aperture; as A_2 goes on, a figure 2 covers the 1, and so on until 5 appears; then $\overline{A}_1$ goes ON, and the figure 6 appears behind the lower window and again is covered as $\overline{A}_2$ goes on and 7 appears, and so on until 0 comes into view.

The disadvantage is that two figures appear at one time. This can be avoided by making the card for A_1 rather more complicated so that whilst A_1 is ON, a mask covers the lower figures, and whilst $\overline{A}_1$ is ON, a mask covers the upper figures (as shown in the figure). Alternatively, a separate mask may be used to switch from the top window to the bottom one.

Another possible method is that which uses one cylinder, A_1, with a movement weight of 1 unit and the other four cylinders with a movement weight of 2 units. The way in which this works may be seen from Table 11.2. The sum of the effects of the movement of the cylinders is shown for

Ternary Decimal	A_1	A_2	A_3	A_4	A_5	Sum
1	1	0	0	0	0	1
2	1	2	0	0	0	3
3	1	2	2	0	0	5
4	1	2	2	2	0	7
5	1	2	2	2	2	9
6	0	2	2	2	2	8
7	0	0	2	2	2	6
8	0	0	0	2	2	4
9	0	0	0	0	2	2
10	0	0	0	0	0	0

each row in the column on the right. Thus if a tape is marked with the appropriate decimal numbers at distances along it according to the right-hand column, the correct number will automatically appear in the viewing window as the quinary pattern is followed by the inputs. A possible mechanical arrangement is shown in Fig. 11.21. Other weighing possibilities exist, and five three-position cylinders may also be used to recognize the 0, 1 or 2 signals and then to operate a read-out.

[11.4] VISUAL DISPLAYS

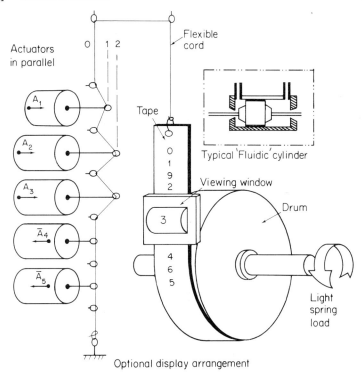

Fig. 11.21 Decoding-display unit for a quinary-code fluidic counter using linear 'fluidic' cylinders in parallel. Figure shows how 11100 (quinary) is decoded to read 3 (decimal)

References

1. Rosenbaum, H. M., and Cant, J. S., A pneumatic tape reader, *C.F.C.* (2), Paper E2.
2. Charnley, C. J., Bidgood, R. E. and Ramanathan, S., The design of a pneumatic tape reader, *C.F.F.* (1), Paper E4.
3. Eisenberg, N. A., A pneumatic tape reader, *H.D.L.*, **III**, 63–92 (1965).
4. British Patent Specification 999528. I.B.M. Company.
5. Bernal, T. W., and Brown, W. R., Development of a pure fluid norgate and a norlogic binary to decimal converter, *H.D.L.*, **III**, 37–61 (1965).
6. Bryan, B. G., A pure fluid binary to decimal converter, *C.F.C.* (2), Paper H2.
7. De Bruyne, N.A., Pneumatic NOR Blocks, *C.F.C.* (1), Paper C3.
8. Bahr, J., The foil element, a new fluid logic element, *Elektronische Rechenanlagen 7*, **No. 2**, 69–78 (1965).
9. Foster, K. and Retallick, D. A., Some experiments on a free-foil switching device, *C.F.C.* (2), Paper D1.

10. O'Keefe, R. F., Fluidic decimal counter for digital control applications, *Advances in Instrumentation*, Vol. 23, pt. 11, Proc. 23rd I.S.A. Conf. 1969, Paper 937.
11. Deason, W. S., A Fluidic Batching Counter and a Fluidic Counter with Visual Decimal Read-Out Unit, *C.F.C.* (3), Paper B1.
12. Cheng, R. M. H., and Foster, K., On some Decoding and Display (D-N-D) Techniques Applicable to Fluidic Counters, *C.F.C.* (4), Paper C3.

12
Shift Registers, Ring Counters and Binary Counters

12.1 Introduction

In Chapter 9, counter design was considered as an extension of the general design method applied to sequence circuits. Following the logic of this approach it was convenient to discuss the T Flip-Flop as a particular case of a sequence circuit, and from this point it is logical to move on to the design of serially operated binary counters. It is only because of the switching delays and consequent spurious signal hazards occurring in serial counters that parallel counters are considered, and from this step it is an easy transition to pass from the four-bit binary-coded decimal arrangement to the configurations of the ring or twisted ring counters that require more bits. The twisted ring counter may also be regarded as a unidirectional shift register, and as a means of understanding the operation of this general class of circuit it is convenient to examine the shift register in some detail first, because once this is understood it is relatively easy to follow the design of a T Flip-Flop and its serial counter derivatives.

Shift registers and counters are basic 'building blocks' in digital computers, but in the less sophisticated world of the machine shop the terminology is strange, and this fact alone prevents a mechanical engineer becoming familiar with their uses. A shift register is used to put 'information' (a number) into a store 'serially' so that it can be read in 'parallel'. For example if a punched tape is being read one row at a time, only one digit of a complete number is being read at that time. However, for use in a machine or computer, the number must be available in its entirety, each digit must be stored in a row in the order in which it arrives. This is achieved by feeding the digits into a shift register which moves all the previous digits along one space as each new one arrives, until the whole number is assembled. (Fig. 12.1.)

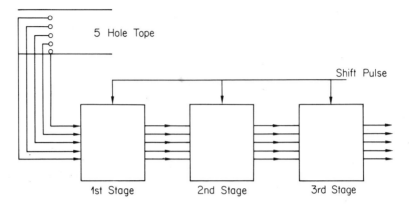

Fig. 12.1 A 5-bit shift register

A similar situation occurs in a domestic telephone; when a number is dialled, the user carries out only one operation at a time, so that the information goes to the exchange serially. The exchange needs the information in blocks before it can act and so requires it in parallel form. The counter may be regarded as a special case of a shift register.

12.2 The Shift Register

12.2.1 Shift Register using Auxiliary Storage

A shift register may be constructed from a number of Flip-Flops, arranged to accept information from, or deliver information to, the Flip-Flops on either side of it. This would constitute a bi-directional shift register. The design of such a device is complicated by the fact that most types of Flip-Flops cannot simultaneously accept and deliver information. Thus the shift registers must operate either by (*a*) starting from the end towards which information is to be shifted, clearing each stage in turn, and transferring its contents to the adjacent previously cleared state; or (*b*) simultaneously clearing all stages and storing their contents temporarily in elements which then discharge the information to the adjacent stage to right or left as desired; or (*c*) by simultaneously clearing all stages by means of a pulsed gating signal and allowing the outputs to go to the next stages on either one side or the other through delay lines so that the gating pulse has disappeared before the signals have arrived at their destination.

The first type would generally be too slow in operation, and the third type will be discussed later in this chapter in the section on twisted ring counters. A typical one of the second type is shown diagrammatically in Fig. 12.2. A four-step process shifts information in the lower register one

[12.2] THE SHIFT REGISTER

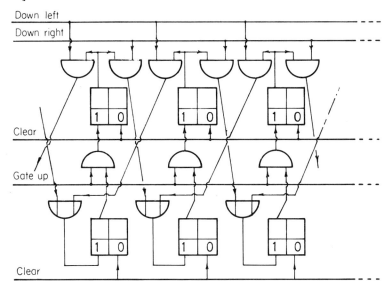

Fig. 12.2 A bi-directional shift register using AND gates and flip-flops

position right or left, the upper register being used as a temporary store. These steps are as follows:
(a) The contents of the lower register are gated directly to the upper register.
(b) All lower-register stages are reset to zero.
(c) The contents of the upper register are gated down to the lower register either one position to the right or one position to the left as desired.
(d) The upper registers are reset to zero.

An alternative way of operating that avoids clearing each row between movement of the information is one in which both outputs of the main Flip-Flop are gated to the intermediate storage Flip-Flop inputs and both outputs of the latter elements are gated either to the left or right, as required. Such an arrangement requires double the number of AND gates that were required by the previous circuit and is therefore cumbersome, but if a single direction of shift is required, the circuit becomes more attractive, particularly since the gating signals to the main Flip-Flop are complementary with the gating signals to the intermediate storage Flip-Flop and therefore both may be obtained from the complementary outputs of a single pulse shaper. (Alternatively, for symmetrical operation, the two signals may be obtained from the outputs of a binary divider which is itself operated by a pulse train; such an arrangement is used in the ring counter described in 12.3.1.)

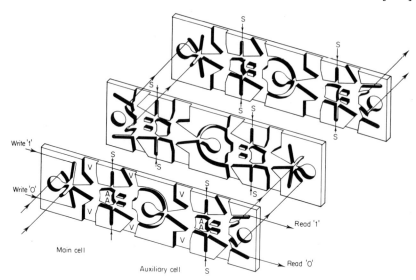

Fig. 12.3 Three sandwich layers of a shift register using wall attachment flip-flops and wall-attachment AND gates (Mitchell, Glaettli and Mueller, *Ref. 1*, Fig. 10)

This latter type of circuit is easily achieved in fluid logic terms by using wall attachment type Flip-Flops and passive AND gates, and arrangement of a uni-directional shift register from the I.B.M. Company[1] is shown in Fig. 12.3. This design is particularly interesting for several reasons. Firstly, the stages, each containing one main Flip-Flop, one intermediate storage Flip-Flop and four AND gates, are stackable so that the connexions from one storage or auxiliary Flip-Flop pass directly into the input channels of the next main Flip-Flop. Secondly the liberal use of vents, marked V in Fig. 12.3, suggests that the large wall-angle Flip-Flops follow the original I.B.M. practice of designing for a high flow gain rather than high pressure recovery. Lastly, the AND gates, labelled A in the figure, are essentially wall attachment monostable devices with one signal passing through what would be the supply channel of a conventional design. This method of providing an AND function has the advantage of high sensitivity, so that little flow is required from the shift pulse, but has the disadvantage of large pressure loss; nevertheless, the design is consistent with the philosophy for a high flow gain throughout the circuit. The operation is as follows: if a write '1' is in the first stage main cell, the jet passes down the lower wall into the lower of the first two AND gates; if no shift signal is present, the output from the main cell switches the jet of the auxiliary cell on to the upper wall and this passes to the read '1' output channel. A shift signal then switches the jet of the auxiliary cell into

[12.2] THE SHIFT REGISTER

the channel leading to the input of the main cell of the next stage; at the same time this signal directs all the main stage outputs via the associated AND gates out through the central vents, thus preventing further passage of the information. The complementary operation of the AND gates associated with the two stages is achieved by using their outputs in opposite ways, rather than by taking complementary outputs from the pulse shaper, but the effect is the same.

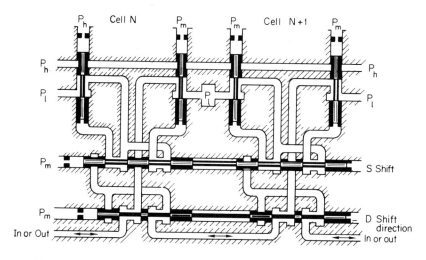

Fig. 12.4 A bi-directional shift register using a spool valve (Mitchell, Glaettli and Mueller, *Ref. 1*, Fig. 4)

Another interesting design from the I.B.M. Company shown in Fig. 12.4 is one using spool valves of the type described in Chapter 8. In the position shown, the information is to be moved from cell $N + 1$ to cell N. The signal from the right-hand piston of cell $N + 1$ passes to the left-hand piston of cell N via the extreme left-hand end of the shift valve and is allowed through to the base of the piston when a shift signal is applied. If the right-hand piston of cell $N + 1$ is up, the high pressure is directed to the left-hand one of cell N, pushing it up; otherwise it stays down. At the same time, the signal is also applied to the base of the right-hand piston of cell $N + 1$ in order to maintain its position—this is an example of memory produced by combinational logic as described in Chapter 9. When the shift signal goes off, the shift piston allows the output signals from every left-hand piston of a cell to be passed on to each right-hand one, whilst at the same time a feedback loop to provide memory is connected around each of the left-hand pistons. The whole assembly is built into an integrated cast epoxy resin block, with the complex interconnecting passages made by the lost wax process.

12.2.2 Shift Register using Pulsed Information

Yet another design of shift-register from I.B.M., reported by Mitchell, Mueller and Zingg[2] uses an AND gate which is a variant on the one used in the earlier pure fluid shift register, and which relies upon the pulsed power jet principle first suggested by Warren[3]. In a symmetrical bistable device, the wall to which a pulse flow will attach can be influenced by a flow asymmetry previously set up in the region downstream of the nozzle. This asymmetry can be provided by a normal control flow, or by an arrangement as shown in Fig. 12.5.

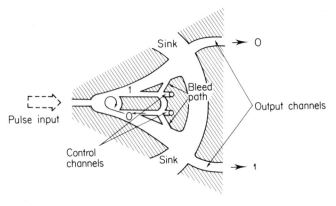

Fig. 12.5 A flip-flop operating with a pulsed supply (Mitchell, Mueller and Zingg, *Ref.* 2, Fig. 4)

Here, the control channels are moved from the position of maximum pressure gradient near the nozzle to a position downstream, which reduces the changes in the static pressure level in the control lines. A flow asymmetry is set up, for example, in the clockwise direction as shown, by a flow from the 'O' channel. A pulse emerging from the nozzle senses the direction of the vortex-like flow, attaches to the upper wall and then exhausts through the 'O' channel. A characteristic of this method of operation is that the control flow required to divert the jet whilst it is being pulsed is very small, but once the jet attachment is established a very much larger control flow is necessary to cause it to switch. If used with a small control flow, the device provides the 'latching AND' function which gives the following facility:

(*a*) It provides the correct output only when the two inputs arrive in the correct sequence.

[12.2] THE SHIFT REGISTER

(b) It retains an output when a particular one of the two inputs is removed.
(c) It ceases to give an output when the remaining input is removed.

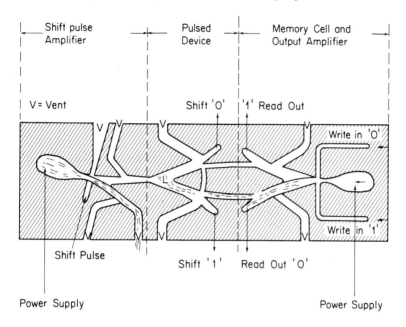

Fig. 12.6 A shift register stage using a pulsed supply element (Mitchell, Mueller and Zingg, *Ref.* 2, Fig. 5(*a*))

The shift register using this principle is shown in Fig. 12.6 and consists of a pulse amplifier, a pulsed-supply device and a memory cell and output amplifier. A shift pulse causes a pulse flow to enter L. If a zero is stored in the memory cell, the pulse is made to attach to the top line of element L and it then passes through the shift 'O' output to the next stage. The write-in signals, and therefore the memory cell state, can be changed whilst the shift pulse is still on without changing the pulse direction in element L, so that the system will accept either pulsed or level inputs. A similar design of shift register is reported by Chadwick[4]. In this case, the pulsed element is more like a conventional monostable device and is shown in Fig. 12.7. Because of the geometrical bias and vent on one side, in the absence of a control signal the pulse will pass down the side close to the control channel, but in the presence of a control signal it will pass down the other output. To be pedantic, this element more strictly provides the latching AND function than does the I.B.M. design. A shift register circuit using this element is shown in Fig. 12.8.

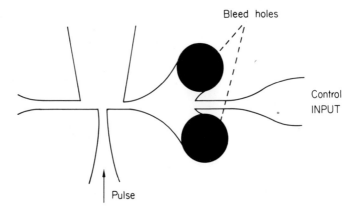

Fig. 12.7 The latching AND gate (Chadwick, *Ref. 4*, Fig. 6)

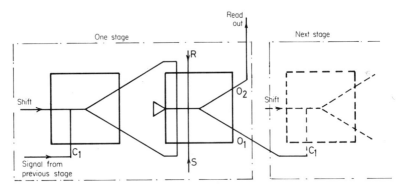

Fig. 12.8 Schematic of a shift register with a pulsed supply element (Chadwick, *Ref. 4*, Fig. 8)

Output 0_1 of one stage is connected to input C_1 of the next stage, whilst output 0_2 is used for the read-out. Channels S and R are used for the set or reset signals. The input pulse or level signal can be provided by a monostable device which may be used to pulse several stages together.

A circuit, designed at the University of Birmingham is based on the conventional AND gate and Flip-Flop[5,6]. The two basic layers of the arrangement, in which stages can be stacked one on top of the other, are shown in Fig. 12.9. The outputs 0 and 1 of the Flip-Flop pass through points a_0 and a_1 to the layer containing the delay lines and thence after a delay to points b_1 and b_0. These two points connect to the next stage in the succeeding layer and are allowed to pass to the inputs of the next Flip-Flop by pulses, the duration of which must be shorter than the delay in

[12.2] THE SHIFT REGISTER

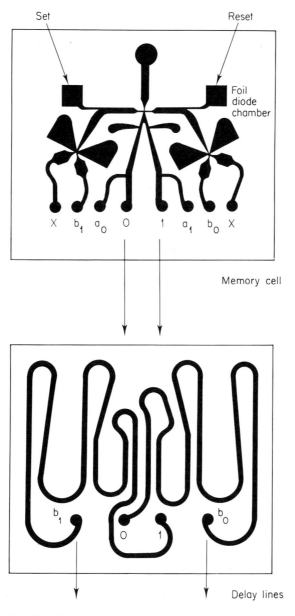

Fig. 12.9 University of Birmingham Shift Register showing inter-stage connexions (Foster and Retallick, *Ref. 6*, Figs. 47 and 48)

the lines, applied at X to the AND gates. Only one pulse shaping circuit is necessary for the gating of up to five stages, providing that the elements of that circuit are large enough to pass sufficient flow to the AND gates. The Set and Reset signals pass to the Flip-Flop through foil diodes as indicated in the figure.

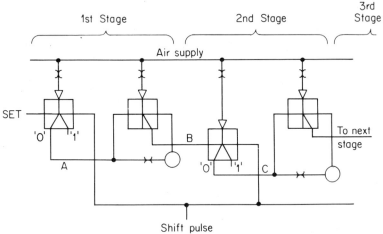

Fig. 12.10 A shift register using wall attachment devices only (Componetrol, Inc. (a subsidiary of I-T-E Imperial Corporation), *Ref. 7*, Fig. 7.16)

Componetrol, Inc., in their book *Fluid Systems Design Guide*[7], suggest a different circuit using only wall attachment elements, and this is shown in Fig. 12.10. Suppose A is initially at '1' and all the other stages at '0'. A shift pulse switches A to '0', which triggers the 'one shot' circuit so that a pulse passes along line B to switch the following bistable to 1. The shift pulse is by now off so that line C does not receive any signal. On receipt of the next shift pulse, the information moves along another stage. Note that this circuit and the previous one shift on receipt of pulses which must be of a limited time duration, whereas the circuits described earlier would accept a rectangular wave input of any duration.

The design of any circuit of this complexity of a shift register requires a compromise between the requirements of several aspects of the circuit. Air consumption, including that of the pulse shapers and the output circuits, must be considered; speed of operation should be kept high and the circuit must be able to provide the outputs necessary for a given application. In the majority of such circuits, the speed of operation is dictated by the length of pulse necessary to give good switching action, and this in turn depends on the size of the element used for pulse shaping—the smaller the element, the shorter can be a well shaped pulse. Of course, several pulse

[12.2] THE SHIFT REGISTER

shapers may be run in parallel, but the problem is simplified by reducing the flow required for switching to as low a level as possible, remembering that a static switching characteristic is not necessarily a good guide to a dynamic switching flow, since at the higher operating speeds the total volume displaced into the attachment region of the element is as important as the rate of flow into the control channel. Thus, the gating circuit of each stage may be seen to be important, and therefore, should be examined carefully; for example, the signal loss in a Greenwood type AND gate is less than the loss in a normal wall attachment element used in a passive mode.

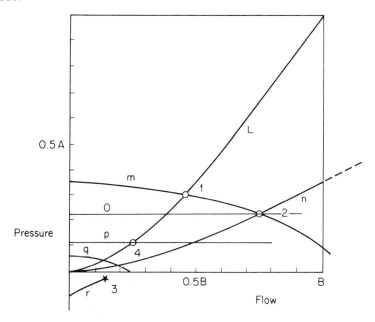

Fig. 12.11 Approximate matching of AND gate, flip-flop and pulse shaper characteristics for University of Birmingham Shift Register (Foster and Retallick, *Ref. 6*, Fig. 42)

As an example of the matching problem, consider a very approximate analysis of the circuit of Fig. 12.9 by referring to Fig. 12.11. This shows the output characteristic of the memory cell as line m, with the switching characteristic as line r. Points A and B represent the pressure and flow to the memory Flip-Flop respectively. Line L represents the characteristic of an orifice, such as one input of the AND gate whilst n shows the characteristic due to an additional orifice in parallel, on the assumption that each output of the memory cell will supply both an AND gate and another element. Therefore, point 2 indicates the delivering pressure to the delay lines. Assume that in the worst case, 50 per cent of this pressure will be lost in the delay lines and line p gives the pressure available at the AND, intersecting the AND gate characteristic at point 4. If the AND gate has two level inputs of pressure magnitude p, then the output characteristic is approximately as shown by line q. This is more than enough to switch the memory Flip-Flop, the switching point of which is given by 3. The flow to each AND gate, given by point 4, is thus 0·25 B. This analysis is very approximate, since it does not allow for the fact that the instantaneous flow into the control channel of the memory cell necessary for fast switching is likely to be much greater than that indicated by point 3. Also, the balance of flow depends upon the output load on the memory cell and the loss down the transmission lines. Nevertheless, in this design the switching flow is generous in order to allow for variation in the circuit.

Table 12.1 at the end of this chapter gives a rough comparison of the comparative characteristics of the various pure fluid shift registers. The estimates of the flow requirements from the pulse shaper are little more than guesses, because the switching characteristics of the elements of the circuits are only known accurately for the circuit of Fig. 12.9.

An attempt has been made to estimate not only the number of elements actually in a given shift register stage, but also the number of elements required for pulse shaping in addition to those in the circuit, together with elements necessary to give output signals suitable for decoding. The extra elements are divided by the number of stages and added to the basic number of active elements per stage in order to arrive at the estimated total number of active elements per stage given in column 6 of Table 12.1. The Shift Registers of Figs. 12.8 and 12.9 require the fewest active elements and there is little to choose between them. The latter has a more symmetrical output availability that might be useful on occasions.

It is not claimed that this type of analysis is necessarily definitive, but a circuit should be examined as completely as possible before any decisions are made on a final configuration and the method adopted may serve as a good guide. To conclude this section, it is worthwhile pointing out a design that may be used for economy of elements in a uni-directional shift register. A device similar to a conventional AND gate may be employed in which not only the A.B output is used but also the A.$\bar{B}$, which is normally a vent. The signal A is the shift pulse input, whilst the signal B is the connexion, through a delay line from the output of the previous stage. If no output comes from the previous stage, i.e. it is an '0', the shift pulse switches the present stage to '0' via channel A.$\bar{B}$. If on the other hand, the previous stage is at '1', the shift pulse guided down the output A.B of the 'AND' gate causes the current memory element to go to state 1. Fig.

12.12 shows the layout. Again, the disadvantage is that a full signal is available from only one of the outputs from each memory element, but with careful design of AND gate, the carry signal between stages may be kept to a minimum value.

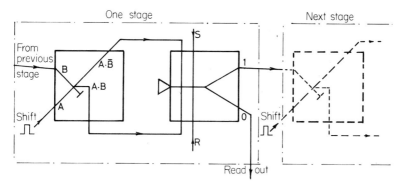

Fig. 12.12 Pulse operated uni-directional shift register using A.B signal and A.B̄ signal of a device similar to passive AND gate

12.3 Ring Counters

12.3.1 Ordinary Ring Counters

A ring counter is a device where a single pulse is moved progressively round a circuit from stage to stage, and at the end of the circuit it is reintroduced into the start again, so that there is in reality no beginning or end, but just a 'ring' of elements. If the output of a unidirectional version of the shift register of Fig. 12.2 is connected back to the input, a ring counter is formed, but a simplification of this circuit is possible because the chain of elements may be regarded as being bottom left, top left, bottom second from left, top second from left and so on. In other words, the intermediate storage elements are used as output elements. The bottom ones are gated by one pulse from a binary divider and the top ones from the second pulse of the binary divider. A further simplification of the circuit may be made because once the '1' signal has passed to a stage, the previous one must be switched off; instead of using an AND gate for the '0' signals, the AND gate that transmits the '1' may be used to also switch off the previous stage so that one AND gate per stage is saved.

A block diagram of such a ring counter circuit is shown in Fig. 12.13 and a silhouette of one stage in Fig. 12.14. Considering Fig. 12.2, an output '1' at the first stage (with all other stages at '0') moves to the second stage, switching the first stage to '0' when the line b of the binary counter gates the first AND gate. The third stage cannot be switched because the second AND gate is not gated. When the binary counter switches to a, the signal is allowed to pass to the third stage, switching the previous one off again, and so on. An alternative arrangement using NOR logic is shown in Fig. 12.15.

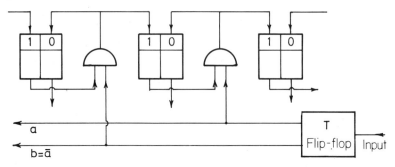

Fig. 12.13 Schematic of a ring counter using AND gates and flip-flops (Foster, Mitchell and Retallick, *Ref. 8*, Fig. 3.1(*a*))

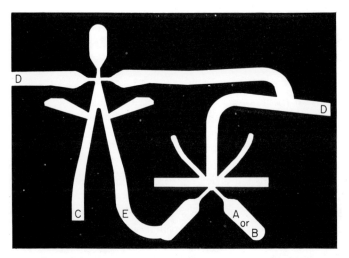

Fig. 12.14 Silhouette of one stage of ring counter with AND gate logic (Foster, Mitchell and Retallick, *Ref. 8*, Fig. 3.2)

A disadvantage of this arrangement of a ring counter is that a spurious pulse can cause a stage further along the ring counter to switch to '1' so that two signals circulate. Once the arrangement has broken down, it is not self-correcting, and may become worse on the next cycle, when a further one or more elements may be switched jncorrectly. Such a breakdown is reported in Ref. 8, where the original design of AND gate allowed

[12.3] RING COUNTERS

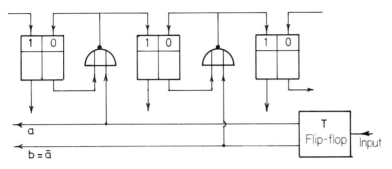

Fig. 12.15 Schematic of ring counter using NOR elements and flip-flops (Foster, Mitchell and Retallick, *Ref. 8*, Fig. 3.1(b))

a spurious pulse to pass through when a new signal arrived from the memory element, even though no gating signal was present. Careful redesign of the vents of the AND gate was found to be necessary in order to prevent this occurrence, but after correction, the output of one stage of the ring counter at 250 Hz was as shown in Fig. 12.16.

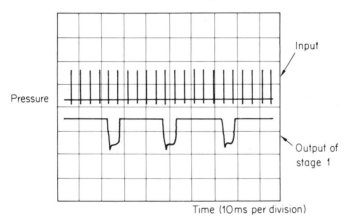

Fig. 12.16 Output of ring counter operating at input pulse rate of 250/sec (Foster, Mitchell and Retallick, *Ref. 8*, Fig. 3.6)

12.3.2 Twisted Ring Counters

The twisted ring counter in decade form consists of a five stage shift register which is connected head to tail, but with a twist in the connecting leads as illustrated in Fig. 12.17.

Also shown is the quinary code in tabular form, since this is the code adopted in the twisted ring arrangement. The mode of operation is explained in the table, the shift register stages successively filling up with

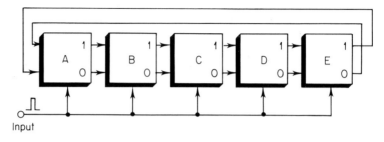

	A	B	C	D	E	Coding
0	0	0	0	0	0	$\bar{A}.\bar{E}$
1	1	0	0	0	0	$A.\bar{B}$
2	1	1	0	0	0	$B.\bar{C}$
3	1	1	1	0	0	$C.\bar{D}$
4	1	1	1	1	0	$D.\bar{E}$
5	1	1	1	1	1	$A.E$
6	0	1	1	1	1	$\bar{A}.B$
7	0	0	1	1	1	$\bar{B}.C$
8	0	0	0	1	1	$\bar{C}.D$
9	0	0	0	0	1	$\bar{D}.E$

Fig. 12.17 The twisted ring counter (Foster and Retallick, *Ref.* 6, Fig. 39)

1's and emptying again cyclically. The decoding logic is also shown in the table and is particularly simple, requiring only five AND gates, or alternatively five OR gates, but it is clear that both the signal and its complement is required from each memory cell.

As described in the previous section, the stage to stage shifting is achieved by feeding shift signals to AND gates on the SET and RESET inputs of each stage and the basic design depends upon the nature of the shift signals. If the shift signals are producible in pulse form, each stage need only consist of one memory cell, two AND gates and two time delays that are longer than the pulse width, as in the design of Fig. 12.9. Alternatively, each stage can be made of two memory cells and four AND gates; one cell and two gates forming a main cell and the other memory and two AND gates forming an auxiliary cell, as in Fig. 12.3. Asymmetric designs, such as that of Fig. 12.8, are less attractive because complementary output signals are required for decoding, although with care taken in matching the output signals to the decoding logic, successful operation is possible. The comments of Table 12.1 still apply and it is clear that circuits using pulsed inputs have the lowest power consumption.

[12.3] RING COUNTERS 485

Therefore, the symmetrical, pulsed-input circuit of Fig. 12.9 was originally chosen for use at Birmingham as a twisted ring counter. Fig. 12.18 shows a decimal output from one decade of counting running at 160 Hz. Trace 1 is the oscillator input, trace 2 the pulse shaper output and trace 3 the output from one decoded decimal bit.

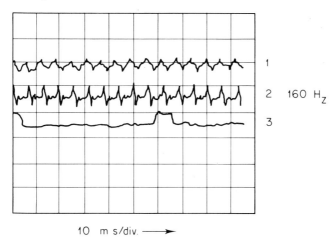

Fig. 12.18 The output from a twisted ring decade counter at an input pulse rate of 160 Hz (Foster and Retallick, *Ref. 6*, Fig. 45)

12.4 Binary Counters. The 'T' Flip-Flop

The work on binary counters in fluidic circuits has been mainly concerned with the design of the T Flip-Flop for use in serial arrangements. The T Flip-Flop is a device in which successive input pulses switch the output alternately from one state to the other; thus it provides a 'divide by two' action, so that by cascading a number of T Flip-Flops a serial binary counter is produced. The basic equations for this device were derived in Chapter 9, and these will now be translated into fluidic designs under the headings I, II and IV used for the discussion of Section 9.7. We shall refer to the circuit diagrams given in Chapter 9 for the following notes.

12.4.1 Case I, Fig. 9.36

This is the simplest design to achieve fluidically, requiring two Flip-Flops and four AND gates, and is a special case of a shift register or single stage twisted-ring counter with level inputs. Feedback from Y_1 steers the input pulse to set the auxiliary memory Y_2. When the input pulse goes off the output from Y_2 is gated through to Y_1 to set it to the complement of its previous state.

12.4.2 Case II, Fig. 9.37

This case has received most attention in the literature and two designs are of particular note, the design due to Warren[9] and that due to Bantle[10]. The latter reference gives a good tabulation of feedback counter circuits. The equations for the operation, i.e.

$$S_{Y_1} = x\bar{y}_2, \quad R_{Y_1} = xy_2, \quad y_2 = y_1\overline{(x\bar{y}_2)} + xy_2, \quad \bar{y}_2 = \bar{y}_1\overline{(xy_2)} + x\bar{y}_2,$$

indicate that for both the S_{Y_1} and R_{Y_1} expressions, an AND function is required and that x, the input, is common to both expressions. This represents a 'steering' function with the input being directed along the S_{Y_1} channel or the R_{Y_1} channel and is achieved fluidically in nearly all the cases by means of a passive bistable element, as illustrated in Fig. 12.19. The signals y_2 and $\bar{y}_2$ now become the steering signals for the input x and may be obtained in several ways.

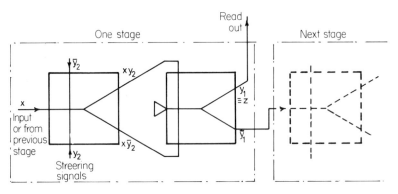

Fig. 12.19 Binary counters case II (Foster and Retallick, *Ref.* 6, Fig. 18)

(*a*) *Case II according to Warren.* This counter is illustrated in Fig. 12.20 and is notable because of the elegant way in which the expressions for $y_1\overline{(x\bar{y}_2)}$ and $\bar{y}_1\overline{(xy_2)}$ are produced by internal feedback.

Assume that the output is in limb z. The low pressure in the separation bubble of Y_1 causes a circulation around a, b in the clockwise sense; thus the steering stage knows that y_1 is present. Even if there is no input signal, the circulation means that y_2 exists and satisfies the switching function $y_2 = y_1\overline{(x\bar{y}_2)}$. The xy_2 part of the original expression for y_2 is not necessary because the passive bistable possesses memory when activated. When the input x arrives, the circulation is sufficient to steer the incoming signal into

[12.4] BINARY COUNTERS. THE 'T' FLIP-FLOP

channel a and give the $R_{Y_1} = xy_2$ signal, thus resetting the Flip-Flop Y_1. Due to the presence of xy_2 in channel a, the $\bar{y}_2$ expression cannot be realized, that is to say that the suction in channel b when the memory element has switched is insufficient to pull the input pulse into channel b.

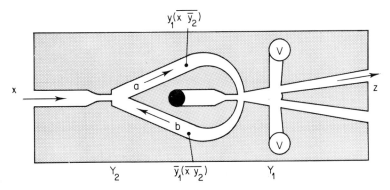

Fig. 12.20 Binary counters case II (Warren, *Ref. 9*, Fig. 2)

However, when x ceases, the circulation in a and b does become anti-clockwise and the $\bar{y}_2 = \bar{y}_1\overline{(xy_2)}$ function can now be realized ready for the next input pulse, which will pass down channel b and set the Flip-Flop Y_1.

Warren counters have been made to operate successfully, but suffer from two conflicting design requirements:
(a) For stability, a bistable element needs high control port impedance or a modified geometry.
(b) For establishment of good circulation flow, low control port impedances are needed. A geometry modification to regain stability usually results in a loss of gain, and the alternative is to closely control the load impedances which then limits the fan out capability.

The solution to the problem must be a fine compromise between these factors and for this reason, various designs have been produced which use a more positive and reliable switching and steering action. Westerman, Richards and Depperman[11] suggested using a Warren type counter with supplementary feedback from the bistable bleeds to improve steering, and Deason[12] has used this approach successfully. Bantle however, devised a fundamentally different approach using external positive feedback and subsidiary feedback in the steering unit.

(b) *Case II according to Bantle*. This design is illustrated in Fig. 12.21 and it may be seen that the y_1 and $\bar{y}_1$ signals are given by the external feedback lines, whilst the passive stage is stabilized by the xy_2 and $x\bar{y}_2$ functions which are fed back to the passive control inputs. These functions override the y_1 and $\bar{y}_1$ signals whilst the input x is still present so that:

$$y_2 = y_1\overline{(x\bar{y}_2)} + xy_2$$

$$\bar{y}_2 = \bar{y}_1\overline{(xy_2)} + x\bar{y}_2$$

as required. When x disappears, the feedback signals are allowed to reset the steering stage ready for the next input.

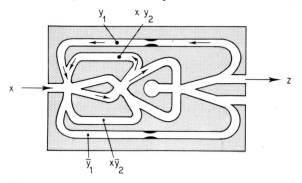

Fig. 12.21 Binary counters case II (Bantle, *Ref. 10*, Fig. 7)

(*c*) *Case II according to I.B.M.* Another design from I.B.M. uses a pulsed input steering stage and this is shown in Fig. 12.22.

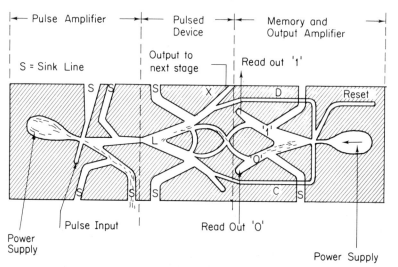

Fig. 12.22 One stage of a binary counter using a pulsed supply device (Mitchell, Mueller and Zingg, *Ref. 2*, Fig. 5(b))

[12.4] BINARY COUNTERS. THE 'T' FLIP-FLOP

It may be seen that it is very similar to the shift register, but with feedback between the steering stage and the memory output. In this design, a pulse input is indicated, but with correct design of pulsed-supply device, the length of the input pulse will not be important, because once the input signal is established in region L, a change in output of the memory cell will not change the direction of the input.

(d) Case II according to Retallick. To supplement these existing designs, a further external feedback circuit was conceived using external feedback but with inhibit gates to realize the $y_1\overline{(x\bar{y}_2)}$ and $\bar{y}_1\overline{(xy_2)}$ functions.

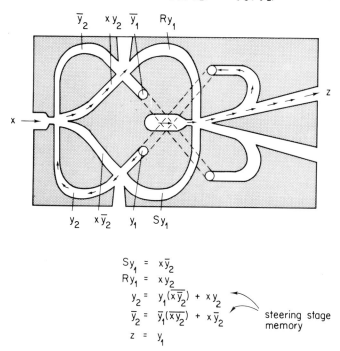

$$Sy_1 = x\bar{y}_2$$
$$Ry_1 = xy_2$$
$$y_2 = y_1\overline{(x\bar{y}_2)} + xy_2$$
$$\bar{y}_2 = \bar{y}_1\overline{(xy_2)} + x\bar{y}_2$$
$$z = y_1$$

steering stage memory

Fig. 12.23 Inhibition counter (Foster and Retallick, *Ref. 6*, Fig. 26)

Consider the counter of Fig. 12.23 when the input x has been steered down channel xy_2 and has switched the output to channel $\bar{y}_1$ so that feedback appears in channel $\bar{y}_2$. The only constraint necessary for correct operation is that the feedback signal $\bar{y}_2$ should not be allowed to steer the input to channel $x\bar{y}_2$. One way is to stop the input signal before the feedback has arrived by using a pulse shaper. Another way is to prevent the feedback from reaching the steering unit until the input signal has ceased;

in other words to inhibit the feedback, a function that can be performed by means of an EXCLUSIVE–OR gate. The mode of operation is as follows:

(a) x appears at the steering unit and is steered down xy_2.
(b) Main jet switches to $\bar{y}_1$.
(c) Feedback $\bar{y}_1$ reaches EX–OR gate and whilst $\bar{y}_1 = xy_2 = 1$ the feedback is inhibited.
(d) Signal x ceases, $\bar{y}_1$ is allowed through to the steering stage and the counter is ready to accept the next input.

A disadvantage with the design is that the feedback signal to EXCLUSIVE–OR gate has to be connected externally.

12.4.3 Case III, Fig. 9.38, Counters using only Combinational Logic

This case is of particular note because no bistable memories are used, either for the steering stage or for the output stage. Consequently, the counter must be designed using NAND or NOR logic and represents counters constructed of turbulence amplifiers, providing the AND gates are replaced by NOR logic.

J. Bulk[13] discusses counters using turbulence amplifiers in some depth, and whilst he considers circuits similar to that of Fig. 9.41, he finally uses a delay line in order to eliminate some of the logic and adopts the circuit shown in Fig. 12.24.

Thus the preferred circuit falls between classes III and IV. Bulk reports that the output from the T.A. contains a certain amount of 'jitter' which results in a variation of the pulse time of a periodic signal. A 12 bit binary

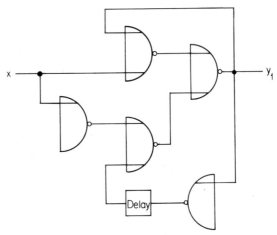

Fig. 12.24 Case III, binary counter stage (Bulk, Ref. 13, Fig. 12)

[12.4] BINARY COUNTERS. THE 'T' FLIP-FLOP

counter and a 3 decade counter were tested up to 100 pulses per second with several different lengths of input pulse, and the figures quoted for the average number of dropped pulses per 10^6 counts are quite small. Not much indication of the reasons for missed counts is given.

12.4.4 Case IV, Fig. 9.39, Counters with Time Delay in the Feedback Path

This counter shown in Fig. 12.25 simplifies the equations for y_2 and $\bar{y}_2$ because the delays in the feedback lines ensure that the input pulse has gone off before the feedback signals arrive at the steering stage. When the feedback signal is established, the circuit is ready to receive the next pulse. The design has been reported by Shooh, Chen and Reader[14]. The design may be made asymmetric by using a monostable element for the steering stage with only one feedback line in a manner similar to that suggested for a shift register in Fig. 12.12. An alternative design described by Parker and Carley[15] uses a G.E. half adder as the steering stage.

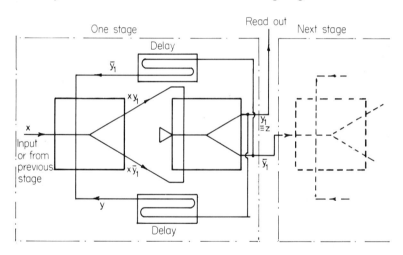

Fig. 12.25 Case IV, binary counters (Foster and Retallick, *Ref. 6*, Fig. 21)

12.5 Up-and-Down Counters

An up-and-down binary counter may be made by using the fact that if there are two binary numbers A and B, then A − B can be obtained by adding the complement of B to A and complementing the result. For down counting, in order to add one to the complement of the number, the complements of the output of each state of the counter must be connected to the input of the next. This necessitates AND gates on each of the two outputs of each stage, so that a gating signal may be applied to one set of AND gates to gate the signal 'up' or 'down'.

A difficulty is that confusion may arise when the gating is changed over, because a series of spurious pulses may pass through the counter. This may be avoided by using only pulsed outputs on each stage (Fig. 12.26), and

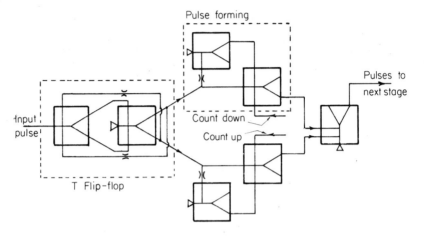

Fig. 12.26 Schematic of an up-and-down binary counter stage

the change over is made at the end of an output pulse. The use of 'one shot' Flip-Flops in each output leg, however, means a total of three active elements per stage with treble the air consumption of the basic Warren type counter.

An up-and-down ring counter is simply a bi-directional shift register and is therefore simple to make but requires twice as many AND gates as the equivalent uni-directional one. Section 12.2 gives enough information on this point.

12.6 Pulse Shapers

For most of the counter designs pulse shaping is necessary and this can be achieved in various ways, the simplest being as shown in Fig. 12.27.

In this method the input pulse is divided and directed to the two control nozzles of a bistable wall attachment element as shown. The restriction reduces the magnitude of the first signal arriving at the Flip-Flop to switch it ON, so that the second signal arriving via the delay line, being unrestricted, switches the Flip-Flop back OFF. This method tends to be unreliable due to the quasi-stable condition of the Flip-Flop which is subject to two control signals simultaneously, one being slightly greater

[12.6] PULSE SHAPERS

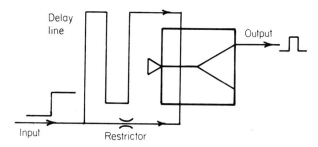

Fig. 12.27 A pulse shaper (one-shot flip-flop)

than the other; changes of operating pressure level will change the difference between the two and an adjustment of the restrictor may be necessary. Also by having flow into both control ports, the basic wall attachment phenomenon is less effective, and the element may be operating more as a beam-deflexion amplifier.

Most other methods of shaping involves at least two elements and a delay line as indicated in Section 9. If two active elements are combined with a delay of 0.3 m s^{-1}, which is about the minimum possible due to the rise and decay time that occurs in the delay line, the minimum pulse width obtainable is 0.6 m s^{-1}. Therefore, assuming a Mark-to-Space ratio of one, the maximum frequency of operation is 830 Hz. This method of using an active element as the output of the shaper is, therefore, not suitable for frequencies of 1000 Hz. The same remarks apply to a pulse shaper designed according to Fig. 12.27 and a trace of a typical minimum

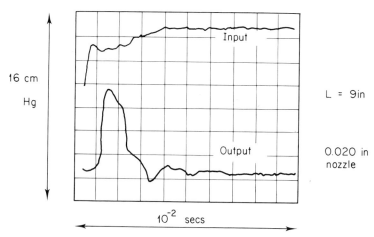

Fig. 12.28 Output pulse from a typical 'one-shot' pulse shaper (Foster, Mitchell and Retallick, *Ref. 8*, Fig. 3.13)

length pulse is given in Fig. 12.28, showing a pulse length of a little over 1 ms for a delay line length L of 9 in.

To obtain a shorter pulse, use may be made of the fact that the operating time of a beam-deflexion device is short compared to that of a wall attachment element. Such a design is shown diagrammatically in Fig. 12.29.

As both the OR and NOR outputs are used, the AND gate output, as shown by Fig. 12.30 is purely a function of the delay, the AND gate itself

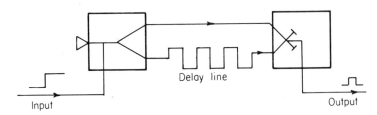

Fig. 12.29 Pulse shaping circuit

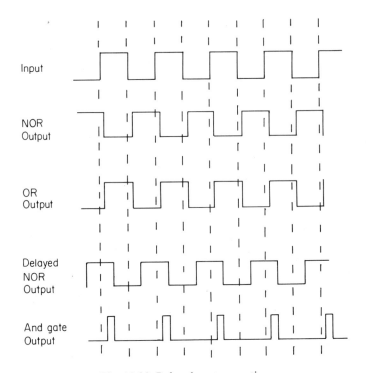

Fig. 12.30 Pulse shaper operation

[12.6] PULSE SHAPERS 495

contributing very little to the total delay. The shaped pulse is triggered by the leading edge of the input pulse as it is the NOR output of the Monostable that is delayed as shown in Fig. 12.30. The rise and decay time of the delay lines may produce an overlap on the trailing edge of the input pulse thereby producing a spurious signal, and it is this phenomenon which limits the minimum pulse width. Pulse lengths as short as 0·34 ms may be achieved and a typical result is given in Fig. 12.31.

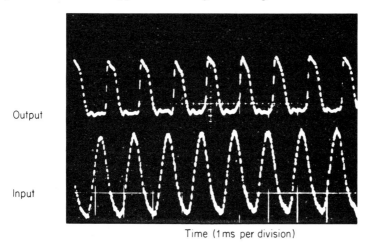

Fig. 12.31 Pulse shaper of Fig. 12.29 operating at an input frequency of 920 Hz (Parker and Carley, *Ref. 15*, Fig. 15)

For larger flow outputs, it may be necessary to use an active element on the output, but these remarks give an indication of the relative merits of the designs.

12.7 Binary-Coded Decimal Counters

Binary-coded decimal means that the number is split up into groups representing numbers up to 10, up to 100, up to 1000 and so on. Each group representing a number up to ten requires a four bit binary number. Since four bits will give numbers up to fifteen, some of the information is superfluous. Because there is some superfluous information, there is a choice of information which may be used or left out, and a number of possible coding arrangements exist. Whatever the arrangement, a logic circuit must be used to recognize when the number 10 appears and the counter must then be reset to its initial state.

In a straight bindary-coded decimal number represented by digits

$a_4a_3a_2a_1$, the number 9 is 1001, so that if an AND gate is connected to a_4 and a_1 a signal will be produced when nine is reached and this may be used to give a reset pulse to all counters when the next input pulse to the counter appears. Other coding arrangements may be used and similar resetting logic provided. Chapter 9 gives more basic information on designs using various weights for the four bits of the binary-coded-decimal.

12.8 Unit-Distance Binary-Coded Decimal Counters

One difficulty of using counters in fast operating systems is that a false signal may be read from the counter because of the delay in any 'carry' signals passing down the counter. Similar difficulties occur in encoding discs, subtractors and other digital blocks. Because of this, a large number of codes have been evolved in which only one of the binary digits representing the number changes its state on the receipt of the pulse. These are called Unit-Distance codes (or Gray Codes) and circuits may be arranged to count in these codes. See Chapter 9.

K	=	A	B	C	D
0	=	0	0	0	0
1	=	0	0	0	1
2	=	0	0	1	1
3	=	0	0	1	0
4	=	0	1	1	0
5	=	1	1	1	0
6	=	1	1	1	1
7	=	1	1	0	1
8	=	1	1	0	0
9	=	1	0	0	0

K_1	=	A_1	B_1	C_1	D_1
0	=	0	0	0	0
1	=	0	0	0	1
2	=	0	0	1	0
3	=	0	0	1	1
4	=	0	1	1	0
5	=	1	0	0	0
6	=	1	0	0	1
7	=	1	0	1	0
8	=	1	0	1	1
9	=	1	1	1	0

K_2	=	A_2	B_2	C_2	D_2
0	=	0	0	0	0
1	=	0	0	0	1
2	=	0	0	1	0
3	=	0	0	1	1
4	=	0	1	0	0
5	=	0	1	0	1
6	=	0	1	1	0
7	=	0	1	1	1
8	=	1	0	0	0
9	=	1	0	0	1

Fig. 12.32(a) A specific unit-distance Lippel code, K; the standard 5–2–2–1 code, K_1; and standard 8–4–2–1 code, K_2.

[12.8] UNIT-DISTANCE BINARY-CODED DECIMAL COUNTERS

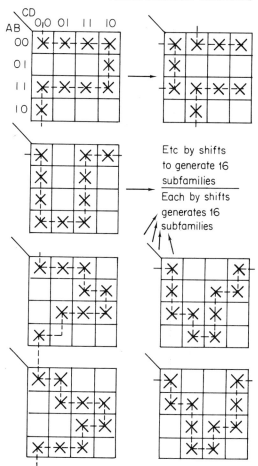

Fig. 12.32(b) Representation of unit-distance codes on Karnaugh maps

An example of a unit-distance code is shown in Fig. 12.32(a) and compared with a straight binary-coded decimal. Below this figure, several Karnaugh maps are given in Fig. 12.32(b), showing how one may arrive at a unit-distance code. Ten out of sixteen spaces are to be used up successively so that only one digit changes at any time with the result that the entries make a continuous path on the map. It may be seen that a large number of possibilities are simply mirror images or rotations of the few basically different patterns. The final choice of code will probably depend on simplicity of decoding, since this varies from code to code. Ref. 16 gives useful information on this topic.

TABLE 12.1

Figure in which shift register is shown	Number of Active Elements per basic shift register	Pulse Shaper requirements	Pulse Shaper requirements expressed as number of active elements per stage	Outputs available	Total number of active elements per stage including a guess at number of extra output elements required	Remarks
12.3	2	To supply four passive AND wall attachment elements	0.8	'0' and '1' after passive element, i.e. flow available but pressure only 10% of supply. Hence extra element required for output	3.8	The circuit must be run at a high enough pressure for the passive wall attachment element to work, so that the element preceding it must be run at perhaps 3 times the minimum possible pressure
12.6	2	To supply only one wall attachment element	0.2	'0' and '1' available directly from active element	2.2	The same remarks as for previous case, but a good pressure is available for the read-out
12.8	1	To supply one active element through a passive wall attachment one	0.2	'1' output only available but a half flow '0' output might be possible	1.2	The pulse shaper must be run at a high enough pressure for wall attachment to be possible in the steering element.
12.9	1	To supply two sides of a wall attachment element through an AND gate	0.5	Half flow is available from both '0' and '1' outputs	1.5	This circuit will work at low supply pressure. The output pressure and flow are just about sufficient for a read-out. Will only work from a pulsed input
12.10	2	To supply one active element per stage	0.2	'1' output available but a half flow from the '0' output might be possible	2.2	This circuit will work only at low supply pressure and from a pulsed input

References

1. Mitchell, A. E., Glaettli, H. H., and Mueller, H. R., Fluid logic devices and circuits, *Trans. Soc. Inst. Tech.*, **26** (1963).
2. Mitchell, A. E., Mueller, H. R., and Zingg, R. H. W., Some recent developments in the design of fluid switching devices and circuits, *Fluid Power Int. Conf.* (London), 1964.
3. Warren, R. W., Fluid pulse converter, *U.S. Patent*, No. 3,001,698, (1960).
4. Chadwick, V. J., A method of using wall attachment devices in an ultra sensitive mode, *C.F.C.* (2), Paper B1.
5. Foster, K., Parker, G. A., Jones, B., and Retallick, D. A., A digital fluidic point-to-point positioning system, *C.F.C.* (3), Paper A1.
6. Foster, K., and Retallick, D. A., Fluidic Counting Techniques. Fluidics Feedback. Aug. 69, Dec. 69 and Feb. 70. British Hydromechanics Research Association.
7. *Fluidic Systems Design Guide*, Componetrol, Inc. (a subsidiary of I-T-E Imperial Corporation). U.S.A. (1966).
8. Foster, K., Mitchell, A. E., and Retallick, D. A., Fluidic circuits used in a drilling sequence control, *C.F.C.* (2), Paper H4.
9. Warren, R. W., Fluid amplification No. 3, Fluid Flip-Flops and a counter, *D.O.F.L. Rep. U.S. Dep. Commerce*, AD285572, (1962).
10. Bantle, K., Counters with bi-stable fluid elements, *C.F.C.* (2), Paper H5.
11. Westerman, W. J., Richards, E. R., and Depperman, W. B., A Miniature fluidic oscillator and pulse counter, *Soc. Automotive Engrs., Proc. Aerospace Fluid Power Systems and Equipment Conf.*, 282–94 (1965).
12. Deason, W. S., A fluidic batching counter and a fluidic counter with visual decimal read-out unit, *C.F.C.* (3) Paper B1.
13. Bulk, J., Counting circuits with turbulence amplifiers, *C.F.C.* (3), Paper K8.
14. Shooh, T. A., Chen, T. F. and Reader, T. D., Fluid Amplification No. 12, Binary counter design, *H.D.L. Rep. U.S. Dep. Commerce*, AD617699.
15. Parker, G. A., and Carley, J. B., A pure fluid oscillator and frequency divider suitable for an industrial timer circuit, *Brit. Hydromechanics Res. Assoc.*, Res. Rep., RR967.
16. Tompkins, H. E., Unit distance binary-decimal codes for 2-track computation, *I.R.E. Trans. Elect. Comp.*, **E.C.5**, 139 (1965).

13
Digital System Encoders

13.1 Introduction

In electronic technology, the use of digital as opposed to analogue techniques has proved to be immense value in many applications for increasing the accuracy and resolution of control systems[1]. For instance, digital circuits for the control of linear movement and angular rotation are widely used for positioning machine tool tables with great precision. A digital positional control circuit is shown schematically in Fig. 13.1.

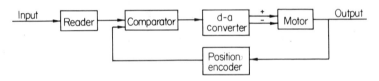

Fig. 13.1 Digital position-control system (Parker, *Ref. 5*, 18)

Input information, in binary coded form, is supplied from a reader accepting, for example, punched cards or tape. In simpler cases this may be manually selected. Feedback information of the actual physical position is generated in coded form by an encoder and compared with the coded input. Any error between the two binary numbers is changed to an analogue signal in a converter and amplified before directing the motor so that the machine moves towards the desired position. This chapter is concerned with the feedback encoder part of the circuit while summing junction and converter units are considered in the next chapter.

The resolution of a digital control system is dependent on the number of discrete intervals into which a given physical quantity may be subdivided and the accuracy is determined by the repeatability of measuring these intervals. For most fluidic systems requiring a resolution of the order of 0·002 in. to 0·001 in., the scales used by the encoder may be manufactured

[13.1] INTRODUCTION 501

relatively cheaply. The main limitation on accuracy and resolution therefore occurs in the fluidic device used for sensing the scale position. The methods of sensing invariably utilize mechanical modification of either two-dimensional or circular jet flowfields. Accuracy is reduced when the sensor is severely scaled down, and also the miniature nozzles required by the sensor are difficult to manufacture.

The accuracy discussed here is concerned with the ability of the encoder to provide feedback information in the steady state or static condition. Under dynamic conditions, the overall accuracy of the encoder is also dependent on the signal delay times through the logic networks associated with the receiver nozzles which may become a dominant factor. This point will be discussed further later in the chapter and also in Chapter 14.

13.2 Classification of Encoders

An encoder subdivides the physical motion, either linear displacement or angular rotation, into incremental motions and converts it to digital form by determining the number of increments contained in a particular motion. Two general types of encoder exist which are classified as Incremental Pattern and Coded Pattern. They are illustrated in Fig. 13.2 for a fluidic angular encoder.

In an Incremental Pattern encoder, all signals generated between successive position increments are identical and, in the simplest configuration, pass along a single line in the form of a series of pulses to a binary counter. Here all the pulses are counted and a binary output equivalent to the physical movement is produced. The counter must be frequently checked against some fixed datum as the encoder does not provide any absolute position signal.

In contrast, a Coded Pattern encoder uses position increments which are coded or numbered so that the number corresponding to any position can be determined directly from the encoder output. The Coded Pattern encoder is analogous to a scale, and no further counting or coding operations are required, although the code itself may have to be changed to another form for use in the control system.

For a pneumatic system, the most convenient method of translating physical motion into an incremental position signal is to sense a recessed or slotted sheet, joined to the moving part, which blocks or allows the passage of air. For an Incremental Pattern encoder only one row of slots is required, but for a Coded Pattern type the encoder has a row of slots corresponding to each digit of the binary number.

Generally speaking, Incremental Pattern encoders use a minimum number of logic elements to generate the desired digital output signal but

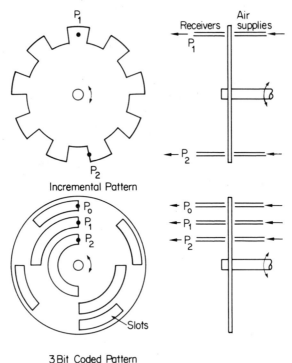

Fig. 13.2 Types of angular-position encoders (Parker, Ref. 5, 19)

sacrifice speed and possibly reliability in comparison with Coded Pattern encoders, which have more complex circuitry associated with them. Each type will now be considered in more detail.

13.3 Incremental Pattern Encoders

As all position increment signals are identical, a simple encoder of this type transmits information concerning the number of increment position changes which have taken place in a given time but is unable to determine the direction of motion associated with change.

To overcome this problem, one possible circuit suggested by Jones[2] is shown in Fig. 13.3 in which an increasing or decreasing binary number is controlled in the counter by signals from a direction discriminator. This consists of a wall attachment bistable element with control ports connected to a flapper valve, designed for switching on atmospheric pressure. The flapper arm is connected to the disc shaft through a slipping friction disc pad so that it moves with the shaft only between the limit stops.

[13.3] INCREMENTAL PATTERN ENCODERS

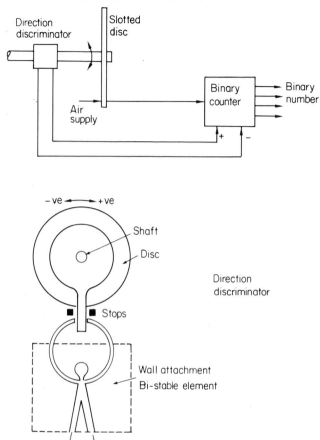

Fig. 13.3 Simple incremental pattern encoding system (Parker, *Ref.* 5, 20)

Direction indication can be achieved by using two receivers at the disc which are displaced from each other for correct phasing by an odd multiple of half the slot width. Fig. 13.2(a) shows a typical receiver configuration and this has been used by Taft and Wilson[3] for a direction interpreting circuit shown in Fig. 13.4.

A pulse train from pickup P_2 passes through a flip-flop and pulse forming circuits to generate two pulse trains, α and β, 180° out of phase with each other. Two signals P_1, and its negation $\bar{P}_1$ generated from a flip-flop attached to the other pick-up, provide correctly phased gating to divert α and β pulses into the appropriate direction output line and then on to the counter.

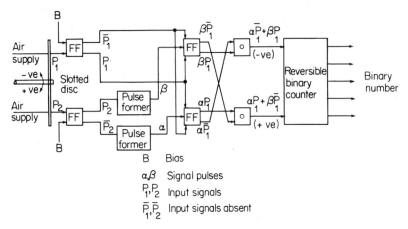

Fig. 13.4 Incremental pattern encoder with interpreting circuit (Parker, *Ref. 5*, 21)

The primary advantage of Incremental Pattern encoders is that they use the minimum number of logic elements for the generation of digital signals. However, this simplicity is only achieved in practice if the disadvantage of measuring increments relative to a reference position can be tolerated. If one increment signal is lost, all subsequent data is in error unless correction is provided. It is possible to introduce error detecting and correcting circuitry at the expense of increased complexity in the system, so that the important feature of simplicity tends to be lost.

13.4 Coded Pattern Encoders

With this type of encoder each increment position is numbered using a binary code so that it is uniquely defined, which dispenses with the need to refer relative displacements to a fixed datum position. Also direction indication is inherent in Coded Pattern encoders as the binary number output merely increases or decreases depending on the sense of the movement.

The two most common codes used for numbering position increments are the Natural Binary (N.B.C.) and Gray (G.C.) which have been discussed in Chapter 9. The distinctive feature of the Gray Code is that only one binary digit changes during one increment change of the code, and is a member of a class of codes, called Unit Distance Codes, which all have this property. For example, the change from decimal 7 to decimal 8 involves changes in all four least significant digits using N.B.C. whereas with G.C. only the fourth digit requires to change its sense.

Because of imperfections in the construction of an encoder, it is im-

[13.4] CODED PATTERN ENCODERS

possible to guarantee that all required receivers change their signals at the transition points between digits. Since the receivers cannot change exactly simultaneously there occur narrow intervals of partial transition where some, but not all, receivers have changed. The numbers corresponding to these partial transition conditions do not necessarily interpolate between the pair of numbers represented before and after the transition. For example Fig. 13.5 shows the ambiguous signals generated by a row of receivers with the most significant digit receiver (P_3 in (a) and P_4 in (b)) misaligned by one segment for both Natural Binary and Gray Codes. It will be noted that errors in the output signal at transition may be very great with N.B.C. whereas with G.C. the error is small.

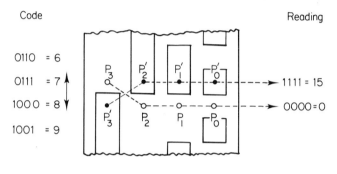

(a) Natural Binary Code

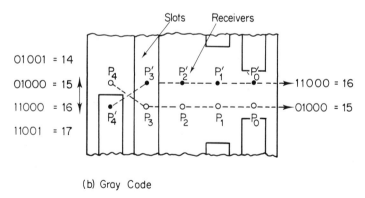

(b) Gray Code

Fig. 13.5 Coded pattern reading ambiguity (Parker, *Ref. 5*, 22)

If the problem of ambiguous reading of N.B.C. is severe, it may be overcome by adopting a more complex arrangement of signal receivers using 'V-scanning', so called because of the 'V' pattern of the receiver configuration. V-scanning calls for two receivers per row, placed a maximum of half the row segment width apart symmetrically about the index or reading line, as shown in Fig. 13.6.

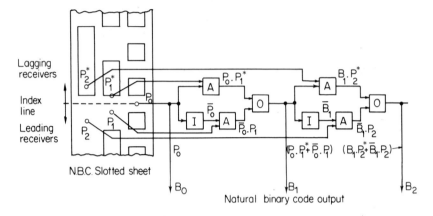

Fig. 13.6 V-scanning logic circuit for reading N.B.C. (Parker, *Ref.* 5, 23)

The receiver displaced from the index line in the increasing binary number direction is termed a leading receiver, whilst the other receiver in the row is displaced in a decreasing number sense and so may be called a lagging receiver. These receivers provide an error free reading for a given row when they are controlled by a signal from the adjacent row in the least significant direction. A logic circuit selects one of the two receivers according to the following rules:

(*a*) When the controlling receiver changes from 1 to 0 generating a carry, select the leading receiver in the next row.

(*b*) When the controlling receiver changes from 0 to 1, select the lagging receiver in the next row.

This may be implemented by the logic components shown in the figure, which consists of a circuit containing two AND's one OR and one INVERTER for each binary digit, except for the least significant digit receiver, P_0. This receiver is read at a transition, all others being read in the middle 50 per cent of the segment widths.

A further twin nozzle read-out configuration may be used called 'U-scanning' which is in some ways simpler than the V arrangement. Again an N.B.C. disc is used on which two parallel rows of nozzles are arranged one digit apart, as shown in Fig. 13.7. Each row is selected for reading according to the least significant digit reading i.e., '0' directs signals through leading nozzles and '1' through lagging nozzles.

The basic binary read-out logic circuit required to interpret a U-scan

[13.4] CODED PATTERN ENCODERS

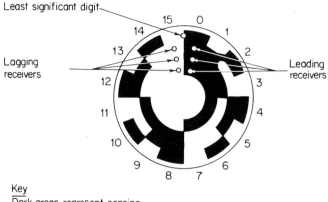

Key
Dark areas represent sensing nozzles switched off.

Fig. 13.7 U-scanning N.B.C. encoder (Ramanathan, Aviv and Bidgood, *Ref. 6*, 49)

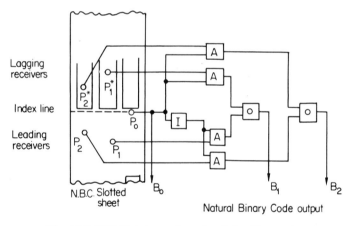

Fig. 13.8 N.B.C. interpreting circuit for U-scanning

configuration, using AND–OR logic units, is shown in Fig. 13.8 for a four row encoder. It is somewhat faster in reading than V-scanning because either all the lagging, or all the leading receivers are activated at a time rather than alternate lead and lag receivers in the latter case.

Due to the unambiguous nature of the Gray Code, a single receiver for each binary digit is used which is situated on the index line, as shown in Fig. 13.9. However, the Gray Code is unsuitable for general use in digital control circuits so logic conversion circuits must be used to generate

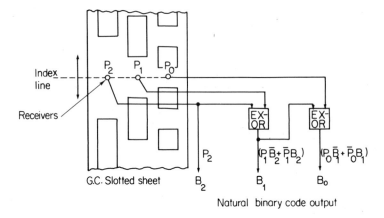

Fig. 13.9 N.B.C. interpreting logic circuit for Gray-code encoders (Parker, *Ref.* 5, 24)

N.B.C., or some other convenient code, at the encoder. The figure shows the logic required for conversion from G.C. to N.B.C. which involves the use of a cascade of EXCLUSIVE–OR elements.

Again the conversion logic circuit may introduce relatively large time lags, in the worst cases, before the N.B.C. output can be transmitted to other units in the system. This is because any digit position output is dependent on a gating signal from the preceding higher digit position which may have to be propagated through all the stages before the true output can be achieved. This raises the question of the most useful design of encoder and associated logic. It is of little practical use to use an obscure, but simple, code on the disc if a complex, slow conversion network is required to produce a standard code for use in the rest of the control system. Generally speaking, for computation in a circuit a form of N.B.C. is required although, as has been shown above, this leaves something to be desired in the encoder itself. A compromise must be reached between encoder simplicity and code conversion complexity.

The codes discussed so far split the linear or angular scale up into increments of 2, 4, 8, 16, 32, ... etc., which is not always convenient for a control system. Normally, of course, distance is measured in decimal units so that it is usually a requirement to have some sort of decimal visual display of the actual control system position. Taking machine tools as an example, a decimal visual display is required to check the component drawings the machine is working to, as these are normally in decimal form. Under such circumstances it is preferable to code the decimal number into a Binary-Coded Decimal number (B.C.D.). These codes

[13.4] CODED PATTERN ENCODERS 509

have been described in Chapter 9 from which it will be recalled that there are numerous possibilities for such codes the commonest of which is the so-called 8–4–2–1 code for each decimal digit.

```
              Example
   Decimal    1     3     7
   B.C.D.    0001  0011  0111
```

It is seen that we can code decimal numbers directly if a minimum of four receivers per decimal digit are used. Again the difficulty of ambiguous reading occurs with the Binary Code Decimal (B.C.D.) code shown above. However, a few B.C.D. codes are also Unit Distance Codes, so that only one increment changes at a time.

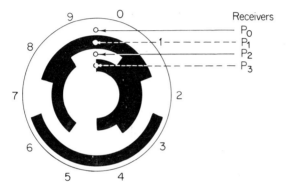

Fig. 13.10 Petherick coded disc for one decimal-decade (Ramanathan, Aviv and Bidgood, *Ref.* 6, 51)

Fig. 13.10 shows one of these codes, the Petherick Code, used on an angular encoder to code one decimal decade. It is similar to the Gray Code, except that ten combinations are made use of, instead of sixteen, to provide the decimal digits 0 to 9. It has the provision for changing from 9 back to 0 with only one bit alteration in the code as well as the usual one bit change for each decimal digit change. Thus an angular encoder using this code will read one decimal decade unambiguously during one revolution and has interesting uses in conjunction with a counter recording the total number of revolutions. By this means the encoder scale may be extended indefinitely depending on the capacity of the counter. For example, if the encoder codes a linear measurement of 1 in. and the counter has a capacity of 32 counts, then the linear scale may be read with the same accuracy over the range 0 to 32 in. as it can over the range 0 to 1 in. provided the counter is faster in operation than the reading speed of the encoder. However, a major disadvantage of a Petherick Code would appear to be the relatively large code converting circuit to change to binary form. Fig. 13.11 shows how this is implemented using OR–NOR logic.

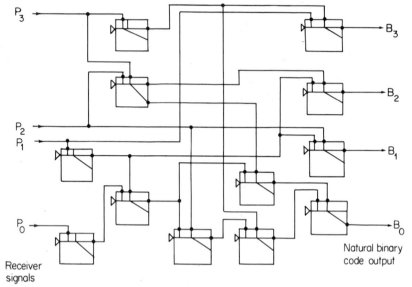

Fig. 13.11 Fluidic Petherick to binary code conversion circuit

A further way of extending the number of an encoder is to gear two or more units together. Normally a fast encoder, providing the fine resolution, is coupled to the physical motion and a second slower encoder, providing coarse resolution, is geared down from the fast unit. Thus, if the gear ratio is N and the encoders use N.B.C. the slow unit has segments $(1/N)$ as wide as they are on the fast unit for the first and last rows respectively.

An encoder configuration in which the physical movement is coupled to the coarse encoder and step-up gearing used for the fine encoder shaft should be treated with caution. This is because the inertias of the rotating masses in the system may generate large forces at the input shaft. Since inertias are reflected by the square of the gear ratio, the total inertia reflected at the input shaft A in Fig. 13.12, assuming both encoder discs are geometrically similar, is

$$I_{\text{total}} = I_1 + I_e + N_1^2[I_2 + I_e] \qquad (13.1)$$

The inertia of a gear wheel is proportional to the fourth power of its diameter, so that approximately

$$I_2 = I_1/N_1^4 \qquad (13.2)$$

[13.4] CODED PATTERN ENCODERS

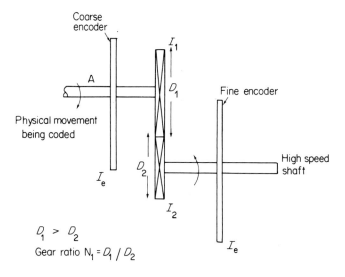

Fig. 13.12 Inertia effect of step-up gearing

Combining equations 13.1 and 13.2.

$$I_{\text{total}} = I_1 \left[1 + \frac{1}{N_1^4} \right] + (1 + N_1^2)I_e$$

or

$$I_{\text{total}} \simeq I_1 + N_1^2 I_e \tag{13.3}$$

This shows that the encoder inertia is the dominant factor of the reflected inertia as it is magnified by the square of the gear ratio.

V- or U-scanning techniques may be used without modification, but unit distance code arrangements may produce errors at the transition between one encoder and the other due to mechanical misalignment, gear backlash, etc. Although the number of the encoder is increased the resolution is not improved. Increased resolution is only obtained by providing step up gearing to the encoder so that each segment of the fast unit represents a smaller increment of the physical scale. For fluidic systems the main advantages in using a multi-speed encoder are that the manufacturing tolerances on each encoder unit are less severe, and also larger numbers may be processed without difficulties in reading the slot width of the least significant digit segment. Against these must be set the disadvantage of mechanical and gear backlash problems mentioned above.

13.5 Practical Angular Encoders

All pneumatic pure fluidic encoders use either back pressure or in-line receiver sensing to determine the presence or absence of a slot or groove in the scale representing the physical dimension.

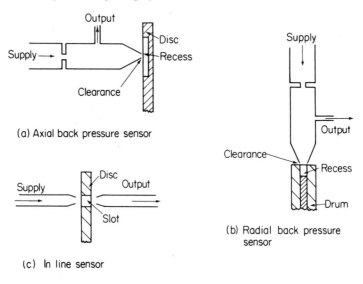

Fig. 13.13 Pneumatic encoder sensors

Fig. 13.13 shows the general arrangement of sensors and encoder scales. In (a) a back pressure sensing device is used to determine the position of a recess in a rotating circular disc or encoder. It is essential with such an arrangement to ensure that the clearance between the downstream nozzle and the disc remains constant otherwise an unwanted sinusoidal signal is superimposed on the output pressure wave-form. The figure shows an an alternative back pressure sensing arrangement in (b) where the sensor nozzle is positioned radially relative to the rotating encoder disc or drum. By this means, although it still senses recesses in the drum, the clearance can be more easily held constant.

An in-line receiver senses signals from a supply jet issuing from a nozzle in line with the receiver. By arranging the encoder in the form of a slotted disc between the supply and receiver nozzles it can interrupt the jet thereby signifying the position of the encoder. The clearance between both nozzles and the disc is not critical although, for good pressure recovery, the gap between the nozzle and receiver should not be too large. The gap to nozzle diameter ratio should be in the range 2 to 5, to minimize losses by operating

[13.5] PRACTICAL ANGULAR ENCODERS

with the receiver impinging on the jet core region. Some tests by Parker and Jones[4] also suggest that a larger receiver than the supply nozzle might result in an improved pressure recovery when the receiver is passing flow.

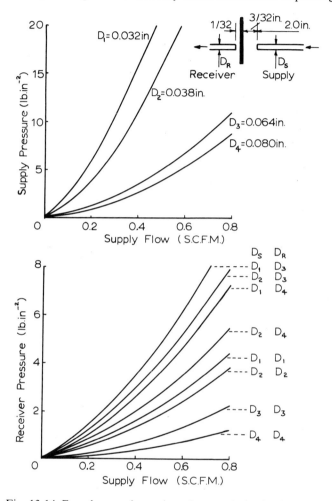

Fig. 13.14 Encoder supply receiver characteristics (Parker and Jones, *Ref. 4, 21*)

Fig. 13.14 shows some of the results obtained for tubular supply nozzles and receivers separated by a distance of about $\frac{1}{8}$ in. in the configuration illustrated. The receiver load was a 0·020 in. × 0·040 in. nozzle. It indicates a maximum pressure recovery with a receiver diameter roughly twice that of the supply nozzle.

A further practical difficulty encountered in using fluid jets for encoder design is the finite width of the receiver nozzle. Ideally, it should be infinitely narrow so that whatever the direction of disc rotation the pressure in the sensor is exactly the same for a given disc position, as shown in Fig. 13.15.

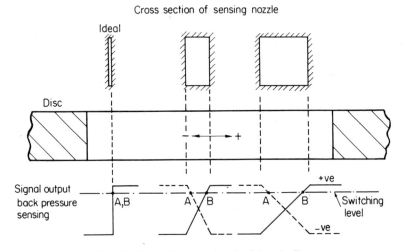

Fig. 13.15 Sensing nozzles dead-band effect

If a pressure signal of a fixed magnitude is required to switch an element downstream of the sensor, the switching would always occur at exactly the same position of the encoder disc irrespective of the approach direction. However, the figure also shows the effect of a finite width for a rectangular nozzle and indicates two switching points, A and B, getting further apart as the width increases. Each switching point is identified with the two directions of motion. This means that there is a dead-band produced by the sensor that cannot be resolved by adding extra interpreting logic into the system. The only simple method of removing the problem of receiver width is to arrange the encoder to approach the desired position in one direction only.

Apart from making the nozzle width as narrow as possible, the problem may be lessened by using two interconnected geared encoder discs so that the slots of the disc coding the smallest distance increments are effectively amplified. This means that relative to the slot width the sensor nozzle appears smaller, thereby producing a sharper signal profile. In any case, the slot width to receiver width ratio should not be less than about 4, otherwise the wave-forms for a slot will approximate to sinusoidal rather than rectangular shape.

[13.5] PRACTICAL ANGULAR ENCODERS

A profile of one stage of a fluidic V-scanning N.B.C. circuit which has been developed by Parker[5], is shown in Fig.13.16. It interprets one row of receivers and provides a natural binary output. The carry signal B_{N-1}, from the adjacent less significant row is passed into a monostable wall

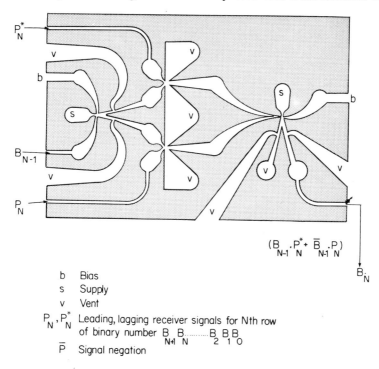

- b   Bias
- s   Supply
- v   Vent
- P_N, P_N^* Leading, lagging receiver signals for Nth row of binary number $B_{N+1} B_N \ldots\ldots B_2 B_1 B_0$
- $\bar{P}$ Signal negation

Fig. 13.16 One row of a fluidic V-scanning N.B.C. circuit (Parker, Ref. 5, 25)

attachment element which conveniently generates B_{N-1} and its negation, $\bar{B}_{N-1}$ with signal amplification. The lead and lag receiver signals, P_N and P_N^* are gated with $\bar{B}_{N-1}$ and B_{N-1} respectively in passive AND elements before passing into a two input OR wall attachment element. The output signal, B_N, is used as the binary output signal for the row and is also fed into the logic circuit for the next row.

The circuit was etched in Dycril plastic from a photographic negative which was produced by a 5:1 reduction of an ink drawing. All nozzles were made 0·020 in. wide × 0·040 in. deep.

Fig. 13.17(a) shows a typical trace of pressure signals generated in a single interpreting circuit operating with a signal frequency of about 300 Hz. A time delay of approximately 1·5 ms between the generation of B_0 and B_1 was found at these frequencies due to the time taken to switch the logic elements. The circuit time delay may be serious as it is accumulative from row to row. Thus it might be expected, for example, that the seventh row digit signal B_6, is delayed by 10 ms, compared with B_0, giving a long transition time before the output may be read. This type of

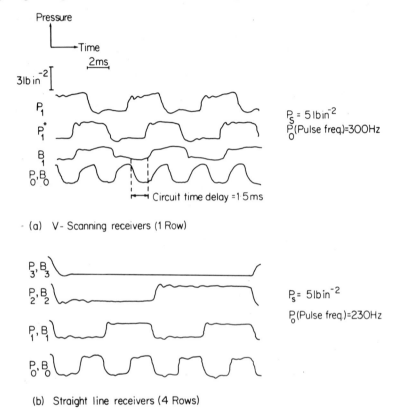

Fig. 13.17 Pressure traces from experimental N.B.C. encoders (Parker, Ref. 5, 26)

problem is typical of fluidic digital circuits and occurs due to the accumulation of delays from individual switching elements and also the interconnecting channels. As far as possible logic circuit arrangements have to be chosen to provide simple parallel flow of information or alternatively operate synchronously in order to minimize these effects. The V-scanning interpreting circuit is not capable of further logic reduction to remove the cascade feature so that the only possibility is to use synchronous operation to produce a correctly phased binary output signal. This involves gating all

[13.5] PRACTICAL ANGULAR ENCODERS 517

the binary output channels with a synchronizing pulse, which is generated after all the logic delays have occurred, thereby giving a true output reading in all channels at the same time. The problem also arises in connexion with adder and subtractor circuits.

Fig. 13.17(b) shows the pressure signal outputs from four rows of a straight line receiver configuration under similar test conditions to the previous experiment. The timing of each row was slightly random due to unavoidable imperfections in receiver alignment but the results were surprisingly good as the greatest deviation of any row compared with the least significant signal P_0, was less than 1 ms. However, at very low angular velocities spurious reading would be expected to give trouble as timing differences between receivers are accentuated.

An angular fluidic encoder based on the Petherick Code has been introduced by Bidgood, Ramanathan and Aviv[6]. The authors used back pressure sensing and found difficulty in controlling the clearance between the sensors and a rotating encoder disc. To overcome this problem, they positioned the sensor to read the disc radially at its periphery and used one disc for each row of the encoder number i.e., four discs for each decimal decade of the number (see Fig. 13.13(b)).

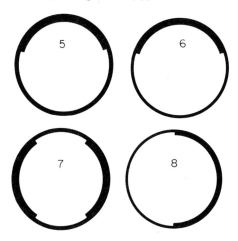

Fig. 13.18 Four discs required for a Petherick code drum encoder (Ramanathan, Aviv and Bidgood, *Ref.* 6, 59)

Fig. 13.18 shows the profiles of four discs required for a Petherick Code drum encoder which may be compared with the same code generated on one disc illustrated earlier in Fig. 13.10. The big advantage with the modified form is that all the discs may be mounted as a drum, and the periphery accurately machined without difficulty to the required sensor clearance. The areas marked black on the profiles are where material is removed from the disc so that when the back pressure sensor is opposite these regions, the sensor is vented giving an 0 output signal and when restricted over the rest of the disc periphery the output is a 1 signal.

One advantage of angular encoders is that it is relatively easy to provide mechanical gearing between the encoder disc and the actual physical scale to match the two scales together. This effectively provides mechanical amplification between the encoder scale and the actual scale. However, the total number of increments of the encoder may be too large for one disc without very precise, expensive machining. In this case it is advantageous to split the code between two encoder discs with a suitable gear ratio linking them together. Bidgood *et al.* used gearing between two encoders to increase the total number capacity and, by suitable design, mounted the unit directly on a 5 t.p.i. lead screw for a machine tool table positioning system. There were four decades in inch units; namely units, tenths, hundredths and thousandths. The hundredths and thousandths codes were repeated twice in one encoder drum to give a total measurement up to 0·20 in/rev, which allowed it to be coupled directly to the lead screw. The units and tenths decades made up the other drum which was geared down 50:1 relative to the first drum to provide coarse measurement over the full range 0 to 10 inches.

13.6 Practical Linear Encoders

When linear encoder scales are used, the physical dimension being measured is commonly split up into the total number of increments to provide the resolution required. In positioning systems, such as machine tools, attractive fluidic applications may be found if a resolution in the order of 0·001 in. may be achieved. In terms of cost and miniaturization techniques, it is not economic to consider linear encoder scales with slots and recesses of 0·001 in. width. A more practical lower limit for such widths seems to lie between 0·005 in. and 0·010 in.

One method of increasing the resolution beyond the slot width has been developed by Sharp, Maclean and McClintock[7] and consists of positioning accurately a diagonal line of 10 back pressure reading nozzles across one scale pitch of the linear encoder, as illustrated in Fig. 13.19.

These reading nozzles act as an interpolating scale and require control logic to ensure that only the nozzle actually at the edge of the scale groove is reading. This may be achieved by the logic circuit shown in Fig. 13.20 and is based on a tree arrangement of OR–NOR wall attachment elements. An indicator is only selected when one nozzle has triggered its

[13.6] PRACTICAL LINEAR ENCODERS 519

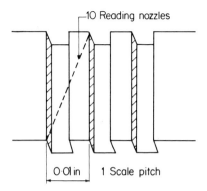

Fig. 13.19 Interpolating reading nozzles for improving resolution (Sharp, Maclean and McClintock, *Ref.* 7, 32)

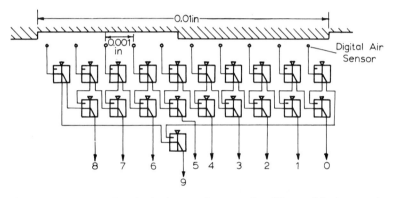

Fig. 13.20 Interpolation logic for reading nozzles (Sharp, Maclean and McClintock, *Ref.* 7, 32)

switching element while the adjacent one has not. A scale reading repeatibility of 0·00025 in. is reported allowing for a 10 per cent supply pressure variation.

Instead of staggering the nozzles across the scale, it is possible to stagger the scale itself in a similar manner. Gant[8] has obtained a resolution of 0·002 in. using linear scales with machined grooves of 0·010 in. width, offset 0·002 in. from each other. Fig. 13.21 shows the arrangement. This requires two rows of four reading nozzles in the lowest decade: the rows being spaced 0·020 inches apart so that the first row reads 0·002, 0·004, 0·006, 0·008 in. and the second row 0·012, 0·014, 0·016 and 0·018 in. For the higher decades, because of the larger linear scales, only 5 nozzles are needed in this design for each decade. Apart from the lowest decade, a

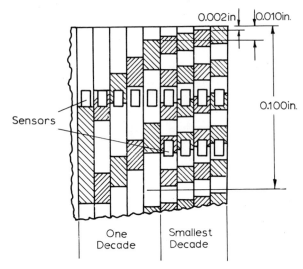

Fig. 13.21 Offset scale linear encoder (2 decades) (Gant, Ref. 8, 10)

straight line of receivers is used which give an unambiguous output. However, all the decades require fairly complex circuits to give a decimal output display.

References

1. Suskind, A. K. (Ed.), *Notes on Analog-Digital conversion techniques*, Techn. Press, M.I.T. (1957).
2. Jones, N. S., A study of the characteristics of turbulent re-attachment control devices, *Ph.D. Thesis* (Birmingham) 1965.
3. Taft, C. K., and Wilson, J. N., A fluid encoding system, *H.D.L.*, **4** (1964).
4. Parker, G. A., and Jones, B., A fluidic subtractor for digital closed-loop control systems, *C.F.C.* (2), Paper H1.
5. Parker, G. A., Digital fluid position encoders, *C.F.C.* (1), Paper E2.
6. Ramanathan, S., Aviv, I., and Bidgood, R. E., The design of a pure fluid shaft encoder, *C.F.C.* (2), Paper E4.
7. Sharp, R., Maclean, R., and McClintock, A., A fluidic absolute measuring system, *C.F.C.* (2), Paper E3.
8. Gant, G. C., A fluidic digital displacement indicator, *C.F.C.* (2), Paper E1.

14
Summing Junctions, Converters and Signal Hazards in Digital Systems

14.1 Introduction

The previous chapter dealt with the methods of signal generation in the feedback path of a digital closed loop control system. Input devices were also discussed in Chapter 11. The remaining important units are the summing junction for the input and feedback signals, which provide the digital error signal, and the digital-analogue (D/A) converter which gives an analogue error signal to the amplifiers and prime mover. A variety of output power devices are used in conventional digital control systems and these have a bearing on the type of summing junction, shaping networks etc. in the forward loop of the system. Broadly speaking, they may be divided into two categories depending on whether they are digital or analogue in action. Digital electric and electro-hydraulic motors are available but it is only comparatively recently that work has been reported on pneumatic digital motors[1-3] which would be compatible with fluidic summing junctions. All analogue output devices require a conversion of the summing junction digital error signal into analogue form for compatibility.

Various degrees of sophistication may be designed into the digital summing junction depending on the system mode of operation. If the control circuit only has an 'on-off' action, the summing junction need only give a signal when the input and feedback signals are equal, i.e. zero error signal. Such a function is provided by a comparator. The usefulness of comparators may be increased by also providing an indication of which of the two binary numbers representing the signals is the greater at any arbitrary position so that the system may be driven in the appropriate direction. In general, when a comparator is used to compare two binary numbers, A and B say, it gives an output signal when $A > B$, a second signal when $B > A$, and a third signal when $A = B$. Additional facilities, such as signals for slowing down the system as it approaches the desired position, may also be incorporated.

Digital control systems in which it is desirable to have continuous modulation of the output require either Adder or Subtractor circuits for the summing point. For negative feedback systems, the coded output signal is subtracted continuously from the input value, the difference being the digital error signal. An Adder would be used in a similar fashion but for a positive feedback circuit. It is also possible to construct a Subtractor using basically Adder circuits and this arrangement is sometimes preferred.

Addition and subtraction of binary numbers can be carried out by a simple process of counting. This is achieved by adding the number of signals counted to one number while at the same time subtracting the same number of counts from another number; the sum of the two numbers is obtained when the second number is reduced to zero. This method of addition is often used when only one line is used to transmit information signals in the form of timed pulses. Such a circuit configuration is said to be serial operation and has the advantage that a minimum amount of equipment is needed to interpret the signals. The drawback to this mode of operation for fluidic circuits is that it is relatively slow in operation; only one signal pulse may be processed at a time. Because of the inherent slowness of operation of fluidic circuits compared with their electronic counterpart, only circuits which allow information transference with a minimum of time delay are practicable. Faster addition and subtraction is achieved by carrying out these computations with digits of corresponding orders separately. For example, if two ten digit binary numbers are added together each digit position has a separate line making ten in all. This is known as parallel operation as individual pieces of information from the complete coded number are all processed simultaneously thereby reducing computation time to a minimum. In this chapter we shall only be concerned with addition and subtraction of the parallel type.

14.2 Adder Logic

If two binary digits A and B in corresponding orders of two numbers are added together a 'carry' of 1 into the next higher order occurs when both A and B are 1. This is because we cannot admit the number '2' in a binary system. The rules for the binary addition of digits is most easily seen by tabulating all the combinations of A and B.

[14.2] ADDER LOGIC

TABLE 14.1

A	0	1	0	1
B	0	0	1	1
Sum digit	0	1	1	0
Carry digit	0	0	0	1

A logic element block which accepts two inputs (A and B in this case), and produces two output signals (Sum and Carry), conforming to Table 14.1 is called a Half Adder. This block is not a Full Adder because no provision is made for also adding in a carry signal supplied from the preceding lower order digit position. The rules for adding in this carry signal are exactly the same as for A and B and so a second Half Adder may be used for this purpose.

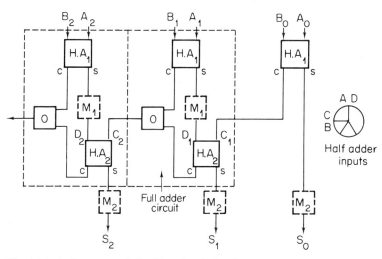

Fig. 14.1 A three-stage full adder circuit (Parker and Jones, *Ref. 5*, Fig. 6)

Fig 14.1[4] shows the simplest form of circuit for adding two numbers $A = \ldots A_2 A_1 A_0$ and $B = \ldots B_2 B_1 B_0$ to produce a sum $S = \ldots S_2 S_1 S_0$. The Half Adders adding the input signals together are marked HA_1 and the other Half Adders, marked HA_2, add in the carry signals C. As there are two sources of carry signal from each stage, they are passed through a two input OR device denoted by 0. The figure also shows elements marked M_1 and M_2 which are signal amplifiers which may be required in fluidic circuits and are discussed further in Section 14.3.

A further simple Full Adder circuit can be obtained by summing the carry signal into the first Half Adder of the next higher order stage instead of the second. Fig. 14.2 shows the resulting logic diagram with the same notation as the preceding figure. This raises the question of the speed of operation of adder circuits and which is likely to be inherently faster.

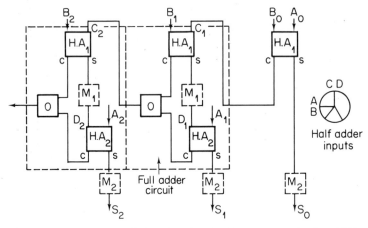

Fig. 14.2 A three-stage full adder circuit with carry to the first half adder (Parker and Jones, *Ref. 5*, Fig. 7)

Although the adder circuits illustrated operate in parallel, they are interconnected by the carry signal which means that the final sum output signals can only be read when the carry signals have passed across the complete chain of elements. It is therefore seen that it is critically important that the path of the carry signal should be as simple as possible to achieve high speed. Now comparing the two Full Adder circuits so far discussed, it is seen that the second one illustrated in Fig. 14.2 is inherently slower in propagating the carry signal than the first. In the worst case, when $B = 0$ and $A = C = 1$ in any stage of the adder, the propagation path passes through both Half Adders and the OR. By comparison the carry propagation path in the first case is through the second Half Adder HA_2 and the OR.

At first glance it may appear that two 'n' digit binary numbers can be added together in parallel paths 'n' times as fast as a single or serial path. That this is not so is due to time delays occurring between the reception of the three inputs to a stage and the formation of the sum and carry outputs. Although two of the inputs are from external sources and can be applied simultaneously to all stages, the carry input to a stage is dependent upon the condition of the previous stage. There are two ways in which a carry can be produced at the outputs of a stage; firstly generation of a carry within a stage which occurs independently from a carry input to it, and secondly, propagation of a carry through a stage.

[14.2] ADDER LOGIC

The Boolean algebraic expressions for the sum S and carry C for the Half Adder can be derived from Table 14.1 given earlier as

$$S = A\bar{B} + \bar{A}B \qquad C = AB \qquad (14.1)$$

If we consider the general nth stage of a Full Adder, it can be deduced that the condition for the generation of a carry C_{n+1}^G can be formally stated as:

$$C_{n+1}^G = A_n B_n \qquad (14.2)$$

Similarly, the necessary condition for propagating a carry is:

$$C_{n+1}^P = (A_n \bar{B}_n + \bar{A}_n B_n) \cdot C_n \qquad (14.3)$$

Hence, the general condition for the formulation of a carry from the nth stage is:

$$C_{n+1} = A_n B_n + (A_n \bar{B}_n + \bar{A}_n B_n) C_n \qquad (14.4)$$

It is evident that the conditions for C_{n+1}^G and C_{n+1}^P are mutually exclusive and that a stage generating a carry does not propagate a carry.

It is clear that the outputs of a multi-stage Full Adder operated in the parallel mode do not appear simultaneously, but depend upon the carry propagation time. The worst case occurs when a carry generated in the least significant stage has to be propagated through all the remaining stages. For example, when 10101 is added to 01011.

In both the configurations discussed so far the carry is propagated from one order to the next in these instances where a carry output from a stage creates a carry in the next highest order. It is possible, by increasing the circuit complexity, to handle all the carry signals more or less simultaneously thereby increasing the overall operating speed. This is shown for three stages in Fig. 14.3 and involves the additional use of AND elements.

Consider the highest order, that is, the one on the left-hand side of the figure. A carry from this adder stage to be sent to the next higher order can be generated in any one of three ways. First, the sum of the digits B_2 and A_2 may be 1; second, the sum of these two digits may be 1 and a carry from the next lower order may be present; and third, the sum may be 1, the sum of A_1 and B_1 may also be 1, and a carry from the lowest order may be present. The two AND elements and the extra input line to the OR provided for the carry generation described in the second and third case. It will be noted that one of the AND's in the highest order position has three inputs; the next highest order not shown would require a four input AND and so on. These multi-input AND elements cannot be simply implemented using pure fluid elements so the method is only applicable to two stages of a fluidic adder.

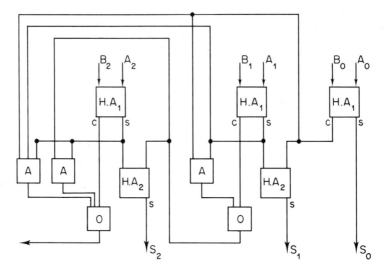

Fig. 14.3 A full adder using simultaneous carry (Parker and Jones, Ref. 5, Fig. 8a)

14.3 Fluidic Adder Circuits

As the Full Adder circuit uses Half Adder circuits for basic building blocks, it is the latter configurations which form the basis for different adder designs. A Half Adder may be constructed in several ways using pure fluidic components. In Chapter 1 the type proposed by Greenwood and also the cusp type suggested by G.E. Co. have been discussed. At the present stage of development, the latter type would appear to be more promising particularly as dynamic operation has already been achieved Parker and Jones[5]. Using an element with nozzle sizes of 0·020 in. wide × 0·040 in. deep satisfactory dynamic operation has been obtained for frequencies up to about 400 pulses/second. A sample of these results using the notation of Fig. 1.16 are shown in Fig. 14.4 for the effect of three different pulsed input combinations on the sum output. Although correct logic operation is achieved, it will be noted that spurious signal generation is possible at the transition of control from one input to the other. Test (a) is possibly the most stringent as the sum output should be continuously present when signals A and B are pulsed with the same frequency exactly out of phase. Due to the finite rise times of the signals and the dynamic response of the element spurious interruption of the sum signal occurs at every control signal change-over. Subsequent tests have shown that the

[14.3] FLUIDIC ADDER CIRCUITS

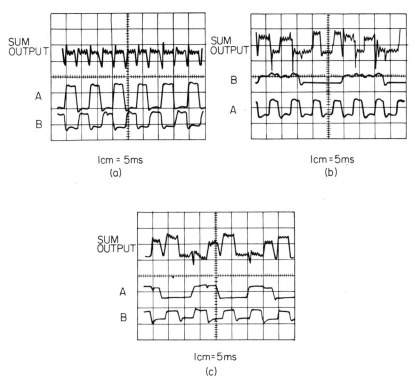

Fig. 14.4 Dynamic operation of a cusp type half adder sum output (Parker and Jones, *Ref. 5*, Figs. 27a, 27b, 27c)

element performs rather better in sizes scaled down to about half the original size.

Because the Half Adder discussed above is passive, signal amplification may be required between the two Half Adders of each stage (drawn as M_1 in Figs. 14.1 and 14.2) and gives rise to the problem of air consumption in a fluidic circuit. It is apparent that some circuit configurations require fewer active elements than others, while in turn, other circuits may be faster in operation. The optimizing of a circuit to meet requirements such as these involves compromises. This can well be illustrated by an investigation by the above authors of various adder circuit combinations when one of the inputs is supplied from a feedback encoder disc. Both the air consumption and signal propagation time are considered to be critically important.

The basic adder circuit originally shown in Fig. 14.1 is denoted as Type 1 without the interstage amplifier M_1 and as Type 1A with the amplifier.

Similarly the second configuration shown in Fig. 14.2 is Type 2 and Type 2A with the amplifier. It is not possible to completely implement the simultaneous carry adder circuit shown in Fig. 14.3 using fluidic components but if the technique is applied to two stages at a time the circuit shown in Fig. 14.5 is obtained.

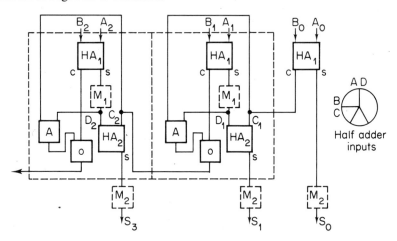

Fig. 14.5 Three-stage simultaneous carry full adder circuit (Parker and Jones, *Ref. 5*, Fig. 8b)

This is denoted at Type 3 or Type 3A with the amplifier. 'Bread-board' versions of each circuit were assembled from the basic logic elements manufactured in photosensitive Dycril, all nozzles being approximately 0·020 in. wide × 0·040 in. deep.

A comparative assessment is made by testing each configuration under various conditions of loading on the sum output of one stage. The load consists of a monostable turbulent re-attachment device (M_2) operating over a range of supply pressures from 1 to 5 lb. in^{-2}. The air consumption of one stage of each type, including the encoder supply, is determined and shown graphically in Fig. 14.6.

It may be concluded from these curves that the configurations that include amplifications consume less air than do their counterparts without amplification; this is a direct result of the significant reduction in the level of the input signals from the encoder to the stage. Of these configurations Type 1A consumes the least quantity of air, whereas the circuit with the highest air consumption is Type 3A due to the monostable amplifier having to supply a fan-out of two passive elements.

Simple tests may be carried out on the dynamic carry propagation time of a single stage of each configuration with the following combination of

[14.3] FLUIDIC ADDER CIRCUITS 529

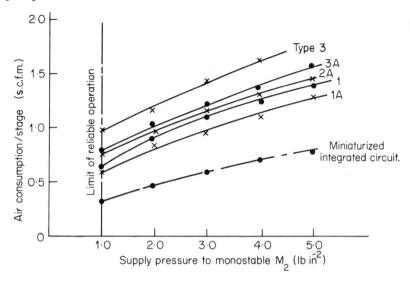

Fig. 14.6 Air consumption characteristics of full adder configurations (Parker and Jones, *Ref. 5*, Fig. 10)

input signals. Input 'B' supplied with a steady '1' signal while A is held at zero for Types 1, 1A, 3 and 3A; the situation is reversed for Types 2 and 2A. A step input is supplied to the carry input to the stage and the time delay through the carry propagation path is typically:

$$\begin{aligned}
\text{Types 1 and 1A} &= 2 \text{ ms} \\
\text{Type 2} &= 3 \cdot 0 \text{ ms} \\
\text{Type 2A} &= 3 \cdot 5 \text{ ms} \\
\text{Types 3 and 3A} &= 1 \cdot 9 \text{ ms}
\end{aligned}$$

Evaluation of the results of the static and dynamic tests on the broadboard version indicate that Types 2 and 2A have by far the slowest propagation times as expected. Of the remaining types the configurations with amplification, Types 1A and 3A, have lower overall air consumption requirements than their counterparts without amplification. Carry propagation times are essentially the same for both types and, since Type 1A consumes less air than Type 3A, it is considered to be the optimum design.

Thus, what might appear superficially to be increasing the air consumption by the additional use of an interstage amplifier, does in fact reduce the overall stage consumption. This is because the amplifier acts as a buffer thereby allowing substantial reductions in the air consumption at the encoder disc supply. The approach of optimizing as much of the complete system as possible is always beneficial in fluidic circuit design. An integrated circuit version of the Type 1A circuit stage is shown in Fig. 14.7 and has been operated successfully with all the input signal combinations at frequencies over 200 Hz with good signal definition.

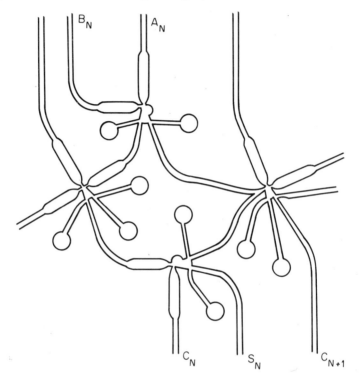

Fig. 14.7 Integrated full adder circuit (Type 1A) (Parker and Jones, *Ref. 5*, Fig. 20)

A further Full Adder circuit has been proposed by Glaettli[6] based on the EQUIVALENCE element described in Chapter 1. Fig. 14.8 shows the logic arrangement and Fig. 14.9 illustrates a practical circuit based on this diagram which uses two active and two passive elements per stage.

It has been constructed in several sizes to assess the effect of geometric scaling on the circuit speed of operation, as illustrated in Fig. 14.10. The propagation time per stage achieved for various circuit sizes based on the nozzle width would suggest comparable operating speeds to the preceding circuit.

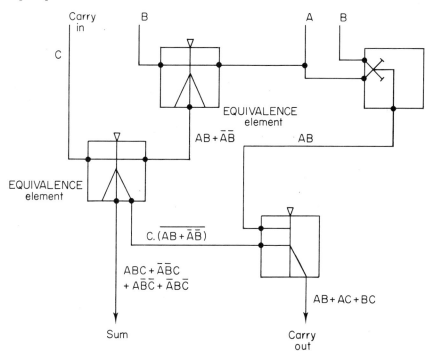

Fig. 14.8 Fluidic EQUIVALENCE elements used in a full adder circuit (Glaettli, *Ref. 6*, Fig. 1)

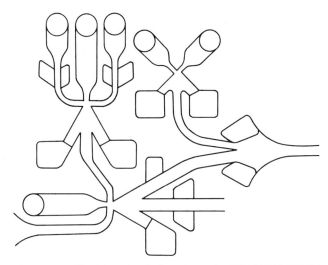

Fig. 14.9 Full adder circuit based on the EQUIVALENCE element (Glaettli, *Ref. 6*, Fig. 2)

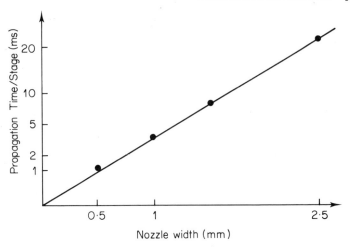

Fig. 14.10 The variation of stage propagation time with geometric scaling for the full adder (Glaettli, *Ref. 6*, Fig. 4)

14.4 Subtractor Logic

When subtracting two binary digits A and B, one from the other, it is possible that the difference between the magnitudes is less than zero. For example, if B is subtracted from A and B = 1 while A = 0 then the difference would read '−1'. However, this is not possible with binary codes so that a 1 must be borrowed from the next higher order. This procedure is referred to as direct subtraction and may be summarized for two binary digits A and B as in Table 14.2.

TABLE 14.2

A	0	1	0	1
B	0	0	1	1
Difference digit	0	1	1	0
Borrow digit	0	0	1	0

A device which conforms to this table is called a Half Subtractor and is analogous to the Half Adder used in adder circuits. The Boolean logic expression for the two output signals from a Half Subtractor may be stated as

$$\text{Difference} = A\bar{B} + \bar{A}B$$
$$\text{Borrow} = \bar{A}B \tag{14.5}$$

[14.4] SUBTRACTOR LOGIC

and the logic diagram for a Full Subtractor is very similar to Figs. 14.1 and 14.2 for Full Adder configurations.

Whenever the complete binary number B being subtracted from number A is the larger number, then the output difference is negative and the subtractor presents the information in what is called '2's complement' form. For example, if the true difference were 1101 the subtractor would give 0011. To convert from 2's complement form to the correct value involves the following procedure:

(a) Subtract 1 from the 2's complement (giving 0010 in the example quoted above.

(b) The number is now in what is called '1's complement' form. To convert to the true number invert all the digit values (thus 0010 becomes 1101, which is the true value in the example).

This process of altering the form of the difference output of the subtractor depending on whether the magnitude is positive or negative means extra logic complexity compared with an adder circuit. However, the borrow signal from the highest order digit position changes sign when the difference magnitude changes sign, and so is used to gate the output signals to provide the correct difference.

It is often more convenient to carry out the subtraction function by the addition of the complement representation of numbers instead of through the use of a subtractor. Either the 1's complement or the 2's complement may be used; the former has the advantage of simplicity of conversion back and forth between the true and complement forms, and is the only type considered in this discussion.

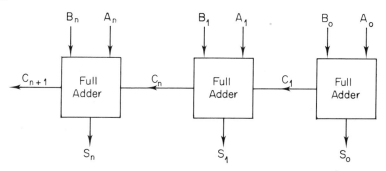

Fig. 14.11 Functional diagram of a full adder circuit (Parker and Jones, *Ref. 5*, Fig. 1)

Consider an Adder circuit shown in functional form in Fig. 14.11. The inputs $A_0, A_1, \ldots, A_n$ to an 'n' digit binary full-adder, represent the 1's complement form of the subtracting number $a_0, a_1, \ldots, a_n$. Inputs B_0,

$B_1, \ldots, B_n$ represents the number to which they are to be added. In performing the addition of two binary numbers, A_0 and B_0 are first added to give S_0, the least significant digit sum, and C_1, the carry digit. Each succeeding stage therefore requires 3 input positions and 2 outputs.

The Boolean expression for the nth stage of such an adder is given as:

$$S_n = \bar{A}_n \bar{B}_n C_n + \bar{A}_n B_n \bar{C}_n + A_n \bar{B}_n \bar{C}_n + A_n B_n C_n \qquad (14.6)$$

and

$$C_{n+1} = B_n C_n + A_n C_n + A_n B_n \qquad (14.7)$$

The table representation of the expressions is given in Table 14.3:

A_n	B_n	C_n	S_n	C_{n+1}
0	0	0	0	0
0	0	1	1	0
0	1	0	1	0
0	1	1	0	1
1	0	0	1	0
1	0	1	0	1
1	1	0	0	1
1	1	1	1	1

When the process of subtraction is to be performed the sum output from the Full Adder must be handled differently depending upon the relative magnitude of the binary words.

The three possibilities may be considered as follows:

(a) $a_0, a_1, \ldots, a_n > B_0, B_1, \ldots, B_n$. The difference is negative and will be represented in the 1's complement form at the output of the Full Adder. In this instance a carry will not be produced in the most significant stage.

(b) $a_0, a_1, \ldots, a_n < B_0, B_1, \ldots, B_n$. The difference is positive and under these conditions a carry will be produced in the most significant stage, will be one digit less than the true difference.

(c) $a_0, a_1, \ldots, a_n = B_0, B_1, \ldots, B_n$. Zero difference is represented in the 1's complement form, i.e. 1111 ... No carry signal is produced by the most significant stage.

For the conditions where $a_0, a_1, \ldots, a_n \geqslant B_0, B_1, \ldots, B_n$, it is necessary to complement the Full Adder outputs in order to obtain the difference in true form. The presence or absence of a carry signal from the most significant stage is a ready means of indicating the sign of the difference

[14.4] SUBTRACTOR LOGIC

that can be used to gate the complementing logic. As stated above, the output of the Full Adder, for the case when the difference is positive, will be one digit less than the true output. In some control systems an error of one digit in the difference is unimportant while the system is moving to its desired state. The basic requirement is that zero difference is represented correctly.

14.5 Fluidic Subtractor Circuits

There is very little information concerning subtractor circuits because of the added complexity of interpreting positive and negative differences discussed in Section 14.4. A subtractor has been constructed by Parker and Jones[5] based on the use of a 1's complement adder circuit. The complementing in this case (which involves changing the magnitudes of all the individual digits in the number) is not too difficult to achieve in fluidic circuits.

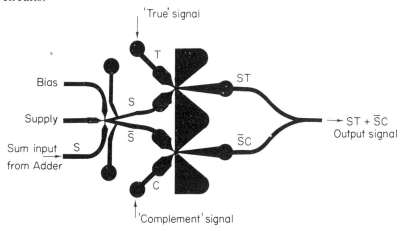

14.12 Simple fluidic complementing circuit

Fig. 14.12 shows a simple complementing circuit using fluidic components. A monostable wall attachment element provides at its output both a signal and its negation, which are both necessary for complementing. If these two outputs are gated with passive ANDs and then joined together in a passive OR arrangement, the output reads the true output or its complement (i.e. the opposite value) depending on which of the controlling signals marked 'true' or 'complement' is present. The controlling signals themselves may be generated from the carry of the highest order digit position of the Adder circuit. A block diagram of the complete circuit for a six-stage subtractor is shown in Fig. 14.13.

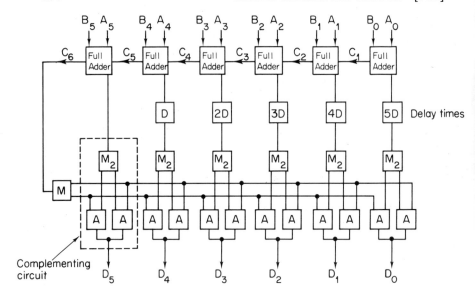

Fig. 14.13 Six-stage subtractor circuit diagram using adder complementing (Parker and Jones, *Ref. 5*, Fig. 9)

In order to construct a multi-stage subtractor using any of the configurations discussed so far, it is necessary to synchronize the sum outputs of the Full Adders. The simplest solution to this problem in fluidic circuitry is to delay the sum outputs of each stage relative to each other by time intervals equal to the carry propagation time through a stage, with the least significant stage having the greatest delay. The delays are marked with the letter 'D' and the approximate magnitude of the delays are as shown in the delay blocks. Although this method gives approximate matching, exact matching is not possible as the carry propagation time varies according to the number combination being processed at any instant of time.

A more precise method of obtaining the correct timing of all the output signals is to arrange for asynchronous or synchronous operation. In the former case, each operation is started only when a signal is received which indicates that the previous operation has been completed. This is usually achieved by the use of further logic networks to generate the signal within the system. On the other hand, a synchronous mode of operation occurs when the output is controlled by means of an external signal. Commonly a clock generator is used for this purpose so that its output may control certain of the logic elements and allow their operation at known time intervals.

14.6 Comparator Logic

Fig. 14.14 shows one possible logic arrangement for a comparator requiring to generate 'greater than' or 'less than' signals as well as the equality signal.

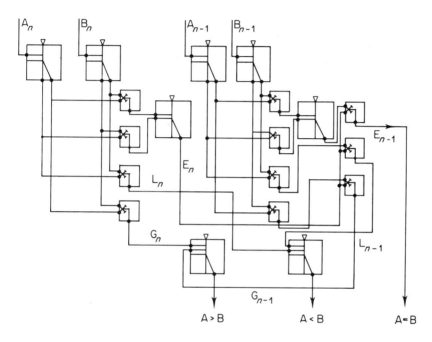

Fig. 14.14 Fluidic comparator with A > B, A < B, A = B outputs

The two highest stages of an 'N' stage comparator are illustrated in the figure. Comparison of the two most significant digits, A_n and B_n in this case, must be made first and the appropriate signal generated depending on their magnitudes. Hence if $A_n > B_n$ a signal G_n is generated which produces an output through a multi-input OR element. Similarly if $A_n < B_n$ the stage output signal L_n results signifying a reversal in magnitude. Clearly, if the most significant digits do not compare any comparison of lower order stages is misleading so the equality signals E_n, E_{n-1}, etc. are arranged to gate lower stages in cascade as shown in the figure. Only when $A_n = B_n$, producing the equality signal E_n, is a similar comparison allowed in the $(n - 1)$ stage. In general, the three output signals in the $(n - 1)$ stage will occur when the following conditions are fulfilled.

$$E_{n-1} = E_n \cdot (A_{n-1} \cdot B_{n-1} + \bar{A}_{n-1} \cdot \bar{B}_{n-1}) \text{ equality} \quad (14.8)$$

$$G_{n-1} = E_n \cdot A_{n-1} \cdot \bar{B}_{n-1} \qquad A_{n-1} > B_{n-1} \quad (14.9)$$

$$L_{n-1} = E_n \cdot \bar{A}_{n-1} \cdot B_{n-1} \qquad A_{n-1} < B_{n-1} \quad (14.10)$$

One disadvantage in the method of producing the 'greater than' and 'less than' signals is that a high fan in to OR elements is required, the actual figure depending on the number of comparator stages.

14.7 Fluidic Comparators

There are several possible circuits for providing the function $A = B$ required by a simple comparator which is comparing two binary numbers A and B. Three of these circuits are shown in Fig. 14.15 and represent single stages of multi-stage comparators.

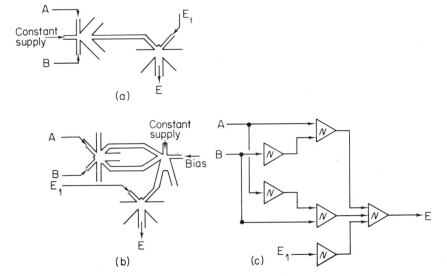

Fig. 14.15 Various comparator circuits giving outputs when $A = B$ (Mitchell and Foster, *Ref. 7*, Fig. 2)

Fig. 14.15(a) illustrates the use of an active EQUIVALENCE gate supplying a passive AND element. If the two inputs are either both 0 or both 1, a signal passes to the AND gate. If also a carry is present from the previous stage, a carry is also produced by this stage. When the complete numbers A and B are the same an equality signal appears at the final output. Fig. 14.15(b) shows an alternative form using an EXCLUSIVE-OR

[14.7] FLUIDIC COMPARATORS 539

function. If A is present but not B, or vice versa, the turbulent reattachment element will switch away from the final AND gate so that no equality signal may be transmitted. However, if A and B are both 0 or both 1, the signal passes to the AND gate and if E_1 is present from the preceding stage, E is produced. The major drawback of these two circuits is that a passive AND gate appears at the output of each stage. If a large number of stages are required, the strength of the equality signal is reduced significantly at each stage and occasional interstage amplification may be necessary. However, they have the advantage that only one active element is required per stage.

For comparison, a circuit using turbulence amplifier elements is shown in Fig. 14.15(c). This produces the same function, but requires six amplifiers, giving a total air consumption for each stage comparable to the other designs.

Economically, the turbulent reattachment element circuit would appear to offer advantages over the latter circuit. In addition, the turbulence amplifier circuit has the disadvantage of significant time delay because an input signal must pass through three turbulence amplifiers, each of which is comparatively slow, before an output appears.

A further circuit designed by Mitchell and others[7] overcomes the objection to gating the carry signal at the output from a stage by gating before an active element.

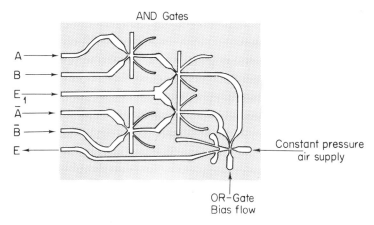

Fig. 14.16 Silhouette of a comparator circuit (Mitchell and Foster, *Ref. 7*, Fig. 4)

Fig. 14.16 shows the silhouette of the comparator circuit stage developed by Mitchell but omits the two monostable wall attachment elements used to generate the two pairs of input signals A, $\overline{A}$ and B, $\overline{B}$. It is seen that the equality signal E, from the preceding higher digit position divides and supplies two passive AND gates which individually gate signals AB and $\overline{AB}$ from two further passive AND gates. The two output signals ABC, and $\overline{AB}$C, then pass into an active wall attachment OR element to provide the stage equality signal E. Thus the equality signal is always amplified between stages and is sufficient to provide a fan-in of 2 to the next AND elements. Against this, the circuit arrangement is more complex and uses three active elements per stage.

Comparator circuits generating A > B and B > A signals as well as equality signal are more complex but several fluidic designs have been reported. Barker and Parker[8] developed one circuit based on the EQUIVALENCE gate, the silhouette of which is shown in Fig. 14.17.

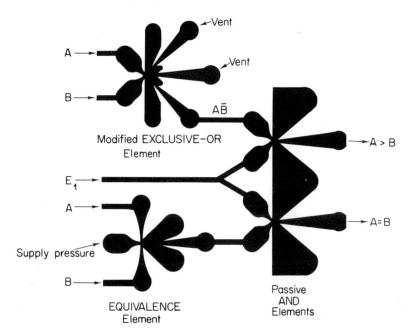

Fig. 14.17 Silhouette of a comparator stage giving A > B and A = B

The equality signal E is generated in an identical manner to the first simple circuit illustrated in Fig. 14.15 using the EQUIVALENCE gate. The A > B signal is obtained by using one channel of a passive EXCLUSIVE–OR to give A$\overline{B}$ and then gating this signal with the higher order equality signal E_1. The stage only requires one active element but the authors experienced difficulty in isolating the inputs to the EXCLUSIVE–OR from the active element.

[14.7] FLUIDIC COMPARATORS

More complex comparator circuits have been investigated by Aviv, Bidgood and Ramanathan[9]. Using commercially available OR–NOR wall attachment elements they constructed a four stage binary comparator as shown in Fig. 14.18.

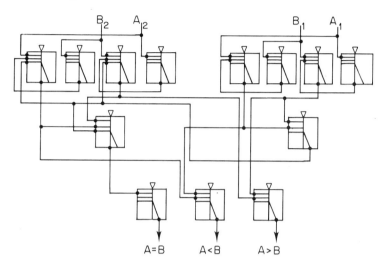

Fig. 14.18 Two stages of a comparator giving $A > B$, $A < B$, $A = B$, based on three input OR–NOR wall re-attachment elements

Each stage operated satisfactorily but was limited to an operating frequency of 200 Hz partly due to the length of interconnecting channels. They conclude that the comparator, with 20 active elements, had an unacceptable power consumption.

Their second more promising approach was to develop a new type of inhibited EXCLUSIVE–OR element for use in the comparator which reduced the number of elements per stage from five active to one active and one passive. Fig. 14.19 illustrates the silhouette of one comparator stage of their final design.

The left-hand element is somewhat similar to a Greenwood Half Adder except that the normal AND function is vented to atmosphere and an additional input acting in the reverse direction provides inhibition. In the design adopted, an inhibition signal is sufficient to prevent any signal combination of the binary inputs A and B appearing in the EXCLUSIVE–OR channels. In addition, the same inhibition signal is sent to a three input OR–NOR wall attachment element, the other two inputs being from the EXCLUSIVE–OR. The OR output provides inhibition for the next less significant stage while the NOR output when present signifies

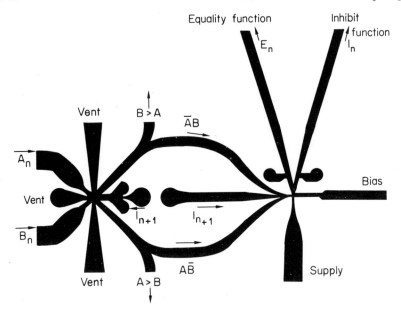

Fig. 14.19 One stage of a comparator based on the inhibited EXCLUSIVE–OR (Aviv, Bidgood and Ramanathan, *Ref. 9*, plate 2)

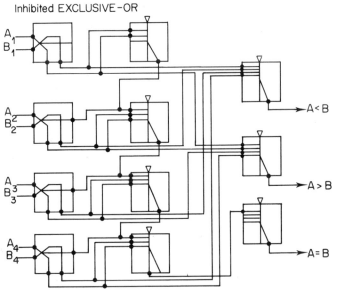

Fig. 14.20 Circuit diagram of a four bit binary comparator using the inhibited EXCLUSIVE–OR

equality. For multi-stage comparators every stage but the most significant must be inhibited as long as the preceding stages are not equal.

Fig. 14.20 shows the logic arrangement of a four stage comparator using the stage logic described above. It is evident that the greater circuit simplicity, as well as appreciably reducing power consumption, should be significantly faster. The authors report a maximum operating frequency of 700 Hz and a time delay for signal propagation through the four stages of 1·3 ms.

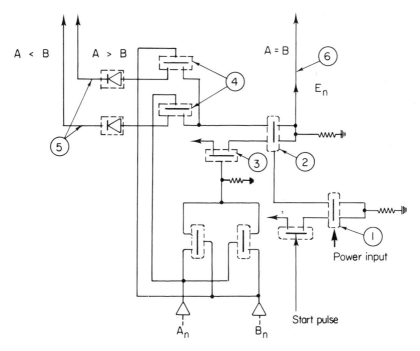

Fig. 14.21 Comparator using foil elements (Bahr, *Ref. 10*)

In contrast, a comparator based on moving part foil devices has been described by Bahr[10] which is illustrated in Fig. 14.21. Binary digits A_n and B_n of the nth stage enter an EXCLUSIVE–OR gate and a start signal, which has to be present during the whole operation, operates the two staged power amplifier marked 1 and 2. The state of the inverter gate 3 determines the position of the foil in 2. If $A_n \neq B_n$, 3 is closed and foil 2 moves to the right allowing an output in one of the channels marked 5 via inverter gates 4. These outputs give the $A > B$ or $A < B$ signals. If $A_n = B_n$, then 3 is open and foil 2 moves to the left generating an equality signal E_n in the form of a pulse through channel 6. No information is available concerning the dynamic capabilities of this circuit although it is probably somewhat slower than the pure fluid counterparts.

14.8 Digital-to-Analogue Converters

14.8.1 General

As yet, a wide range of fluid prime movers compatible with fluidic control systems do not exist, particularly for digital systems. Conventional linear fluid control components, with some modification, are the best available solution at the present time so that a compatible fluidic Digital-to-Analogue (D/A) converter becomes essential for digital closed loop control systems.

The design of a D/A converter has differing performance criteria depending on the type of control system for which it is intended. Two potential areas of application are readily apparent: numerically controlled machine tools and process control systems. For machine tools, the input information rates to the converter are required to be higher than for process control. In addition, the information capacity requirements of a machine tool control system is usually high as the resolution must be retained over large operating ranges. Process controls, on the other hand, operate at lower frequencies and tend to require only moderate information capacity. Typically digit word sizes involved are 10–17 bits for machine tools and 5–9 bits for process controls.

A D/A converter functions to provide a single continuous linear output signal whose amplitude is the direct decimal equivalent of a binary combination at the input to the device. In a fluidic D/A unit the analogue output signal selected could be either a pressure or a flow, dependent on the type of load into which the signal must pass. As in the case of electronic systems, pneumatic D/A converters may be broadly divided into two categories: firstly, resistance or capacitance networks[11] for non-moving part systems and, secondly, force balance systems involving mechanical components.

14.8.2 Pure Fluid Converters

Two techniques which are commonly used in electronics are the current and voltage source ladder networks, as shown in Fig. 14.22[11].

The current source ladder network is widely used for converting a natural binary number to an analogue voltage. The current sources are assumed to have infinite impedance when switched on or off and the load at the output of the network, Z_L, must be large compared to the resistances, although still passing current. Each current source is switched on or off by its corresponding weighted input binary bit so that the output

[14.8] DIGITAL-TO-ANALOGUE CONVERTERS

voltage, e_L, across the load is

$$e_L = \frac{IR}{3 \times 2^{n-1}} \cdot [d_n 2^n + d_{n-1} \cdot 2^{n-1} + \cdots + d_1 \cdot 2^1 + d_0] \quad (14.11)$$

where the value of d is 1 or 0 when the corresponding current source is switched on or off respectively.

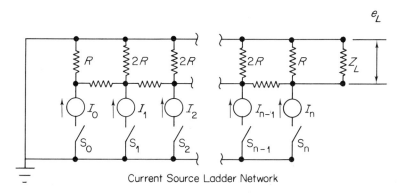

Current Source Ladder Network

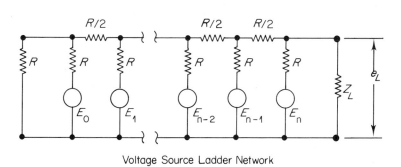

Voltage Source Ladder Network

Fig. 14.22 Electronic ladder networks for D/A converter (Susskind, *Ref. 11*)

A fluidic equivalent of a current source network is not feasible as it is not possible to obtain a fluidic digital device with infinite impedance to reverse flow when switched off. Also the fluid resistances would have to be the capillary type for linearity but designed with a large diamater so that the load impedance was much larger. This would lead to excessively long resistance tubes and large flow rates.

The electrical voltage source network assumes that the voltage sources have zero internal impedance and that the load impedance is infinite. Referring to Fig. 14.22 again, the output voltage, e_L, across the load in this case is

$$e_L = \frac{E}{2^{n+1}} \cdot [d_n \cdot 2^n + d_{n-1} \cdot 2^{n-1} + \cdots d_1 \cdot 2^1 + d_0] \quad (14.12)$$

where the value of d is 1 or 0 when the corresponding voltage source is switched on or off respectively. A digital pure fluidic device as a constant pressure source (analogue of the voltage) is feasible in the switched on condition but generally the impedance is low to reverse flow when the device is in the off condition. This could be improved by the use of a foil element in conjunction with a wall reattachment device provided some degradation in performance is acceptable.

The simplest form of pure fluid D/A conversion, first used by Turnquist[12] is a flow network utilizing the pressure summation of parallel resistance discharging into a downstream volume, as illustrated in Fig. 14.23.

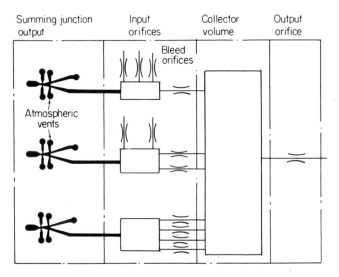

Fig. 14.23 Three stages of a pure fluid D/A converter (Parker and Jones, *Ref. 13*)

Each resistance is associated with a particular flow rate depending on the particular weighing of the binary number it represents. Thus, successively from the least significant digit position the relative flow magnitudes through the resistances are 1, 2, 4, 8, 16, etc. Provided the relationship between pressure and flow is always linear across a restriction, the downstream chamber pressure is the analogue of flow resistance code being used, i.e. the digital signals controlling the flow through the weighted

[14.8] DIGITAL-TO-ANALOGUE CONVERTERS

resistances is directly related to the analogue downstream pressure. Although Turnquist used a low impedance load at the output of the D/A the same circuit can be used for high impedance loads. This is particularly important as the servo-amplifiers required by prime movers present this type of load condition to the converter.

The linearity of the converter will depend upon the type of resistances used in the input and output stages. For example, a design employing linear resistances at the input and output will necessarily have an overall linear characteristic. However, resistances with linear characteristics can only be achieved at low flow Reynolds numbers by using, for example, long narrow capillary tubes. The two chief disadvantages of this arrangement are that dynamic signals are severely attenuated thus impairing the action of the converter and also large numbers of capillary tubes are physically very space consuming. Possible ways of overcoming these difficulties are to use either non-linear orifices, or some combination of linear and non-linear orifice, which must be compromised with the overall converter linear performance.

When either linear capillaries or non-linear orifices are used in the input stages to a converter the binary weighting is best achieved by grouping numbers of identical orifices together in the ratio's 1, 2, 4, 8, etc. In this way uniformity of the pressure-flow characteristic is maintained. Normally the subtractor unit providing the digital error signal to the D/A converter would have wall reattachment output elements operating at constant supply pressure. This implies that the input pressure to each converter stage would vary, dependent on the load impedance presented by the resistances to the flow.

A method of equalizing the impedances of the converter input stages is easily achieved by providing bleed resistances which discharge to atmosphere. The bleed resistances are coupled in parallel to the input resistances such that the impedance of each stage is constant. Providing that the output load impedance of the converter is low the pressure drops across the bleed resistances approximates to that across the inputs. Alternatively, if the load impedance is high, the bleed resistances may discharge through a similar impedance rather than directly to atmosphere.

When the input resistances are not receiving flow from the subtractor unit they are subjected to reverse flow from the downstream chamber. Provided the input resistances are linear and the upstream wall re-attachment elements have adequate venting, the converter output remains linear. However, when non-linear orifices are used, forward and reverse flow characteristics will vary widely adding further to the difficulties of providing a linear output.

Reverse flow may be prevented by providing free foil diode isolaters at the input to each stage of the converter. This may improve the output characteristics of the converter when non-linear orifices are being used or, alternatively, when the wall reattachment elements supplying flow to the converter have badly matched characteristics.

In practice, fluidic D/A converters may be expected to be limited in usage by dynamic considerations but also by the engineering complexity of providing large numbers of identical orifices in a reasonable space. It is clear that, for example, a converter of 5 bit capacity using identical orifices requires 31 input orifices and 49 matching bleed orifices, which represents a reasonable practical limit above which matching and packaging becomes a problem. For a fluidic converter to be used in machine tool applications a limitation on the proportional operating region must be accepted. Process control requirements, on the other hand, may well be met with further development of converters, particularly if compromises are possible in the number of identical input orifices required. A further factor in favour of the latter application is that dynamic requirements are very much reduced, a maximum total count frequency of 1 Hz being typical.

Parker and Jones[13] considered the general case when the resistances may be linear or non-linear and foil diodes used to prevent reverse flow. Table 14.4 shows the possible resistance combinations.

TABLE 14.4 *Possible D/A resistance combinations*

Type	Input resistance	Output resistance	Diode
1	Linear	Non-Linear	No
2	Linear	Non-Linear	Yes
3	Linear	Linear	No
4	Linear	Linear	Yes
5	Non-Linear	Non-Linear	No
6	Non-Linear	Non-Linear	Yes
7	Non-Linear	Linear	No
8	Non-Linear	Linear	Yes

Types 7 and 8 are special cases of Types 5 and 6 yielding similar characteristics and so need not be considered further. Each of the Types 1 to 6 may be analysed, on a flow basis as shown in Fig. 14.24, where T is the total count capacity of the converter and N is the actual count.

Theoretical curves for each of the six system types for a 5 bit converter ($T = 31$) based on an upstream pressure of $P_1 = 0.25$ lb. in^{-2} are shown in Figs. 14.25, 14.26 and 14.27 for widely differing load conditions.

In all cases the blocked load conditions is illustrated while for low

[14.8] DIGITAL-TO-ANALOGUE CONVERTERS

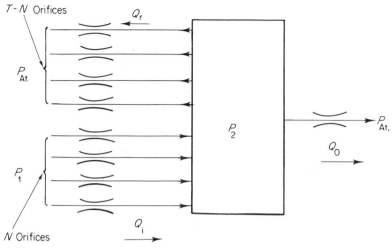

Linear input flow, $Q_i = K_i(P_1 - P_2)$
Non Linear output flow, $Q_o = K_o \sqrt{P_2}$
Non Linear input flow, $Q_i = K_A \sqrt{P_1 - P_2}$

Fig. 14.24 Flow-diagram for a converter (Parker and Jones, *Ref. 13*)

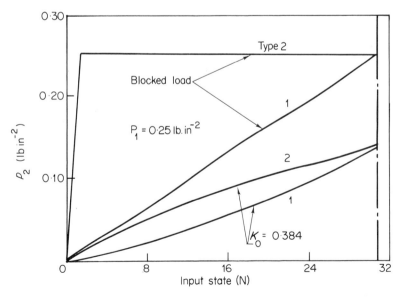

Fig. 14.25 Linear input, linear output (Parker and Jones, *Ref. 13*)

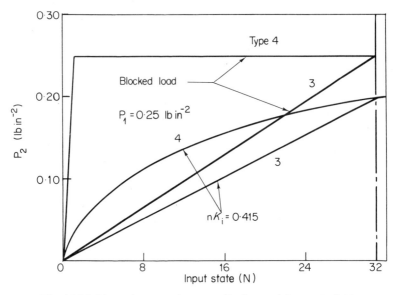

Fig. 14.26 Linear input and output (Parker and Jones, *Ref. 13*)

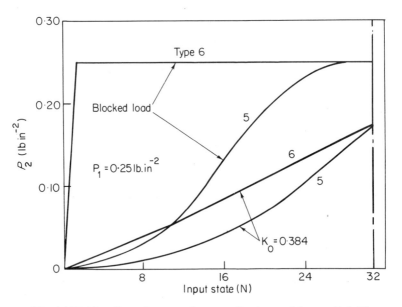

Fig. 14.27 Non-linear input and output (Parker and Jones, *Ref. 13*)

[14.8] DIGITAL-TO-ANALOGUE CONVERTERS 551

impedance loads in the non-linear orifice condition an 0·081 in. diameter orifice with an experimental value of $K_0 = 0.384$ was taken. An arrangement of 10 standard capillaries were used for output loading in the linear case giving $nK_1 = 0.415$. A continuous curve has been drawn for these functions whereas, of course, they should have a 'staircase' form due to discontinuities between each number input.

The curves predict that whenever a linear input resistance is used a linear output should be possible over a wide range of load conditions. Under such conditions the use of diode isolators is detrimental. However, the situation is significantly altered when the input resistances are non-linear, as an approximately linear characteristic is only achieved using diode isolators (Type 6) under lightly loaded conditions. In practice the load could be increased from this value by also loading the matching bleed resistances in a similar manner, as described earlier. Such modifications were not derived in the theoretical treatment, the extension being straightforward.

An experimental 5 bit D/A converter, using linear input resistances manufactured in Dycril plastic, was operated under identical conditions to those used in the theoretical predictions. Not surprisingly, difficulty was experienced in matching different sets of weighted capillaries to give exactly the same resistance characteristics. As the input count N changes, different combinations of capillary are used giving rise to output pressure discontinuities when matching is poor. This is particularly noticeable when changing, for example, from decimal 7 to decimal 8 as there is a flow redistribution from three banks of capillaries to one larger one.

Fig. 14.28 illustrates the correlation obtained between theory and experiment for Type 1 system. As anticipated, the largest output discontinuities were experienced in the experimental converter in changing from 7 to 8 and also 15 to 16. It may further be added that the foil diodes used were not ideal in that they allowed a small reverse flow. This was due to the very low operating pressures of the system.

The authors found that the dynamic performance of the D/A converter configurations tested were compatible with existing subtractor information transfer rates, the maximum operating frequency in the least significant digit position being 200 Hz.

Good linear fluidic converter static characteristics are possible over a wide range of load conditions if linear resistance capillaries are used for flow weighting without diode isolators. The important disadvantages in practice are that the large numbers of identical capillaries required are very difficult to match exactly and packaging becomes a problem. For most applications a 5 bit converter would seem to be the largest capacity unit which could be used at the present state of the art.

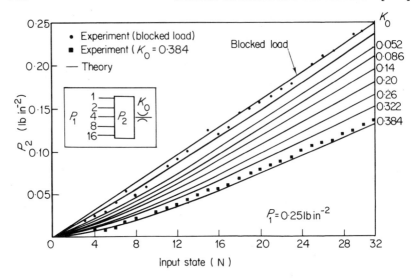

Fig. 14.28 Characteristics of a type 1 pure fluid D/A converter (Parker and Jones, Ref. 13)

A promising approach to overcome these disadvantages appears to be to use non-linear orifices in conjunction with diode isolators to obtain approximately linear characteristics with substantially fewer resistances. Load matching of the vent orifices would also be required.

14.8.3 Moving Part Converters

Because of the severe limitations in accuracy and construction difficulties of the pure fluid converters, a moving part device with built-in geometric weighting to the digit values would appear promising. Two types of moving part force balance converter are used. Firstly, systems which generate digital displacements applied to a spring, thus producing in it a force proportional to a digital value. Secondly, systems which generate a digital force proportional to the input number. In both systems the force generated is balanced with an analogue pressure signal.

Jakubowski[14] has considered converters of both types, suitable for process control applications with resolutions of better than one part in one thousand (i.e. ten stage converters). By using a switchable constant supply pressure to diaphragms with relative area ratios of 1, 2, 4, 8, etc. the required force weighting for a force balance device can be achieved. The author found that a stacked single acting diaphragm capsule, as shown in Fig. 14.29, was the best form of construction.

Although the static accuracy achieved was less than ± 0.2 per cent, due

[14.8] DIGITAL-TO-ANALOGUE CONVERTERS

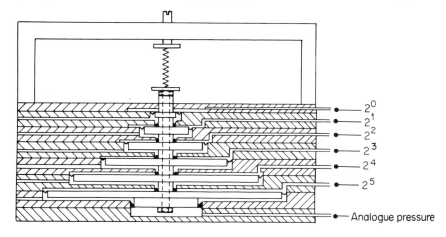

(a) The single-acting diaphragm capsule

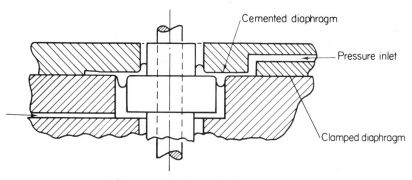

(b) Enlarged detail showing formed diaphragms

Fig. 14.29 Stacked diaphragm force balance capsule (Jakubowski, *Ref. 14*, Figs. 2a and b)

to mechanical imperfections, improvements in design could improve this to ± 0.1 per cent. Fig. 14.30 shows the way in which the force balance device is used in the D/A converter to form a simple closed loop control system. The register holds the binary number which is to be converted to an analogue pneumatic signal in the range 3–15 lb. in^{-2}. When a digit is switched on the reference pressure of 1 lb. in^{-2} is applied to its particular weighted area diaphragm which generates an axial force on the piston assembly thereby displacing it relative to the capsule body. This displacement decreases the flapper nozzle opening, increases the back pressure which in turn is amplified by the relay. The output pressure is fed back into the analogue pressure connexion of the capsule (Fig. 14.30) which produces a force balance between the digital input and analogue output.

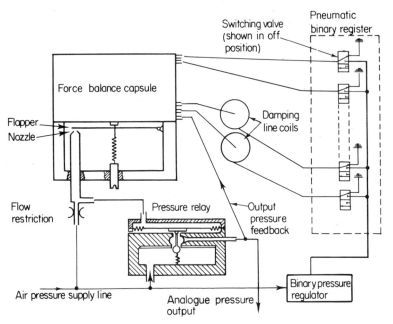

Fig. 14.30 D/A converter using a force balance capsule (Jakubowski, Ref. 14, Fig. 3a)

The same author gives details of a digital displacement system operating against a spring in which an electrical stepping motor provides the incremental spring displacements.

Fig. 14.31 shows the converter which contains a digital rather than analogue, feedback loop. The digital input information is compared with the pneumatic output of a mechanical binary counter driven by the stepping motor. When the binary input is greater than the counter number the stepping motor is driven in the direction of increasing counter total. This increases the spring load on the balance beam, decreases the flapper nozzle opening, and increases the relay output pressure. Another loop feeds back the analogue output pressure to the bellows which restores the linkage force balance. Jakubowski found that the resolution possible with this system was greater than with the stacked diaphragm arrangement, 11 to 12 bit conversion being possible. The speed of conversion with the stacked diaphragm arrangement is limited by the pneumatic relay with typically a response time of 0·1 s to a 10 per cent digital input

[14.8] DIGITAL-TO-ANALOGUE CONVERTERS

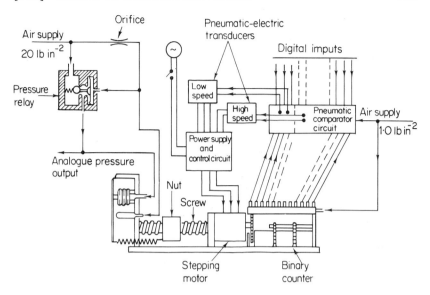

Fig. 14.31 Hybrid D/A converter based on a stepping motor (Jakubowski, *Ref. 14*, Fig. 8)

signal change. The stepping motor converter depends upon the maximum counting frequency of the mechanical counter, which in the system described was 70 Hz.

14.9 Static Hazards in Fluidic Circuits

14.9.1 Introduction

Earlier in the chapter it was shown that two fundamentally different modes of operation are possible for digital switching systems; namely the asynchronous, or free-running, mode in which the speed of operation is governed by the circuit characteristics and the synchronous mode, in which the speed is governed by a 'clock' or synchronizing pulse. If the asynchronous mode is used the steady-state behaviour of a circuit is adequately described by the Boolean representation of the logic used. However, during the transient state when the system is responding to a change in input variables the Boolean algebraic expression may be invalidated, since no account is made of the finite time delays due to switching and signal propagation. During this transition period the actual output may be quite different from the steady state value and 'hazardous' operation is said to have occurred.

The effect of such hazards on overall system performance depends upon whether the circuit has memory. Hazards may not have a permanent effect upon the steady-state output in combinational circuits, although if the circuit is of a sequential nature, i.e. one which acts as if it had memory due to feedback connexions, then the system may adopt an improper stable state.

In the comparatively new field of fluidics the problem of hazardous operation is likely to occur, particularly in complex circuits, as it does in other logic media. Although the theory of switching circuits is well established for steady-state behaviour[15], the theory of transient performance is less well known. In particular the identification and elimination of the various types of hazard is described in only a few recent papers[16,17] relating to advanced electronic and relay switching systems. With the exception of elementary hazard problems there is no straightforward guide to enable the designer of engineering switching circuits to identify and eliminate hazards.

In certain asynchronous circuit applications where the input information sequence is between adjacent states, i.e. where only one input variable changes, complete hazard elimination is possible by introducing additional redundant logic or time delay. However, in multi-variable input applications complete hazard elimination is not possible and this may necessitate the use of a synchronous mode of operation.

The need for suitable methods of designing for a hazard-free operation has been demonstrated in relation to asynchronous binary subtractor logic[5] where hazards result in spurious output transients. An attempt has been made to identify and eliminate such hazards in general and develop formal design techniques for hazard-free operation. Emphasis is given to the types of hazard considered to be most prevalent in fluidic circuits, although the methods are perfectly general.

14.9.2 Types of Hazard

In asynchronous switching circuits there are four main types of hazard that may be generated:

(*a*) Static Hazard.
(*b*) Dynamic Hazard.
(*c*) Essential Hazard.
(*d*) Critical Race Hazard

In defining the nature of each of the various types of hazard, the discussion will be limited to those that arise from a single change of input variable. The static hazard[16] occurs in both combinational and sequential circuits and is characterized by a spurious transient output which occurs when a change between adjacent input states takes place which should

[14.9] STATIC HAZARDS IN FLUIDIC CIRCUITS

ideally result in no change in the output state. Similarly the dynamic hazard can occur in both combinational and sequential circuits although its effect is not so great as that of static hazard. The dynamic hazard is revealed as a spurious output transient which occurs when the output state is changing from 1 to 0 or from 0 to 1. It will be characterized by an output change of at least the form 1010 or 0101. The essential difference between static and dynamic hazard is shown in Fig. 14.32. Whereas the static hazard can result in an improper stable state being adopted in a

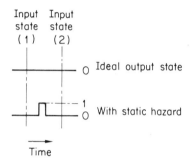

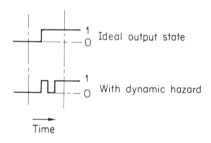

Fig. 14.32 Typical form of hazards on system timing chart

circuit with memory, the dynamic hazard usually will not affect the static output of the circuit. Both the static and dynamic hazard of the single variable type may be eliminated logically or by the insertion of delay in the circuit.

The essential hazard[18] occurs only in sequential circuits and is due to delay introduced in the feedback loops of the circuit. The resulting action is such that an output will adopt an improper stable state when ideally no change should occur. This type of hazard is not due to the particular logic design and is inherent in the structure of certain sequential problems.

Complete elimination is achieved by the insertion of delay but can only be partially diminished logically.

A further hazard peculiar to sequential circuits is better known and is referred to as a critical race hazard. Such a hazard can arise when two or more feedback variables attempt to change simultaneously at a given input change. Due to differences in response one variable may change before the other, resulting in the state of the variables being incorrect. The feedback paths in such a sequential system may either be in the form of a transmission line or as secondary logic elements. If one sequence of response occurs the feedback variables may change through a sequence of unstable states before adopting the required stable state. Alternatively, the sequence of response may result in an incorrect stable condition. Hence a race occurs as to which stable state occurs first and is said to be critical as an incorrect state may occur. The elimination of critical race is readily achieved by the insertion of delay in the appropriate feedback loop, to ensure that a non-critical sequence of response is adopted. Methods of location and elimination of critical race hazards have been fully described by Caldwell[15] and will not be discussed further.

In general the types of hazard most likely to be encountered in fluidic digital systems, are the static and dynamic hazard. Of these two types, the dynamic hazard is not likely to cause system malfunction to asynchronous Full Adder and Subtractor logic.

14.9.3 Static Hazards

A static '1' hazard is defined as a spurious '0' transient which occurs when the output of a circuit should be '1' before and after a change in input state. The inverse of this statement defines the '0' hazard.

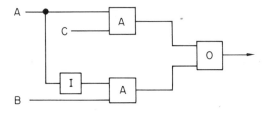

Fig. 14.33 Logic circuit for $\bar{A}B + AC$

The nature of static hazard is shown in the circuit of Fig. 14.33 in which the presence of an inverter gate will introduce a time delay such that the conditions, $A = 1$, $\bar{A} = 1$ or $A = 0$, $\bar{A} = 0$, can exist which are incompatible. Examination of the timing chart of Fig. 14.34 reveals that an input change from ABC = 111 to 011 will cause a spurious transient in

[14.9] STATIC HAZARDS IN FLUIDIC CIRCUITS

the output. Such a hazard is identified as a single-variable static '1' hazard which occurs when a change between adjacent input states takes place.

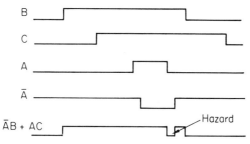

Fig. 14.34 Timing chart for $\bar{A}B + AC$

In order to identify possible hazards in a circuit it is necessary to adopt a method that will describe the transient behaviour of the system. The method adopted is adequate for the location of hazards, although more detailed methods of describing transient circuit behaviour have been developed[17].

In the example of Fig. 14.35, the AND–OR logic circuit has a transmission function f of the form

$$f = ABC + (A + D)(\bar{A} + \bar{C}) \qquad (14.13)$$

which may be expanded into Sum of Products form as

$$f = ABC + A\bar{C} + \bar{A}D + \bar{C}D + A\bar{A} \qquad (14.14)$$

to give the first four terms only in the steady state as the redundant term $A\bar{A}$ is zero. However, this may not be so in the transient state due to different signal propagation delays so that, for example, A and $\bar{A}$ may both be 1 for a short period of time. Thus, in considering the transient transmission function, complementary literals (A and $\bar{A}$) must be used as separate variables and minimization of terms such as ABC to AB are not possible.

If it is necessary to identify signal delays a convenient notation is to number each individual connecting path as shown in Fig. 14.35. In fluidic systems connecting path delay times will normally be greater than or comparable with element switching time so that both delays are assumed lumped together in each numbered path. Using identified paths, the transient switching function of equation (14.14) becomes

$$f = A_{1,8}B_{2,8}C_{3,8} + (A_{4,9,11} + D_{5,9,11})(\bar{A}_{6,10,11} + \bar{C}_{7,10,11}) \qquad (14.15)$$

thus allowing each hazard cause to be identified.

An alternative form for the transient function is obtained if the product of sums form is used. In this case equation (14.13) becomes

$$f = (A + D)(A + \bar{A} + \bar{C})(B + \bar{A} + \bar{C})(C + \bar{A} + \bar{C}) \quad (14.16)$$

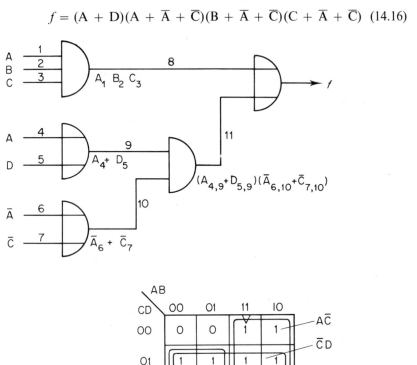

Fig. 14.35 Example of a static hazard network

Having expanded the transient transmission function into the sum of products form of equation (14.14) the individual sets ABC, $\bar{A}D$, etc., may be examined for possible hazards using the map technique of Huffman[16] or a tabular method described by McCluskey[17]. In the sum of products form of the transient function the individual sets are referred to as 'P' sets and will generate a '1' output only if all of the literals in the

[14.9] STATIC HAZARDS IN FLUIDIC CIRCUITS

group are '1'. Similarly the transient function may be represented in the product of sums form and the sets become 'S' sets which will give a '0' output when all of the literals in the set are equal to '0'. The P and S sets do not necessarily correspond to minimized steady state terms as the treatment of complementary literals as separate variables invalidates many of the standard Boolean reduction formulae.

The concept of 'P' and 'S' sets is useful since hazards may be determined from either and in certain circumstances it is easier to obtain the transient function in one form rather than the other. Sets which include a variable and its complement are referred to as unstable sets, e.g. $A\bar{A}$, and those which do not as stable sets, e.g. $A.B.C$.

The general conditions for indicating the presence of a static hazard may be formally stated in the following way.

A static 1 hazard (0 hazard) exists in a network if and only if:

(a) There is a pair of adjacent input states which both produce 1 outputs (0 outputs) in the static case.
(b) There is no stable P set (S set) of the network which covers both the input states of the pair. Since unstable P sets (S sets) cannot cover any input states, they automatically give rise to potential 0 hazards (1 hazards).

If the circuit under examination involves less than 6 variables the map method of Huffman may be used to locate potential hazards. In the example of Fig. 14.35 the transient function given by equation (14.14) gives four stable P sets

$$ABC \quad \bar{A}D \quad \bar{C}D \quad A\bar{C}$$

and on unstable P set

$$A\bar{A}$$

The figure also shows the Karnaugh map for the transmission function in the static case with the stable P sets marked on it. Note that the suffices have been omitted as they only indicate the particular path through which the signal passes in the network. Examination of the map shows that the transient function may give rise to static 1 hazards in the transition between $ABCD = 0111$ and 1111, 1111 and 1101, 1110 and 1100 as in each case there is a pair of adjacent input states not covered by stable P sets which have 1 outputs. The unstable P set $A\bar{A}$ will give a static 0 hazard if there are two adjacent input states which both produce a 0 output and differ only in the value of the A variable. Inspection of the map shows that the pair of input states 0010 and 1010 satisfies these conditions and therefore gives rise to a static 0 hazard.

The same results could have been obtained by expanding equation

(14.16) into the product of sums form to give the equivalent S sets, finding the transition points for the stable S sets to give static 0 hazards, and finally inspecting the unstable S sets for static 1 hazards. However, synthesis by this procedure tends to be slightly more complicated as a whole so that the easily found P set approach is usually preferred.

To eliminate the 1 hazard in the above example, additional redundant logic may be inserted to cover the adjacent states responsible for the hazards. From the map it can be seen that an additional logic block BCD prevents any hazard between states 0111 and 1111 and a logic block AB eliminates both hazards between 1111 and 1101, 1110 and 1100. The static 0 hazard caused by $A\bar{A}$ may be removed by modification of the term to $A\bar{A}B$ as B is always 0 during the change. Insertion of delays may be used instead of redundant logic to remove hazards.

If a fluidic circuit were used to realize the logic function of Fig. 14.35 the connecting channels between the elements may be of various lengths and, as a result, introduce finite different propagation delay times. Assuming the element switching times are negligible compared with the connecting line delay times, the width of the hazard signal is related to the difference in path length through which the two critical signals must pass. An example

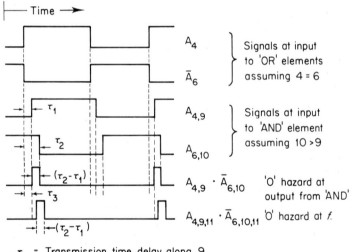

τ_1 = Transmission time delay along 9
τ_2 = Transmission time delay along 10
τ_3 = Transmission time delay along 11

Circuit from fig 14.35 with B = 0, C = 1, D = 0

Fig. 14.36 Effect of transmission time delay on hazard generation

[14.9] STATIC HAZARDS IN FLUIDIC CIRCUITS 563

from the figure illustrates the mechanism. Assuming that $B = 0$, $C = 1$, $D = 0$ a 0 hazard arises due to the changes in variable A, as discussed earlier. The signals along each of the critical lines 9 and 10 can be represented as shown in Fig. 14.36, assuming that 10 is larger than 9.

14.9.4 Static Hazards in Asynchronous Parallel Binary Subtractor Logic

Parker and Jones[5] have described the development of integrated fluidic adder and subtractor circuits. They found that the operation of such a circuit in the asynchronous mode caused spurious output transients to occur due to carry propagation delays inherent in the circuit. On a logic basis the transients may be formally identified as a particular type of hazard.

The behaviour of the nth stage of 'm' stages of subtractor logic, shown in Fig. 14.37, is given by the following equations:

$$S_n = \overline{A}_n\overline{B}_nC_n + \overline{A}_nB_n\overline{C}_n + A_n\overline{B}_n\overline{C}_n + A_nB_nC_n \quad (14.17)$$

$$C_n = A_nB_n + A_n\overline{B}_nC_n + \overline{A}_nB_nC_n \quad (14.18)$$

$$D_n = C_{m+1} \cdot S_n + \overline{C}_{m+1} \cdot \overline{S}_n \quad (14.19)$$

Each of these equations is in the sum of products form and give rise to static hazards even though there are no unstable P sets. In the case of equation (14.17), the variables A_n, B_n and C_n all give rise to possible static hazards, as does equation (14.18) with respect to A_n and B_n, and equation (14.19) with respect to C_{m+1} and S_n. Since more than one variable can cause static hazards and the logic is based on the binary code which involves more than one variable changing at a time, the hazards become multi-variable static hazards. In particular the hazards generated as a result of equations (14.17) and (14.18) have the greatest effect since long duration transients can be produced.

The worst case of static hazard generation is associated with the Full Adder logic in a fluidic circuit due mainly to accumulative delays suffered by the carry signal C as it passes through several stages. The mechanism of hazard generation can best be seen by selecting an example from the first two stages of an adder circuit. Fig. 14.38 shows the circuit operating under the specific conditions of $B_0 = B_1 = 1$. It can be seen that when a time delay of τ_1 occurs in the propagation of the carry signal C_1 from the first to the second stage, static '0' and '1' hazards of the same time period are introduced into the sum signal S_1 and one of the carry signals C_2^*. In following stages, not only are existing hazards propagated but fresh conditions for hazard generation are also created which generate longer duration transients.

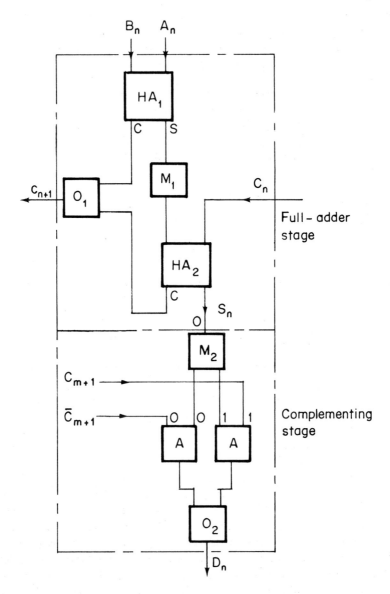

Fig. 14.37 Fluidic binary subtractor stage

[14.9] STATIC HAZARDS IN FLUIDIC CIRCUITS

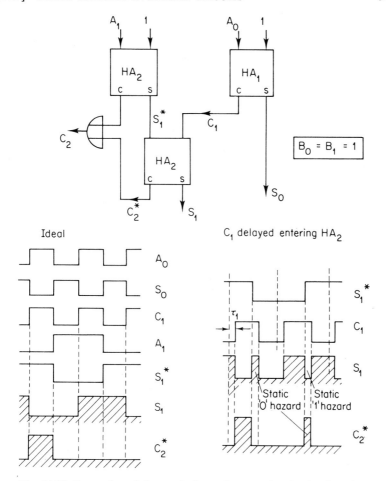

Fig. 14.38 Examples of the static hazards occurring in the first two stages of an adder circuit

Since the hazards are of a multi-variable type, complete elimination either logically or by insertion of delay is not possible and a synchronous mode of operation must be considered. This means that the output signals from the subtractor stages are gated with a clock pulse that only permits read-out when sufficient time has elapsed for all transients to have occurred.

14.9.5 Methods of Synchronous Protection Against Static Hazards

In the case of binary Full Adder logic the signal propagation delay is made up of two components; one due to carry signal propagation delay,

τ_1, and a processing delay, τ_2, in each stage before a difference in output is generated. The maximum delay, $m\tau_1 + \tau_2$, for 'm' stages occurs when a carry is propagated through all stages and determines the period in which hazard can have effect. For this case a set of input information applied to the Full Adder block will produce an output from the least significant stage after a delay τ_2. All outputs are complete when the most significant stage generates an output after a total delay of $m\tau_1 + \tau_2$. In the case when the subtraction function is to be performed an additional delay through the complementing circuitry will be involved denoted by τ_3. The total delay will, therefore, be $m\tau_1 + \tau_2 + \tau_3$ in this case.

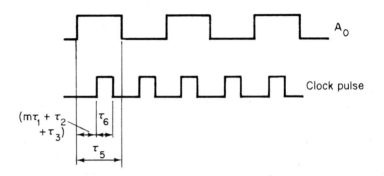

Fig. 14.39 Clock pulse timing

For synchronous operation a clock pulse can be used to gate the outputs, that is lagging behind the input information by a time interval of $\tau_4 = m\tau_1 + \tau_2 + \tau_3$, which covers the worst case. The duration of the clock pulse should be such that it ceases before the input information changes state. Referring to Fig. 14.39 this condition can be formally stated as:

$$\tau_6 \leq \tau_5 - (m\tau_1 + \tau_2 + \tau_3)$$

Thus the general criteria for synchronous operation may be stated as:
(1) The clock pulse should lag behind the input information by an interval of at least $m\tau_1 + \tau_2 + \tau_3$.
(2) The duration of the clock pulse should not exceed the period

$$\tau_5 - (m\tau_1 + \tau_2 + \tau_3).$$

(3) The frequency of the clock pulses should be twice the frequency of the least significant digit of information, so as to provide a clock pulse each time the input changes.

[14.9] STATIC HAZARDS IN FLUIDIC CIRCUITS

For position control systems with low resolutions it is possible to accommodate an extra row of information on the disc to generate the clock pulses for the synchronizing circuit. For high resolution systems, which is more likely to be the case, it becomes difficult to use this method. A more general method adopted involves the use of pulse shaping circuits which can be triggered off the least significant row of the encoder.

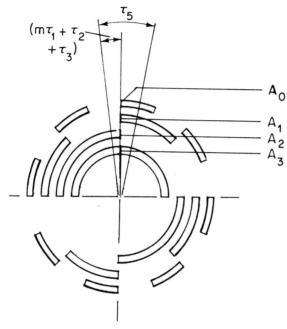

Fig. 14.40 Binary pattern encoder

Since a clock pulse is required for each word of binary information contained on a binary encoder it is necessary to have two additional receivers. The arrangement shown in Fig. 14.40 relates to an interruptable jet type encoder. The physical displacement of one receiver relative to the receiver A_0 will correspond to the delay $m\tau_1 + \tau_2 + \tau_e$, based upon the maximum speed of operation of the encoder. The displacement between the clock pulse receivers corresponds to one least significant digit, A_0, of information.

The outputs of the clock pulse receivers are coupled to pulse shaping circuits, which generate an output of fixed pulse width. The outputs of the pulse shaping circuits are combined in a wall reattachment OR device, producing clock pulses of the required frequency which are used to synchronize the subtractor outputs by gating in a passive AND gate. The resulting synchronized output D'_2 of the 2nd stage output of a 6 stage subtractor, as shown in Fig. 14.41 is a time representation of the difference between input signals over the finite interval of the clock pulse. Since the possible mode of operation involves coupling the subtractor block to a

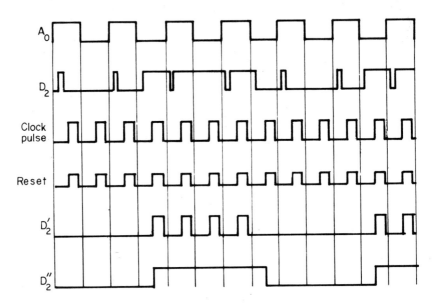

Fig. 14.41 Timing chart for the 2nd stage of a six-stage synchronous binary subtractor

D/A converter, definite advantages are to be gained from restoring the unit pulse width of the synchronized output to that of the least significant digit of input information. This is indicated by D''_2 of Fig. 14.41. One method of restoring the synchronized output is shown in Fig. 14.42 where clock pulses from the two input OR are divided so that a gating signal is passed via a restrictor to provide a reset signal for a bistable wall reattachment device. The amplitude of the reset signal is chosen so as to reset the bistable device in the absence of a synchronized output from the AND gate. The bistable generating D''_2 is set to the '1' state upon reception of a synchronized output D_2 from the AND gate and remains in this state until a reset signal occurs, without the presence of a signal from the AND. Such a restored output is free from the effects of hazard and is delayed relative to the input information by an amount, τ_0, which is composed of individual propagation delays.

[14.9] STATIC HAZARDS IN FLUIDIC CIRCUITS

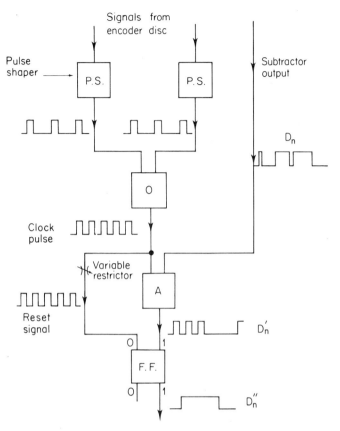

Fig. 14.42 Synchronizing circuit

$$\tau_0 = m\tau_1 + \tau_2 + \tau_3 + \tau_7$$

where τ_7 is the delay through the synchronizing circuit.

This method of synchronous operation has been applied to 6 stages of binary fluid subtractor by Parker and Jones. An undesirable feature of this method of restoring the synchronized output is due to the bistable device being operated in a 'two-level' or 'over-ride' condition which can result in a unstable state. An alternative design is shown in Fig. 14.43 where the reset signal is generated by an EXCLUSIVE–OR device. The dynamic performance of this circuit is satisfactory.

The overall effect of the additional synchronizing circuitry of the type shown in Fig. 14.42 upon the maximum operating frequency is very small. The additional delay introduced is limited to the signal propagation path

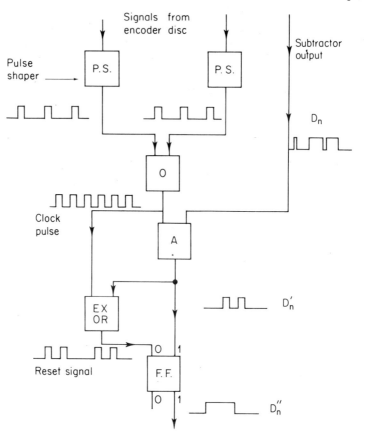

Fig. 14.43 Synchronizing circuit

through a passive AND gate and a bistable device. This delay is of the order of 1·5 ms. With the six stages of synchronous subtractor tested the frequency of operation is of the order of 50 Hz.

Although the foregoing description has been confined to parallel subtractor logic the synchronizing techniques described are equally applicable to other types of logic where hazards are likely to occur. In particular hazard problems are likely to occur with comparator logic, of the type that generates a sense of error and an equality signal, due to inter-stage inhibition signals.

References

1. Blaiklock, P., and Sidel, R., Development of a pneumatic stepping motor system, *M.I.T. Dept. Mech. Eng.*, Summer course notes, 2.73 (1966).
2. Griffen, W. S., and Cooley, W. C., Development of high speed fluidic logic circuitry for a novel pneumatic stepping motor. *A.F.*, **1967**, 402.
3. Martin, R. P., A pneumatic stepping motor, *M.Sc. Thesis* (1965), Pennsylsylvania State Univ.
4. Richards, R. K., *Arithmetic operations in digital computers*, Van Nostrand, 1961.
5. Parker, G. A., and Jones, B., A fluidic subtractor for digital closed loop control systems, *C.F.C.* (2), Paper H1.
6. Glaettli, H. H., Circuits using fluid dynamic components. *C.F.C.* (1), Paper D4.
7. Mitchell, D. G., and Foster, K., The design of a comparator for use in a pure fluid control system, *Mech. Eng. Rep.*, 75, Birmingham.
8. Barker, R., *M.Sc. Thesis* (1965), Birmingham.
9. Aviv, I., Bidgood, R. E., and Ramanathan, S., The design and construction of a pure fluid four-bit binary comparator, *C.F.C.* (2), Paper H6.
10. Bahr, J., The foil element, a new fluid logic element, *Electronische Rechersanlagen*, 7, No. 2, 69–78, (1965).
11. Susskind, A. K., *Notes on Analog Digital Conversion Techniques*, John Wiley & Sons, 1957.
12. Turnquist, R. O., *A fluid state digital control system*, Rep. EDC 7–65–8, Eng. Design Center, Case Inst. of Technology, U.S.A.
13. Parker, G. A., and Jones, B., Fluidic digital to analogue converters, *C.F.C.* (3), Paper K1.
14. Jakubowski, M., Pneumatic Analog–Digital and Digital–Analog converters, *ASME Paper*, 68–WA/AUT–16.
15. Caldwell, S. H., *Switching Circuits and Logical Design*, Wiley, New York, 1963.
16. Huffman, D. A., Design of hazard-free switching circuits, *J. Assoc. Comp. Mach.*, **4**, 47–62 (1957).
17. McCluskey, Jr., E. J., Transients in Combinational Logic Circuits, in *Redundancy Techniques for Computing Systems*, Spartan Books, Washington, 1962.
18. Unger, S. H., Hazards and delays in asynchronous switching circuits, *I.R.E. Trans. Circuit Theory*, **CT-6**, (1959).
19. Parker, G. A., and Jones, B., Protection against hazards in fluidic adder and subtractor circuits. *I.F.A.C.*, Paper B2, London (1968).

15
Graphical Techniques Applied to Elements and Circuits

15.1 General

The fluid mechanics of system components such as resistors, volumes and inductances were discussed in Chapter 2 while analogue and digital amplifiers were described in Chapters 4 and 7 respectively. In many cases the characteristics relating pressure and flow through a device were non-linear and mathematically cumbersome. As in electronics, it is often advantageous in fluidics to use graphical techniques for matching components of this nature together. This is particularly useful for complex circuits as it avoids involved analytic solutions.

Generally speaking, three sets of characteristics completely specify the static performance of a fluid component. These are:

(a) Input characteristic.
   A pressure-flow graph which provides information about control impedance when acting as a load on other elements.
(b) Transfer characteristic.
   An input-output pressure or flow graph giving gain performance.
(c) Output characteristic.
   A pressure-flow graph giving information concerning the output load conditions.

15.2 Digital Element Input Characteristics

Normally, most fluidic devices with the same nozzle configuration will exhibit similar characteristics. Typically, for subsonic flow this follows the

[15.2] DIGITAL ELEMENT INPUT CHARACTERISTICS 573

well-known square law relationship between flow and nozzle pressure drop.

Fig. 15.1 shows the characteristic which may be expected from a rectangular 0·020 in. wide × 0·040 in. deep nozzle configuration. However, downstream of the control nozzle the pressure may vary considerably

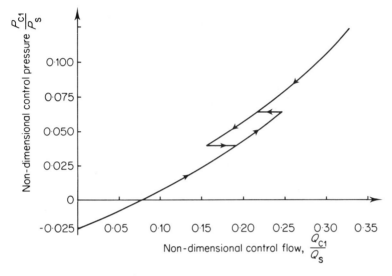

Control C2, Outputs O1 and O2 Open

Fig. 15.1 Control characteristic of a wall re-attachment element [B.A.C. (Stevenage), Commercial Leaflet, IPG 702]

during switching causing a discontinuity in the characteristic as shown in the figure. A wall attachment device, when switched, produces a very rapid change in signal levels, even with nearly static conditions. On the other hand, a momentum interaction digital device, such as the passive AND, has no such switching action so that the input characteristic shows no abrupt change. It is therefore advantageous to consider both classes of device separately.

Consider a wall attachment element which is not necessarily symmetric, so that it may exhibit monostability or bistability. A suitable notation is given in Fig. 15.2.

Normally, pressures and flows in the control channels are measured upstream of the interaction region so that these channel resistances or impedances are not taken into account. If an amplifier is to be designed for different conditions of bias flow to give monostability or bistability the control channel impedances must be known[1,2]. The static impedance

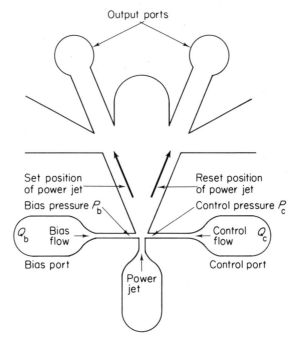

Fig. 15.2 Nomenclature for control characteristics (Wright, *Ref. 2*)

curves, under no output load conditions, at the exit from the control and bias ports may appear as shown in Fig. 15.3.

Looking at the control impedance curves more closely, it is seen that when the jet is attached to the same wall as the control (reset) the bias flow

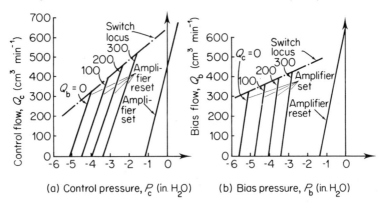

(a) Control pressure, P_c (in. H_2O) (b) Bias pressure, P_b (in. H_2O)

Fig. 15.3 Static impedance curves (Wright, *Ref. 2*)

[15.2] DIGITAL ELEMENT INPUT CHARACTERISTICS 575

level has an appreciable effect on the control characteristic. However, when the jet is attached to the opposite wall (set) the level of the bias flow has no effect on the control characteristic, hence there is only one curve representing this condition. The switch locus signifies the control conditions necessary, over a range of bias flows, to switch from the reset to the set curve. Similarly to switch in the opposite direction the bias impedance curve is consulted. The simplest case to consider is when both bias and control are orifice resistances R_1 and R_2 respectively open to atmosphere as shown in Fig. 15.4.

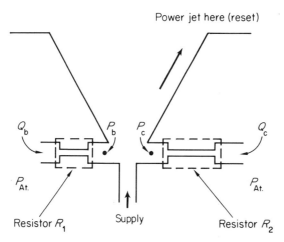

Fig. 15.4 Control orifice configuration (Wright, Ref. 2)

The operating points are found by superimposing the orifice characteristics for R_2 and R_1 on the control and bias impedance curves respectively. They are both shown on Fig. 15.5 as hatched lines passing through the origin due to the upstream pressure being atmospheric (i.e. zero gauge pressure).

If the amplifier is reset, the bias flow is given on the bias impedance curve by the intersection of the control reset and bias resistance curves at point A and has a value Q_{b1}. The corresponding point on the control impedance curve is point B for the bias flow Q_{b1} giving the control flow Q_{c1}.

It is possible that this configuration is also stable in the set position of the jet. In this case the starting point is the control impedance curve which indicates an operating point C giving a control flow Q_{c2}. On the bias curve the operating point may be D, corresponding to the control flow Q_{c2}, with bias flow Q_{b2}. This signifies that the set position too is stable. However,

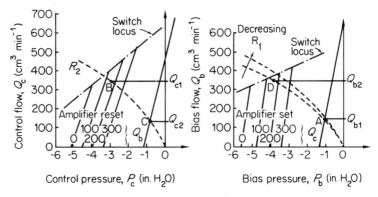

Fig. 15.5 Static operating points (modified Fig. 15.3)

if R_1 had been slightly less, as illustrated, no balance point D would have been possible and the configuration would have produced only one stable state, i.e. monostable in the reset position.

This type of analysis may be extended to more complex input conditions to provide many types of logic function. It does suffer from the disadvantage that no account has been taken of output loading effects which may considerably modify the switch locus.

Momentum interaction digital elements are typified by the passive AND and EXCLUSIVE-OR devices which both use two mutually perpendicular orifices to provide jet interaction. The two signal flows, when both are present, modify their orifice flow characteristics.

Fig. 15.6 shows typical results for 0·020 in. × 0·040 in. rectangular nozzles[3]. It can be seen that a family of nozzle characteristics are obtained for nozzle 1 depending on the constant flow Q_2 from the other nozzle. The upstream pressure P_2 of this nozzle also varies, despite the constant flow through nozzle 2, signifying an increase in pressure in the interaction region of the jets.

15.3 Digital Element Transfer Characteristics

These characteristics show how the input signal affects the output signal and so gives insight into the operating mechanism of a device. If the signals are both pressures or both flows then information concerning the pressure and flow gains respectively of active devices, suitably loaded, may be obtained. For a switching device gain does not have the same meaning as for proportional amplifiers due to the discontinuous nature of the characteristic. Nevertheless, information such as the gain at the point of switching is very useful.

[15.3] DIGITAL ELEMENT TRANSFER CHARACTERISTICS

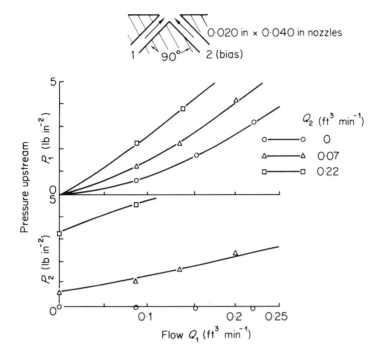

Fig. 15.6 Nozzle characteristics of two impinging jets (Parker and Jones, *Ref. 3*, Fig. 3)

Passive devices do not exhibit gain but the transfer characteristic is important for determining signal discrimination. Each class of device will again be considered separately.

The input-output pressure transfer characteristic of a wall attachment device shows the switching action and hysteresis effect of switching when the control signal is cycled through a range of pressures.

Fig. 15.7 taken from Steptoe[4] shows typical results for a monostable wall attachment element loaded at the output with three identical elements. The switching pressure gain, G_{sp}, may be defined as

$$G_{sp} = \frac{\text{output pressure change after switching}}{\text{input switching pressure change}}$$

which, in this particular case, would give $G_{sp} \simeq 3.5$. Similar flow transfer curves may be obtained although, as with the pressure curves, the load conditions must be carefully specified.

Ideally momentum interaction digital devices should have a definite switching action so that the transfer characteristics of AND and EXCLU-

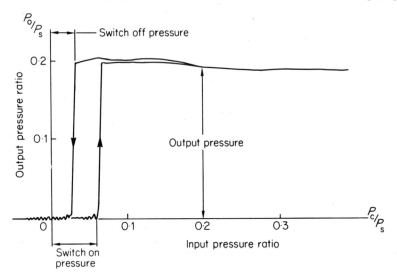

Fig. 15.7 Wall re-attachment amplifier transfer characteristics (Steptoe, Ref. 4, Fig. 1)

SIVE-OR devices should be as illustrated in Fig. 15.8 by the continuous line.

The curves are non-dimensionalized; the horizontal axis is the relative magnitudes of the input flow while the vertical axis is the non-dimensional output pressure. To give maximum signal discrimination between the cases when $Q_1 = 0$ and when $Q_1 = Q_2$, switching should occur at $Q_1 = 0.5Q_2$. This ensures that if the flow signals are badly mismatched correct logic operation is still possible. Unfortunately, as no switching

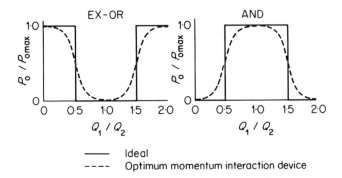

Fig. 15.8 Ideal passive, element transfer characteristics (Parker and Jones, Ref. 3, Fig. 5)

action occurs in practice with momentum interaction, an optimum characteristic may be reduced to a curve similar to the hatched line in the same figure.

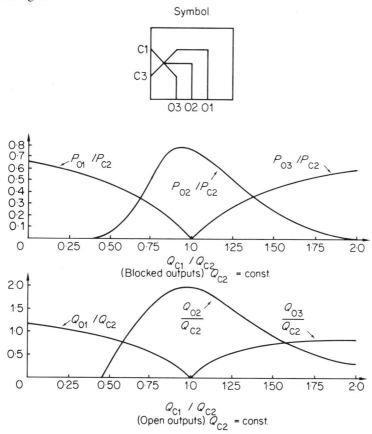

Fig. 15.9 Transfer characteristic of a combined passive AND and EXCLUSIVE–OR element [B.A.C. (Stevenage), Commercial Leaflet, IPG 703]

Fig. 15.9 shows the transfer characteristic of a combined AND and EXCLUSIVE-OR device from Ref. 5 in terms of pressure recovery for blocked load conditions and also flow recovery for no load conditions. Although the AND characteristics would seem to give reasonable discrimination the EXCLUSIVE-OR shows signal levels approaching zero only around the point where $Q_{C1} \simeq Q_{C2}$ and would be more difficult to use statically in practice. Both the curves were obtained by holding one control signal flow constant and varying the other from zero to twice the constant value. This is rather misleading in some ways, as can be seen by examining the AND characteristic more closely.

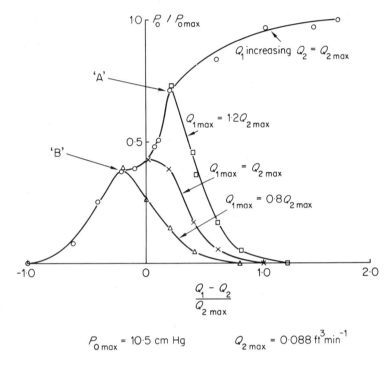

Fig. 15.10 Transfer characteristics of the AND, both inputs variable (Parker and Jones, *Ref. 3*, Fig. 10)

Fig. 15.10 shows a different AND transfer characteristic from Ref. 3 with both inputs variable. The top curve shows that if the input flow through, say, nozzle C2 is kept constant and the input flow through the other nozzle is varied to sweep the resultant jet across the output receiver, then an asymmetric transfer curve is obtained, which does not represent conditions likely to occur in practice. If we assume instead that the input signals both have identical maximum values ($Q_{1\max} = Q_{2\max}$) then the resultant jet, due to the signals impinging on each other, can only be swept across the output receiver in the following way:

(a) Put $Q_2 = Q_{2\max}$ and increase Q_1 from zero to $Q_{1\max}$. At this point the resultant jet is central.

(b) Leave $Q_1 = Q_{1\max}$ and gradually reduce Q_2 from $Q_{2\max}$ to zero. Such a test results in the curve marked '$Q_{1\max} = Q_{2\max}$'. Two other curves are also shown representing the condition when $Q_{1\max}$ is mismatched by ± 20 percent from $Q_{2\text{max}}$, which may be the maximum tolerance to amplitude differences that a fluidic system may stand. Asymmetric conditions are again produced as the maximum output pressures in the two cases (marked A and B) are substantially different.

15.4 Digital Element Output Characteristics

Most pure fluidic digital elements give similar output pressure-flow characteristics which approximate to an inverted nozzle characteristic. By suitable shaping of the vents and the diffuser section of the device, the output curve shapes may be altered. With careful design it is possible to obtain a constant output pressure over an appreciable range of output flow. Practical devices exhibit the nozzle curve shape so that a typical set of characteristics for a bistable wall attachment device may appear as in Fig. 15.11.

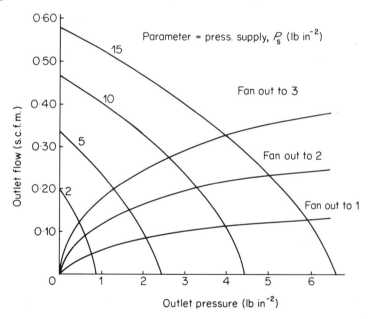

Fig. 15.11 Bistable output and load characteristics (Aviation Electric, Commercial Leaflet)

Each curve is for a different supply pressure. Also shown in the figure is a superimposed set of load lines representing the control impedances of 1, 2 or 3 identical control nozzles. When the output of the bistable element is supplying a signal to between one and three identical elements then the static operating points, for a given supply pressure, are found by the intersection of the appropriate control curve with the correct output curve. This is the basis of most static graphical matching techniques in digital circuits. If the procedure is repeated through all the elements of a complex network all the operating signal pressures and flows may be determined.

If the device is asymmetric, such as a wall attachment OR-NOR function, then it is possible that the two output legs generate different output characteristics, in which case two sets of output curves are required.

For passive devices, the output curves are plotted for constant input signal conditions. In the case of the AND function this is when the two input signals are equal and for the EXCLUSIVE-OR when one signal is present, as shown in Fig. 15.12.

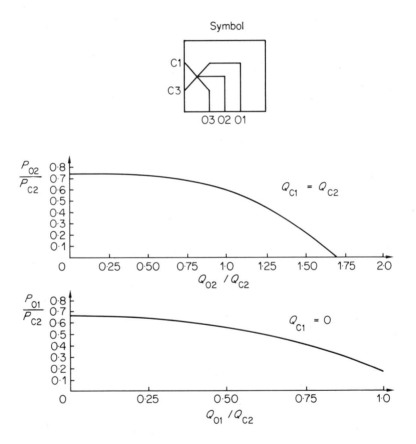

Fig. 15.12 Output characteristics of a combined passive AND and EXCLUSIVE–OR element [B.A.C. (Stevenage), commercial leaflet, IPG 703]

15.5 Combined Digital Characteristics

All the characteristics described above may be combined together, provided consistent values for the variables are chosen between each curve. This is achieved by putting each characteristic in a quadrant of a larger set of axes. This technique has been successfully used to describe a turbulence amplifier static performance[6]. Fig. 15.13 shows the combined characteristics for a planer turbulence amplifier. Quadrant I gives the supply tube characteristic, which is required to determine the power consumption. Quadrant II gives the output pressure-flow characteristic, the same scale being adopted for the supply flow Q_s and output flow Q_o on the horizontal axis. In the same quadrant is also plotted the output pressure against supply flow. Quadrant III gives the transfer characteristic relating the output pressure P_o to the control flow Q_c and quadrant IV gives the input tube characteristic. The maximum pressure recovery is given by the ratio AC/AD and the maximum flow recovery by BO/AO.

The curves can also be used to determine the number of identical elements that may be given input signals from the output of one element to just cause them all to switch (fan-out).

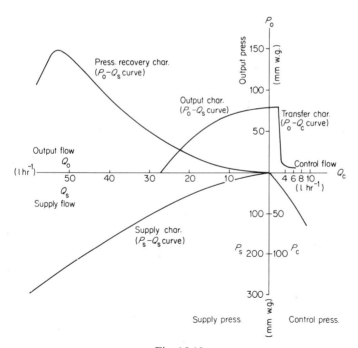

Fig. 15.13

In quadrant III of Fig. 15.13 point P represents the limit of the 'on' state i.e. laminar flow, while point S is the lower limit of the amplifier in the 'off' state. When the upstream amplifier is in the 'on' state represented by P the control pressure P_c is $\leqslant$ 10 mm w.g. (quadrant IV) and the output is sufficient to turn at least one other amplifier 'off'. This means that the output pressure of downstream amplifiers must be smaller than 10 mm w.g. (point T on the transfer curve) and each requires a control flow Q_c = OQ to keep it 'off'. Following the hatched line, the upstream amplifier can deliver a flow of RO at the required pressure and so the fan-out or number of elements supplied is equal to the largest integer of the ratio RO/OQ. Although combined characteristics are attractive they are only applicable to devices which show little or no coupling between supply, control and output ports. Turbulence amplifiers possess this advantage and are thus well suited to this method.

15.6 Two Circuit Analysis Examples

By superimposing output characteristics onto input characteristics of the following element or elements, steady state operating points can be determined. In the case of switching modes of operation the analysis enables minimum operating conditions to be determined i.e. the marginal operating conditions at which switching will or will not occur. First, a simple configuration of a wall re-attachment element supplying an AND device will be considered followed by a more complex example based on a comparator circuit.

15.6.1 Wall Re-attachment Element and a Passive AND

Suppose it is a design requirement to exactly match the signals in the two nozzles 1 and 2 of an AND element being supplied from a wall re-attachment element into nozzle 1 as illustrated in Fig. 15.14.

Assume that the active wall re-attachment element supply pressure is fixed and the flow is always switched to the position shown. The output pressure-flow characteristic of the active element is shown in Fig. 15.15 for a constant supply pressure.

It was shown earlier that the characteristics of both input nozzles are coupled so that the nozzle characteristic of nozzle 1 is modified by the strength of the flow, Q_2, in nozzle 2. A set of nozzle 1 curves for Q_2/Q_1 ranging from 0 to 1·5 are shown in the figure.

Two operating conditions may be determined depending on whether signal 2 is present or absent. When signal 2 is absent the lowest nozzle curve is used giving the operating point where output and input curves intersect. Thus for nozzle 1, the pressure and flow is $P_1 = \text{OD}$, $Q_1 = \text{OC}$ and, of course, the AND output is zero. When signal 2 is present and properly matched, Q_1 is equal to Q_2 and the AND function generated. The appropriate nozzle curve in this case is for $Q_2/Q_1 = 1 \cdot 0$ giving an interaction with the output. Now the pressure and flow for both nozzles is P_1 and $P_2 = \text{OD}'$, Q_1 and $Q_2 = \text{OC}'$.

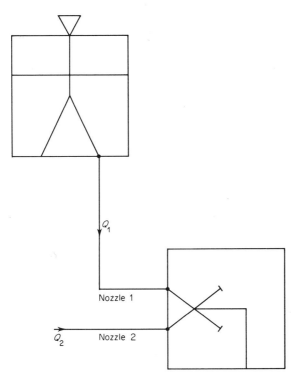

Fig. 15.14 Wall re-attachment element cascaded with a passive AND

15.6.2 *Comparator Circuit*

Fig. 15.16 shows the logic diagram of one arrangement which compares two input signals 'a' and 'b' and gives an output when they are both equal (both '0' or both '1'). A carry signal C_1 provides gating between different stages of the comparator so that two binary numbers are compared in succession from the highest digit to the least digit positions[7].

The method of approach in a complex circuit is to determine operating conditions by working from the last element in the circuit to the first. By

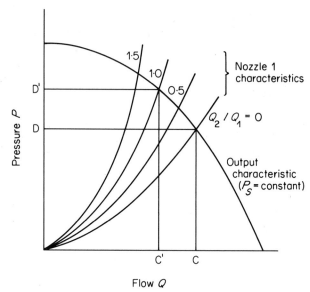

Fig. 15.15 Graphical solution for Fig. 15.14

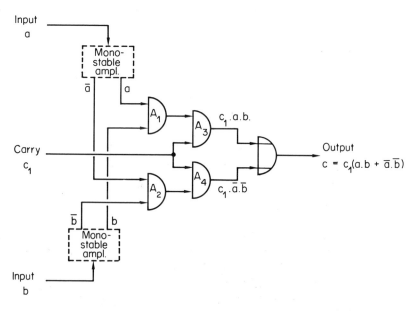

Fig. 15.16 A comparator logic circuit (Mitchell & Foster Research Report No. 75, Mech. Eng. Dept., University of Birmingham)

[15.6] TWO CIRCUIT ANALYSIS EXAMPLES

inspection of the circuit it is seen that the networks associated with signals 'a' and 'b' are identical so that only half the system need be considered, i.e. monostable amplifiers, A_1 and A_2, A_3 and A_4 are identical pairs of elements.

The first step is to match the OR input to AND function A_3 output as illustrated in Fig. 15.17. The output curves represent the A_3 output when signals A_1 and C_1 are both present; three curves are shown for different signal pressure amplitudes. The other curve is the switching characteristic of the OR under specified load conditions. In this instance, the minimum operating point is R so that the pressure signals C_1 and A_1 must be $\geqslant P_2$.

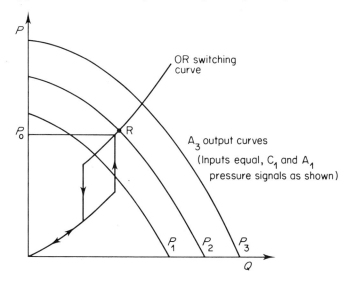

Fig. 15.17 Matching the OR element to A_3

The second step is to match AND elements A_1 and A_3 together (Fig. 15.18). The A_3 input curves are for carry C_1 present and absent. The circuit only operates when carry C_1 is present giving an operating point S corresponding to signal level P_2. In turn this means that the equal input 'a' and 'b' signals to A_1 must have pressures of magnitude $\geqslant P_5$.

The third step involves matching AND element A_1 to the monostable element (Fig. 15.19). The ouput curves are the monostable amplifier output at various supply pressures while the other curves are the AND nozzle characteristics with 'b' present and absent. At the output 'a' of the amplifier must be $\geqslant P_5$, the minimum operating point is T which requires an amplifier supply pressure $\geqslant P_8$.

The last step is to check that the carry signal C_1 generated by an output

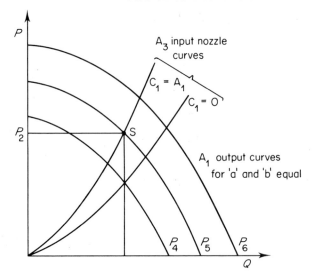

Fig. 15.18 Matching A_3 to A_1

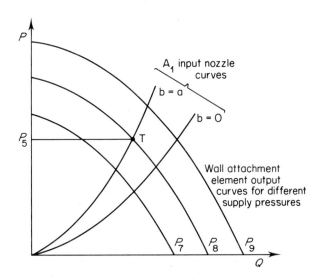

Fig. 15.19 Matching A_1 to a wall re-attachment element

OR is sufficient to supply an identical circuit in the next stage (not shown in figure). This means that signal C_1 must be large enough to fan-out to two identical AND elements A_3 and A_4. Fig. 15.20 shows the output characteristics of the OR and a load of one or two AND input nozzles superimposed on it. As signal C_1 must have a pressure $\geqslant P_2$, point U, the intersection of the two nozzle curves with pressure P_2, must lie inside the output pressure flow curve otherwise the amplifier cannot supply the correct signal level. In the example shown the condition is fulfilled and all the operating pressures and flows for the circuit have been determined.

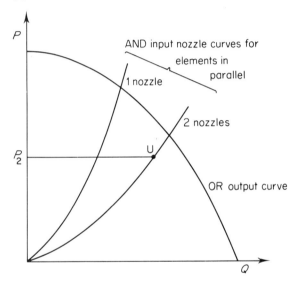

Fig. 15.20 Check for the magnitude of the carry signal C_1

15.7 Interstage Static Matching of Analogue Beam-deflexion Amplifiers

Static Beam-deflexion amplifier characteristics are shown in Figs. 4.8(a) and (b). They consist of an input pressure-flow characteristic, a transfer characteristic between control and output pressures, and an output pressure flow characteristic. As these relationships are often non-linear they are usually presented in graphical form, although an analytic method due to Simson[8] is available for predicting transfer characteristics.

Belsterling[9] has given a rational method for evaluating static gains based on the input, transfer and output characteristics of each active element. The characteristics required are very comprehensive and require a substantial amount of experimental work for their compilation. When applying this technique to, say, the matching of an amplifier to a similar element the output characteristics are superimposed on the input characteristics to see if mismatching is occurring. Should there be mismatching three standard approaches may be used: alter output bias level, alter the second amplifier supply pressure or change series or shunt resistor configurations. The effect of each of these may be seen by trial and error so that matching becomes a series of compromises requiring significant calculations at each stage. When matching has been achieved the change in gain is the final result obtained and there is no intermediate representation of gain change, the most important parameter. Difficulties also arise due to the differential control locus and the bias point locus being frequently very close on the amplifier input characteristics.

A different impedance matching technique may be preferred due to the rapidity with which new configurations may be analysed. It considers the active elements to be linear over the operating range and that their characteristics may be modified by the use of series or shunt linear resistors. Only two characteristics are found to be necessary to fully describe each amplifier, namely the input and transfer characteristics thus avoiding the more complex output characteristic. The correct interstage loading between amplifiers to provide a predetermined gain may be arranged using shunt or series linear resistors as shown in Fig. 15.21.

In general, a shunt resistor R_1 could have a value chosen to make the parallel combination of R_1 and R_i have the same resistance as R_o. In this case

$$R_1 = \frac{R_i R_o}{R_i - R_o} \tag{15.1}$$

provided the input impedance of the amplifier is quasi-linear. Most Beam deflexion amplifiers have a near linear input impedance provided the control flow does not approach zero. This condition may be avoided by the appropriate use of control bias levels. If the input impedance is highly non-linear then Belsterlings method is to be preferred although for such amplifiers it would be difficult to obtain linear proportional operation.

If a series resistor is used then a different technique is required. The blocked load output gain of the first amplifier G_b, is taken and the value of the series resistor R_2 obtained from

$$R_2 = R_i \left[\frac{G_b}{G_i} - 1 \right] \tag{15.2}$$

The loading of the first amplifier is now $(R_i + R_2)$ which has a corresponding gain on the transfer characteristics. If R_2 is not large, this gain will be

[15.7] STATIC MATCHING OF ANALOGUE BEAM-DEFLEXION AMPLIFIERS

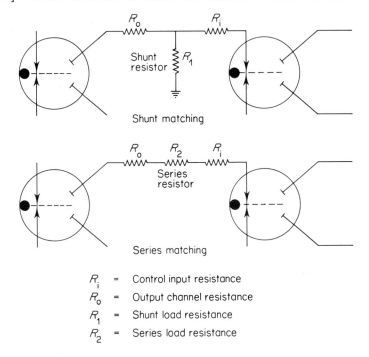

Fig. 15.21 Static matching techniques for beam-deflexion amplifiers

less than G_b requiring the process to be repeated using the new gain value instead of G_b. This gives rise to a short iterative procedure to obtain the correct value for R_2 for a given stage gain G_1.

Although series resistors are frequently used for interchange matching in fluidic circuits they have a number of disadvantages. They give a gain dependent on both the potential divider effect and also the loading of the preceding active element. Also the effects of small variations in the interstage resistors on the interstage gain is more marked for the series than for the shunt configurations. The most important disadvantage in the use of the series resistor, however, is that its phase shift, when placed between two active elements, is much greater than a branched line shunt resistor under dynamic signal conditions.

References

1. Saghafi, H. T., Static design of pneumatic logic circuits, *H.D.L.*, **2** (1964).
2. Wright, C. P., Some design techniques for fluid jet amplifiers, *I.B.M. G.P.D. Rep.*, TR.01.758 (1963).

3. Parker, G. A., and Jones, B., Experiments with AND and EXCLUSIVE-OR passive elements, *C.F.C.* (2), Paper C4.
4. Steptoe, B. J., Steady state and dynamic characteristic variations in digital wall attachment devices, *C.F.C.* (2), Paper B3.
5. Anon, *B.A.C.* (Stevenage), Brochure on Element IPG 703.
6. Verhelst, H. A. M., On the design, characteristics and production of turbulence amplifiers, *C.F.C.* (2), Paper F2.
7. Foster, K, and Mitchell, D. G., The design of a comparator for use in a pure fluid control system, *Mech. Eng. Rep.*, 75, Birmingham.
8. Simson, A. K., Gain characteristics of subsonic pressure controlled proportional fluid jet amplifiers, *ASME Paper*, 64–WA/AUT–2.
9. Belsterling, C. A., and Tsui, K. C., Analysing proportional fluid amplifier circuits. *Control Eng.*, 87, (1965).

Index

Acoustic effects 175
Addition, binary 521–532
  air consumption 527–529
  carry signal 525
  fluidic adder 526–532
  full adder 523, 533, 535, 563
  half adder 523, 526
  logic 522–525
  serial operation 522, 524
  signal propagation 527, 529, 532
Address circuit 423
  letters 401
  logic 430
Air consumption 7, 24, 423, 436–443, 478, 492, 527–529
  see also Power consumption
Analogue devices
  beam deflexion 29–31, 179, 181–195, 224, 230, 493
  confined jet 38
  double leg 35–38, 179
  impact modulator 31, 32–34, 179, 195–206
  knife-edge 153, 315
  vortex 34, 170–174, 179, 206–221
AND, fluidic use of 515, 535, 538, 540, 567, 570, 584, 587, 589
Attachment, *see* Wall re-attachment
Automatic gauging system 430

Back-pressure amplifier, *see* Sensors
Beam-deflexion amplifier
  beat frequency detection 255
  differentiation 247
  hybrid techniques 271
  integration 242
  lag networks 245
  lead networks 248
  matching 589–591

  phase discrimination 258
  signal addition 237
Beat-frequency detection 253, 255
Binary divider, *see* Logic functions, T Flip-Flop
Boolean algebra 274–297
  complement 278–280, 285–287, 396, 412, 413
  dual 278–280
  hindrance function 279
  optional combinations (don't care) 289, 294–296, 380, 383
  prime implicant 290
Boundary layer 94, 103, 117, 133
  in vortex 164, 168–171
  separation 157

Card readers 424, 434, 447–449
Carrier wave modulation 252, 253, 265
Cascade method of design, *see* Sequential systems
Cavitation number 336, 337
Characteristic, of elements 304, 307, 320, 322, 324, 325, 331, 333, 438, 479
Codes 395–402
  binary-coded decimal (B.C.D.) 295, 409–413, 453, 454, 495–497, 499, 508, 509
  block 398
  conversion 441, 453
  excess 3 399
  for N.C. machines 400, 441
  Gray 399, 415, 504, 507
  Lippel 415, 496
  natural binary 395, 399, 403, 433, 462–464, 469, 490, 504, 508
  Petherick 295, 296, 399, 441, 509, 510, 517

Codes—*cont.*
  quinary 417, 444, 464, 465, 483
  unconventional 442
  unit-distance 399, 403, 404, 415, 443, 496, 497, 499, 504, 509
  weighted 396, 397, 509
Combinational circuits 556
Comparator 424, 434, 435, 441, 443, 444, 537–544
Control flow, *see* Switching flow
Converter, *see* Decoder
Cost, of air supplies 436
  relative, of circuits 427
Counter 402–420, 421, 423, 443, 444, 469–499
  batching 468, 499
  B.C.D. 409–413, 495–497, 499
  complementing 412, 413
  controlled circuit (sequence) 429, 431
  decade 408–417, 433, 491
  decimal 468
  down-counting 409–412, 491
  Gray code 415
  Lippel code 415
  natural binary 403, 433, 442, 462, 464, 469, 490, 501
  parallel mode 403, 413–417
  pre-set 409, 441, 443, 444
  ring 416, 417, 448, 449, 469, 481–483, 492
  serial mode 408–413
  spool valve 343, 344
  T Flip-Flop 403–414, 471, 481, 485–491
  twisted ring 417, 418, 433, 464, 469, 483–485
  up-and-down counter 440, 491, 492
  up-counter 409–411
  unit-distance 403, 404, 415, 496, 497, 499
  Warren 408, 486, 487
Curved wall effect 157, 176
Cusped diffuser 172

Decoding 433, 447, 453–468, 484
  binary to decimal 455, 458, 467
  folded tree 456–460
  matrix 453–455, 459, 460, 462
  Petherick to binary 295
  tree 456

De Morgan's Theorem 281, 418, 422
Diffusers 163, 325
Digital Control Systems 500, 521
Digital devices, *see* Moving part digital devices and Pure fluid digital devices
Digital frequency converter 254
Digital-to-analogue (D/A) converter 440, 544–555
  dynamic limitations 548–551
  force balance 554
  moving part types 552–555
  non-linear resistances 548, 552
  pure fluid types 544–552
  static characteristics 548–552
  stepping motor 555
Direction discriminator 502, 503
Displays, visual 433, 447, 462–468, 499
Drilling sequence 426, 446, 499
Dycril plastic 515, 528, 551

Eddy viscosity 148
Edgetones 110–117, 175
Electronics 1, 2, 265
Encoder 297, 424, 434, 443, 444, 446, 500–520
  angular 512
  binary coded 440, 442, 504
  incremental 440, 442, 502
  linear 518, 519
  mechanical design 505, 510, 518
  scanning 505, 506, 511, 515, 516
  sensors 505, 512, 513
Entrainment 26, 131, 140, 299, 309, 312
  parameter, definition 140
EQUIVALENCE, fluidic use of 530, 538, 540
EXCLUSIVE–OR, fluidic use of 538, 540–542, 569

Fan-in 4, 24
Fan-out 4, 10, 305, 333, 437–439, 456–458, 487, 583, 584, 589
Feedback, digital 440
Flip-Flop, preferentially biased 428
  see also Pure fluidic digital devices
Flow, compressible 53, 68, 218
  incompressible 52, 67
Flow diagram, Mealy 376, 377
  Moore 376

INDEX 595

Flow table 376, 378, 393, 402–405
  equivalent rows 378
  Huffman 377, 378
  merged 379, 393
  primitive 378, 381, 393, 418
  pseudo-equivalent rows 378
Fluid networks 64–76
  Helmholtz resonator 75, 256, 264
  ladder 65, 544
  resistance 67
  resistance and capacitance 68
Fluid parameters 47–60
  capacitance 55
  dimensions of 47
  impedance 47
  inertia 58
  resistance 48–50, 52
Fluid transmission lines 76–91
  approximations 86
  characteristic impedance 83, 86
  electrical analogue 81
  large amplitude signals 91
  liquid lines 79–83
  pneumatic lines 84–86
Fluidics, analogue devices 3
  definition of 1
  digital devices 3
Frequency, digital converter 271
  of edgetones 112–117
  modulation 117, 253

Gain 176, 487, 590
  flow 37, 180, 212, 267, 312, 313, 472
  power 303, 305
  pressure 31, 39, 180, 211, 212, 577
  switching 4, 577
Graphical representation 572–591
  analogue matching 589–591
  combined characteristic 583
  input 572–576
  output 581–582
  transfer 576–581

Half adder, use of 526
Hazards 555–570
  critical race 556, 558
  dynamic 556, 557
  essential 556, 557
  static 556, 557, 558–565
  synchronous protection 565–570

  transient transmission function 559–561
Helmholtz resonator, *see* Fluid networks
Hindrance function 279·
Huffman, flow table 377
Hydraulic logic 343, 373
Hysteresis 355, 356

Inertia effects 58, 367–370
Input combinations 377–379
Integrated circuit 426, 428, 530
  spool valves 343
Interlocks 423

Jet deflexion 129–131, 174, 178, 179, 314
Jet, laminar 93–95
  core 122–125, 145
  equations 95
  natural frequencies 104, 106
  neutral stability 98–101, 109, 175
  noise in 110–117, 188
  re-attachment 138
  transition 4, 95–110, 175
  velocity fluctuations 103–105
Jet, spreading rates 127
Jet, turbulent 117–127, 176
  equations 120, 122, 124
  shear stress 125, 126
  static pressure 125, 126

Karnaugh maps 291–297, 453, 454
  for counters 415, 497
  for sequence circuits 379–383

Ladder networks, *see* Fluid networks
Ladder type circuits 384, 394
Lead screw 443, 444
Linearisation 60–64, 191
Load parameter 369, 370
Logic functions 274–297, *see* also Moving part digital devices and Pure fluid digital devices
  AND 274–276, 340, 341, 345, 347, 349, 353, 354, 358, 360, 422, 423
  bistable (Flip-Flop) 342, 345, 351, 358–361, 379–384
  diode 363, 365

Logic functions—*cont.*
  DOUBLE SEQUENTIAL AND 391–394, 427, 428
  EQUIVALENCE 340, 341, 349
  EXCLUSIVE–OR 340, 341, 349
  INHIBITION 340, 341
  IMPLICATION 340, 341, 349, 353
  NAND 340, 341, 349, 357, 358, 490
  NON-IMPLICATION 353
  NOR 340, 341, 349, 353, 357, 358, 362, 373, 422, 423, 427
  NOT (NEGATION, inverter) 340, 341, 347, 349, 353–357, 361–363
  OR 276, 277, 340, 341, 347, 351, 353, 354, 358, 360, 362–365, 422–424
  SEQUENTIAL AND, OR, NOT 389, 391
  T Flip-Flop, *see* Counter
  YES (IDENTITY) 340, 341, 349, 353
Lost wax casting 345, 474
Lucas method 388

Mach number 129, 145, 150, 165, 335–337
Machine tools 544
Management functions on punched tape 398
Manifold 350, 351
Manufacturing errors 314, 505, 517
Mark-to-space ratio 265, 266, 493
Martonair cascade system 385, 386
Matching 201, 437, 479, 590, 591
Miscellaneous functions on punched tape 401, 402
Missile control systems 266, 267
Mixing length 118, 121, 140
Momentum interaction 4, 178, 179
Moving part digital devices 338–373
  air consumption 347, 439
  back pressure devices 338
  ball-valve 339, 358, 361, 362
  decoding with 458–462
  diaphragm-lever 339, 346
  diaphragm with foil 339, 354
  diverter valves 338, 339
  double diaphragm 28, 339, 347–351
  free-foil 339, 362–365, 373, 448, 460–462, 467, 477, 543
  latching spool valve 339, 345

  passive OR element 348, 351
  planar diaphragm 339, 355
  single diaphragm 339, 353, 354, 427
  spool valve 339–346, 384–389, 427, 459, 460, 473
  spring NOR unit 339, 358–360, 434, 458
  triple-diaphragm 339, 352

Natural frequency 369, 370
Navier-Stokes equations 97, 164
Noise, in jet 38, 110–117, 175, 181, 201, 228
  reduction 117
  signal-to-noise ratio 228
Nozzle flow, *see* Flow
Number codes, *see* Codes
Numerically controlled machines 400, 402, 420, 421, 446

One-shot flip-flop, *see* Pulse shaper
Operational amplifiers 224–236
  differentiator 234
  ideal 225, 226
  integrator 231
  signal-to-noise ratio 228
  stability 228
  summation 225
OR/NOR element, use of 510, 515, 518, 528, 541, 567, 584, 587, 589
Orifice flow, *see* Flow
Orr–Sommerfeld equations 98, 105, 108
Oscillator 211, 254, 267, 269, 499
  clock signal 536, 565, 566
Output, assignments 378
  circuits 381
  combinations 377
  excitation 379, 381–383, 393, 404

Paper tape, *see* Punched tape
Parity check 398, 402
Phase discriminator 258
Piston valves, *see* Moving part digital devices, spool valve
Plugboard 424, 434, 447, 448
Position control
  continuous path 440
  point-to-point 439–446, 499
Potential flow 162, 177

INDEX 597

Power amplification 370, 371
Power consumption 127–129
 see also Air consumption for various fluids 128
Preparatory functions, on punched tape, 401, 402
Pressure recovery 130, 301, 302, 311, 313, 319–325
Primary memories 421
Process control 544
Pulse shaping 269, 272, 344, 417, 419, 478, 489, 492–495, 567
 width 493–495
 width modulation 265–273
Pulsed supply 474–476, 488, 489
Punched card 421, 424, 434
 controlled sequence 433
 tape 400, 402, 420–424, 434
 block format 401
Pure fluid digital devices 5–29, 299–337, *see also* Momentum interaction, Wall re-attachment and Turbulence amplifier
 AF relay (AF_R) 431–433
 AND 13–15, 21, 25, 26, 425–428, 437–439, 442, 479–482, 494, 576, 578–581, 582
 AND/NAND 12, 13, 472, 473
 bistable (Flip-Flop) 5–9, 425–428, 431, 437–439, 442, 454, 472, 476, 478–480, 489, 499
 converging wall 332
 diffuser switch 27, 28
 EQUIVALENCE 17, 18
 EXCLUSIVE–OR 15, 16, 490
 half adder 16, 17, 18–21, 491
 latching AND 474–476
 NOR 7, 21–25, 28, 29, 301, 305, 337, 455, 583
 OR/NOR 9–12, 328, 329, 430–433, 442, 491, 494, 495
 tristable 335, 336

Radial drill, circuit 435
Radixes 395
Re-attachment, *see* Wall re-attachment
Receiver, distance 302, 303
 size 186, 303, 304, 320, 326
Rectifier 255
Redundancy 397

Resistors 590, 591
Resonator 114–117, 175
 see also Fluid networks
Restrictors 353, 355
Reverse flow 323–326
Reynolds number 21, 49, 51, 93–110, 123, 125, 129, 135, 164, 165, 299, 335, 336
 critical 96, 127, 129, 300

Secondary assignments 378, 379, 403, 414
 circuit 380, 381
 excitation 379–381,393
 excitation maps 380, 381, 393, 404, 405
 memories 376, 379, 380, 421, 422
 state 377
Sensitivity 305, 313, 333, 472
Sensors, acoustic 117
 back pressure (flapper valves) 41, 71–74, 425, 449, 452, 453, 512
 conical 44
 dead band 514
 digital position 39–44
 interruptible jet 42, 512, 513
 pressure ratio 337
Separation, flow 36
Sequential system (or circuit) 374–395, 402, 421, 446, 469, 556
 asynchronous 374, 378, 421–426
 cascade method 384–389, 394, 420
 flow diagram 378
 fully-specified 376
 input combinations (or state) 377–379
 input signals 376
 ladder type 384
 optional state 379
 optional transition 378
 output combinations (or state) 374, 377–379
 output signal 374, 375
 secondary state 379
 stable state 377, 379, 383
 Stancielescu's logic algebra 389–395, 420
 state map 379, 380
 synchronous 374, 378, 421
 transitions 377

Sequential system—*cont.*
  under-specified 376
  unstable state 379, 383
Setback 309
  effect of 143, 144, 310–315, 319, 320, 330, 333
Shear stress constant 145, 148–150
Shift register 417, 450, 469–499
  hydraulic 343
  latching spool valve 346
  pulsed operation 474–481
Speed measurement 260–264, 269
Speed of response 354, 367–370, 443, 478
  see also Switching speed back pressure sensor 452
Splitter 154, 310, 311, 312, 316–321, 325, 329–331
Spurious signals 443, 469, 482, 483, 492, 495
State diagram 376–379
  internal 377
  map 379, 380, 393
Static friction 369, 370
Static hazards, *see* Hazards
Stepping motor 444, 446
Strouhal number 97, 99, 100, 101, 103, 107, 108, 110, 114, 335, 336
Subtractor, binary 532–536
  carry signal 536
  complementing 533, 535
  fluidic 443, 444, 446, 535, 536
  half subtractor 532
  logic 532–535
  static hazards 563–5
  synchronous operation 536, 555, 556, 565, 566–569
Summing junction 440, 441
Switching flow,
  action 3, 329–337
  dynamic 479
  static 171, 172, 312, 314, 329, 331
  functions, *see* logic functions
  point 355
  speed (or time) 9, 141, 159–161, 305, 306, 333, 335, 347, 359, 364, 469

Tape readers 447, 450–453, 467
Temperature measurement 253–260, 271, 272

Time-delay 501, 508, 516, 527, 529, 536, 539, 559, 562, 565
  delay line 417, 470, 476, 477, 480, 490–495
  hazards 555
  resistance-capacitance 425
Timed-sequence control 434
Timer circuit 499
Timing control 433
  diagram 375, 430
Transformation 162
Transition, of jet 95–110
  of state 377
  optional 378
Transmission function 278
Transport time 334, 335
Truth table 276, 282, 285, 433
  double-diaphragm valve 351
  grinding circuit 432
Tuned cavity, *see* Resonator
Turbulent jet re-attachment, *see* Wall re-attachment
Turbulent shear stress 118
Turbulence amplifier 117, 301–307, 336, 337, 408, 423, 427, 449, 490, 499, 539
  air consumption 436
  characteristics 304, 307
  planar 305
Twist drill flute milling machine 428, 446

Universal logic block 389, 395, 428
'U' scanning 441, 506
  see also Encoders

Velocity profiles (distribution) 299
  of deflected jet 130
  of laminar jet 93, 99
  of turbulent jet 120–125, 146–148, 300
Vent 162, 163, 310, 312, 313, 315, 318, 319–326, 329, 330
Vortex amplifier 34, 35, 170–174, 177, 206–221, 337
  angular rate sensor 177
  control functions 215
  dynamics 221
  effect of 329, 330
  flow 163–174, 216–221

INDEX 599

Vortex amplifier—*cont.*
  geometry 213
  lag networks 245
  lead-lag networks 252
  signal summing 240
  static characteristics 206
  trapped 161–163, 177, 319
  vent 162, 163, 323–325, 336
  with wall re-attachment 326–328
Vortices, in a jet 110–116, 174, 175
'V' scanning 505, 511, 515, 516
  see also Encoders

Wall angle 309
  effect of 136–138, 143, 144, 310–315
Wall re-attachment
  air consumption 436
  bubble pressure 134, 137, 150–152, 175, 309
  characteristic 11, 438, 573–577, 581
  distance (bubble length) 135, 136, 143, 144, 151, 329
  downstream loading 154–157, 176, 315–321, 329, 330, 337
  finite wall 148
  process 4, 131–161, 176
  rate of bubble growth 141, 154, 158–161
  stability of 152–157
  supersonic flow 158, 177
  with foil diode 365
Wavelength, of jet disturbance 110, 112
Weighted codes, *see* Codes